Second Edition

Understanding Symbolic Logic

VIRGINIA KLENK

West Virginia University

PRENTICE HALL, Englewood Cliffs, New Jersey 07632

Library of Congress Cataloging-in-Publication Data

Klenk, Virginia
    Understanding symbolic logic / Virginia Klenk. -- 2nd ed.
        p.   cm.
    Includes index.
    ISBN 0-13-942764-3
    1. Logic, Symbolic and mathematical.   I. Title.
BC135.K53 1989
    160--dc19   88-31179
                                                                    CIP

Editorial/production supervision and
   interior design: Rob DeGeorge
Cover design: George Cornell
Manufacturing buyer: Pete Havens

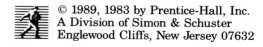
Printed in the United States of America

10   9   8   7   6   5   4   3   2   1

ISBN 0-13-942764-3

PRENTICE-HALL INTERNATIONAL (UK) LIMITED, *London*
PRENTICE-HALL OF AUSTRALIA PTY. LIMITED, *Sydney*
PRENTICE-HALL CANADA INC., *Toronto*
PRENTICE-HALL HISPANOAMERICANA, S.A., *Mexico*
PRENTICE-HALL OF INDIA PRIVATE LIMITED, *New Delhi*
PRENTICE-HALL OF JAPAN, INC., *Tokyo*
SIMON & SCHUSTER ASIA PTE. LTD., *Singapore*
EDITORA PRENTICE-HALL DO BRASIL, LTDA., *Rio de Janeiro*

SUMMARY OF RULES OF CONDITIONAL PROOF (C.P.) AND INDIRECT PROOF (I.P.)

A. Conditional Proof (C.P.)

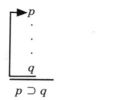

If, given the assumption p we are able to derive q, then we are allowed to infer ($p \supset q$), citing all the steps from p to q inclusive.

B. Indirect Proof (I.P.)

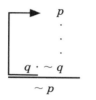

If, given an assumption p we are able to derive a contradiction $q \cdot \sim q$, then we may infer the negation of our assumption, $\sim p$, citing all the steps from p to $q \cdot \sim q$ inclusive.

C. Restrictions on the Use of C.P. and I.P.

1. Every assumption made in a proof must eventually be discharged.
2. Once an assumption has been discharged, neither it nor any step which falls within its scope may be used in the proof again.
3. Assumptions inside the scope of other assumptions must be discharged in the reverse order in which they were made; that is, no two scope markers may cross.

D. General Instructions for Using C.P. and I.P.

1. For both C.P. and I.P., an assumption may be introduced at any point in the proof, provided it is justified as such, that is, provided we label it as an assumption.
2. In using C.P., we assume the antecedent of the conditional to be proved and then derive the consequent. In using I.P., we assume the opposite of what we want to prove and then derive a contradiction. All the steps from the assumption to the consequent (for C.P.) or the contradiction (for I.P.) are said to be *within the scope* of the assumption.
3. The sequence of steps within the scope of an assumption is called a *subproof*.
4. We indicate the scope of an assumption, and set off the subproof, by an arrow (pointing to the assumption) and a vertical line which runs to the left of the subproof and includes every step in the subproof. This arrow and vertical line is called the *scope marker* for the assumption. We also *set in* or *indent* every step in the subproof.
5. There is no limit to the number of assumptions we may introduce in a given proof, and we may make one assumption inside the scope of another.
6. The scope of the assumption ends immediately prior to the step in which we infer the conditional or negation. We say that *the assumption is discharged* at this point. Thus neither the conditional nor the negation, the result of applying C.P. or I.P., falls within the scope of the assumption. We indicate that the assumption has been discharged by cutting off the vertical line (the scope marker) at this point.

**To my mother,
Helen Crooker Klenk**

Contents

2 THE STRUCTURE OF SENTENTIAL LOGIC 18

3 COMPUTING TRUTH VALUES 30

4 SYMBOLIZING ENGLISH SENTENCES 48

5 TRUTH TABLES FOR TESTING VALIDITY 69

PART II   MONADIC PREDICATE LOGIC

10   SINGULAR SENTENCES   *189*

11   QUANTIFIERS   *203*

12   CATEGORICAL PROPOSITIONS   *216*

Preface

This book is intended as a comprehensive introduction to symbolic logic. It presupposes no prior acquaintance with either logic or mathematics, and it includes all the standard topics through relational predicate logic with identity. The book was written in the conviction that any student can master symbolic logic, and it is designed to give the student as much help as possible in attaining that mastery.

The main part of the book is divided into twenty units, each of which has an introduction and a statement of study objectives, so that the student has an overview of what is to come, and knows exactly what is required in order to master the unit. The explanatory material for each unit is divided into several subsections, each of which has a specific function and covers one relatively small, clearly defined topic. The clear separation of topics and the division into easily comprehended small "bites" allow the student to master the material step by step without being overwhelmed by an indigestible mass of information.

One-variable predicate logic is developed, in detail, independently of relational predicate logic, and identity is presented in two separate units. The semantics of predicate logic is also developed in a separate unit, as is the semantics for propositional logic. In addition to the basic material, there are several "extra credit" units, which provide a glimpse into alternative methods of logic and more advanced topics.

I have tried to give as detailed explanations as possible, both for specific techniques, such as drawing up truth tables or constructing proofs, and for the rationale behind these techniques. It seems to me as important for a student to understand *why* things are done in a certain way as to learn the techniques themselves, and in this book I have tried to supply the "why's" as well as the "how's."

The book does, however, supply the "how's" in abundance. Aside from the detailed explanations, there are numerous examples worked out in the text: various types of truth tables, a great many detailed, step-by-step symbolizations, and over fifty fully-worked-out proofs. In addition, there are copious exercises, with answers to fully half of these provided at the back of the book. Problems for which answers are given are indicated by asterisks.

A second edition has given me the opportunity to make some changes that should make the text clearer for students. I have rewritten several sections that seemed to me unnecessarily complicated and have tried to streamline presentations wherever possible. In particular I have tried to simplify and clarify the discussion of singular sentences in Unit 10 and have eliminated all but the most cursory mention of relational predicates and definite descriptions, which are discussed in Part Three. I have also substantially rewritten Unit 23, on proof trees, to include more explicit descriptions of the method.

I have added paragraphs on arguments containing constants in the discussions of invalidity for predicate logic. In many units I have added an explicit set of definitions at the end of the unit. In general, I have been able to update examples and exercises, have added new exercises, and have revised those that I found caused confusion.

I have made some changes in organization. Unit 17 now encompasses all the symbolization for relational logic (without identity), and Unit 18 covers just proofs and invalidity. I have replaced Unit 26, which contained some very general remarks on metatheory, with a unit on the stroke and dagger operators and expressive completeness.

Because of the detailed explanations, the extensive coverage, and the clear division of topics, the book is extremely flexible. It can be used in either freshman courses or upper division courses and is suitable for quarter, semester, or even two-quarter courses. In one quarter, for instance, one might cover just Units 1–14; in a semester course, Units 1–15, 17, and 18; and in a two-quarter course one might cover the entire book, including the supplementary units. Because of the step-by-step approach and the numerous examples and exercises, the book can also be used in self-paced classes. Suggestions on how to structure such a course are included in the Instructor's Manual.

It is a great pleasure to acknowledge at this point my considerable debts to the many people who helped make this book what it is. My greatest debt, both in general and in particular, is to Nuel D. Belnap, Jr., from whom I absorbed most of what I know about logic and much of my interest in pedagogy. In addition to these general contributions, the rule system for predicate logic is a slightly modified version of one of his systems. He also read an earlier version of the entire manuscript and made numerous valuable suggestions, most of which have been incorporated here. Without him the book would not have been written, and without his astute commentary it would not have been as useful as I hope it will be.

I would like to thank Nicholas D. Smith, of the Department of Philosophy of Virginia Polytechnic Institute & State University, for the many excellent comments and suggestions he made in reviewing the manuscript. I would also like to thank numerous colleagues at West Virginia University, and students at both WVU and the University of Pittsburgh for many valuable suggestions and for catching many misprints and outright errors. I would especially like to thank Shirley Dowdy, Henry Ruf, and Patricia Long for their time and expertise. Special thanks go to Myrtle Dodge and Linda Garet for their patient, expert, and good-humored typing of a manuscript which must have been at times enormously frustrating. The Lilly Foundation provided partial summer funding during 1977, and I would like to thank that organization, and Gene D'Amour, for the initial impetus for the book. I would also like to thank the West Virginia University Foundation for making possible in-house publication of an earlier, partial version, which encouraged me to complete the project. Finally, I wish to thank the following for their valuable comments when reviewing the manuscript for the present edition: Ronald Glass, University of Wisconsin, La Crosse; Ron Bombardi, Middle Tennessee State University; Keith Cooper, Pacific Lutheran University; and William DeAngelis, Northeastern University.

Finally, this book is dedicated to my mother, Helen Crooker Klenk, who always encouraged my interest in formal studies.

VIRGINIA KLENK

PART ONE: SENTENTIAL LOGIC

UNIT 1

Introduction to Logic

A. INTRODUCTION

If you have never had a course in logic before, you probably have little idea of what it is all about, and even less idea of what is involved in symbolic logic. You may even wonder what the point is of being "logical"; it sometimes gives the impression of a rather cold, dispassionate approach to life, and doesn't really sound terribly appealing. By the end of the course, however, I hope you will have discovered that the study of logic is not only extremely useful, but can also be fun. I think you will find that many of the procedures you will learn here are intrinsically interesting; you may think of them as puzzles or games, and in the process you will be developing your reasoning ability.

What, then, is reasoning ability, and why should you be concerned with developing it? The ability to reason, or infer, is simply the ability to draw appropriate conclusions from given evidence, and this reasoning ability is extremely important in our daily lives because it is the source of most of our knowledge. Most of our knowledge is inferential; that is, it is gained not through direct observation, but by *inferring* one thing from another. A doctor may *observe*, for instance, that a young patient has a high fever, spots on his throat, and generally looks miserable. What he or she may *infer*, however, is that the patient has a strep infection, and on the basis of this inference will perhaps prescribe penicillin in time to ward off more serious consequences.

Inferring is a process of going from what we do know (the premises) to what we previously didn't know (the conclusion); it is a way of expanding our knowledge, and needless to say, it is very important that we understand how to do this correctly. (If the Pentagon incorrectly infers, for instance, that spots on the radar are enemy missiles, rather than geese, we may all be annihilated.)

Logic is primarily about inferring, about reasoning; in particular, it is the study of what constitutes *correct* reasoning. In this introductory unit you will learn what reasoning is, the difference between deductive and inductive reasoning, a preliminary definition of a valid deductive argument, and the crucial role of the concept of *form* in the study of logic. The last two sections contain a discussion of the role of symbols in logic and a general overview of the discipline of symbolic logic.

B. UNIT 1 OBJECTIVES

    1.   Be able to answer all the study questions at the end of the unit.

    2.   Learn the definitions at the end of the unit.

C. UNIT 1 TOPICS

1. Why Study Logic?

To take first things first, why should you be taking a course in symbolic logic? What (aside from fulfilling a requirement) can you expect to get out of it? Will it really teach you to think logically, and is there any advantage in that? Isn't it better, perhaps, to be spontaneous, intuitive, perceptive? Well, in the first place, there need be no conflict between intuition and logic; intuition and perception are appropriate in certain sorts of situations, for instance, in creating or appreciating a work of art, and logic is necessary in other situations, for instance, in figuring out the instructions on your income tax return. What you may not realize is that you are using logic, that is, *reasoning*, continually. Every time you draw a conclusion on the basis of certain evidence, infer one thing from another, or try to figure out the consequences of a certain course of action, you are using logic. Logic is a matter of *what follows from what*, and the better you are at figuring this out—that is, the better you are at reasoning correctly—the more likely you are to come up with the right decision in practical situations.

Let us look at a few examples. Suppose you can get into law school only if you do well on the LSATs and have at least a 3.5 average. You will do well on the LSATs only if you have a course in logic, but a course in logic will lower your grade point average to below 3.5. Can you get into law school? Or, to take a more complicated example, suppose you are making out your income taxes, and the instructions read as follows: "If you are either single, or married but filing separately, you may take the standard deduction of $2400, but only if your adjusted gross income does not exceed $10,000 and you have no income from capital gains. Otherwise you must compute your deductions from Schedule D, if you are self-employed, or Schedule E if you are not self-employed." Now if you are single, self-employed, and have some income from capital gains, but your income does not exceed $10,000, how do you figure your deductions? Or suppose you are on a jury and the following evidence has been presented: the defendant could not have committed the murder unless he had the cash to hire someone, or was in possession of both a car and a gun. Furthermore, he would not have committed the crime unless he was either drunk or on drugs. But he never touches drugs and he did not have a gun. Is the defense attorney right in asserting that the defendant could not have committed the crime? It is in situations like this that logic is indispensable; we need to reason our way through to a decision, and the better we are at reasoning, the better our decisions will be.

But will taking a course in logic really improve your reasoning ability? It should, in at least two important respects. In the first place, you will learn to recognize and use certain very common forms of correct logical inference, and you will come to recognize and *avoid* certain common logical errors. For instance, if all conservatives *oppose* detente with the Soviet Union, and John is *in favor* of detente, you can correctly infer that John is not a conservative. Suppose David opposes detente. Can you infer that he *is* a conservative? Many people would, but this would be a mistake; the premises don't say that conservatives are the *only* people who oppose detente, just that they *all* do. It is quite possible that there are many others, from all parts of the political spectrum, who also oppose detente. This erroneous inference involves a very common fallacy, and once you become aware of such logical mistakes, you should be able to avoid them.

In the second place, logic should increase your ability to construct extended *chains* of reasoning, and to deal with more complex problems. A good example of this occurs in chess, where success depends in large part on the ability to think ahead, to plan moves several steps in advance. Such situations occur as well in real life; you may have to consider a number of options, consequences of each of those options, and consequences of the consequences. The more clearheaded you can be about these inferences, the better the decisions you will be able to make, and what you learn in logic should help you with this sort of complex reasoning.

In addition to the practical advantages of learning to reason more effectively, however, there is also a theoretical side to the study of logic. You will be learning, not just *how* to reason correctly, but also *why* certain forms of inference are correct and others incorrect. It would hardly be appropriate in a course designed to help you learn to think things through to tell you to take it all on authority; if you really learn to think logically, you will certainly want to be asking *why* things are done in a certain way. You may, in the end, not accept all the presuppositions of modern symbolic logic (logic is not as cut and dried as you may think), but at least you should have a good understanding of why they are accepted by so many logicians.

2. What Logic Is All About

You already know that logic is about reasoning, inferring one thing from another, but we need to be somewhat more precise than this. For one thing, it would not do to say that logic is about the reasoning processes which go on in your head, since these are not directly observable by anyone else, and we have, in fact, no way of really knowing what goes on there. Rather, we must say that logic is concerned with the *verbal expression* of reasoning, since this is the only thing which is publicly ascertainable. The term which we will use for this verbal expression of reasoning is *argument*. An argument, for purposes of logic, is *not* a quarrel or disagreement, but rather a *set of sentences consisting of one or more premises, which contain the evidence, and a conclusion, which is supposed to follow from the premises*. An argument can, of course, be spoken as well as written, and the principles of logic apply equally well to any expression of reasoning, but for obvious reasons we will here be concerned primarily with written arguments. In logic it is customary to write an argument with premises above a line and conclusion below, and the sign '∴' is often used to stand for "therefore." The following is an example of the sort of argument we will be concerned with in this book:

> John will not get an A on this exam unless he studied hard for it.
> John did not study hard for this exam.
> _____
>
> /∴ John will not get an A on this exam.

This is still not enough for a definition of logic, however, for it does not tell us what we are supposed to do with arguments. Just write them down? Count the number of words? Admire the calligraphy or typesetting? Should we perhaps do psychological investigations, such as record the effects of various arguments on the listener? Try to determine which arguments people think are good, or which they use most frequently? Try to figure out the psychological reasons why people argue as they do? None of this is the province of logic. *The only thing logic is concerned with is whether arguments are good or bad, correct or incorrect.* Logic is a *normative* enterprise; its job is to *evaluate* arguments, and this is primarily what you will be learning in this course.

What does it mean, then, for an argument to be good or bad, correct or incorrect? What are the grounds on which we evaluate arguments? The answer is somewhat complex, and you will not fully understand for several more units, but we may begin by noting that, in an argument, a *claim* is being made that there is some sort of evidential relationship between premises and conclusion: the conclusion is supposed to *follow from* the premises, or equivalently, the premises are supposed to *imply* the conclusion. This indicates that the correctness of an argument is a matter of the *connection* between premises and conclusion, and concerns the *strength of the relation* between them. We will evaluate an argument, then, on the basis of whether this evidential claim is correct, on whether the premises do in fact support, or provide evidence for, the conclusion. It is extremely important to realize that the correctness of an argument depends upon the *connection* between premises and conclusion, and *not* on whether the premises are true or not. This connection may be extremely strong even though the premises and conclusion are all false. It is generally the task of other disciplines to assess the truth or falsity of particular statements— biology, for instance, is concerned with the truth of statements about living organisms—while logic explores only the *relationship* between premises and conclusion.

In the most logically compelling kind of argument, a *valid deductive argument*, we have the strongest conceivable kind of relationship; it is absolutely impossible to have all the premises true with the conclusion false, which is to say that under all possible circumstances, if the premises are true, then the conclusion will be true as well. In other words, *the truth of the premises absolutely guarantees the truth of the conclusion.* Before we go into more detail about deductive validity, however, we must mention briefly a kind of argument in which the evidential relationship is not this strong, in which the truth of the premises does not guarantee the truth of the conclusion, but only makes it probable to some degree. Such arguments are called *inductive*, and will be discussed in the next section.

3. Induction and Deduction

The distinction between a deductive and an inductive argument lies in the *intended strength of the connection* between premises and conclusion. Since we cannot always be sure what the intention is, this distinction is not very precise, but for our limited purposes it will be enough to say that *in a deductive argument the premises are supposed to provide an absolute guarantee of the conclusion, while in an inductive argument the premises are only supposed to provide some degree of support for the con-clusion.* The conclusion of an inductive argument is thus often indicated by the term "probably," or some such word.

In inductive logic, for instance, if we have as premises that John gets A's on 95% of the exams for which he studies, and that he did study for this exam, we may conclude that he will *probably* get an A on this exam. Or from the fact that 99.99% of commercial airline flights are completed without incident, we may correctly infer that the next airplane we take will *almost certainly* arrive safely. But it is important to note that in these arguments we cannot be *absolutely* sure of the conclusion even though we know the premises to be true. (This is, of course, why some people refuse to fly.) There is always some "slippage" or "logical gap," between premises and conclusion in an inductive argument. It is always possible for the conclusion to turn out false, even though the premises are all true. This is the nature of induction as opposed to deduction. This does not make inductive arguments bad arguments, however; we are perfectly justified (at least according to most philosophers) in using in-ductive arguments for practical purposes, and in concluding, for instance, that our plane will land safely, even though we cannot really be 100% sure. Most of the arguments we use in everyday life, in fact, are inductive rather than deductive, and we would be unnecessarily restricting our inferential powers if we refused to use them. The use of induction in science and ordinary life seems to be essential.[1]

Nevertheless, there is no doubt that a deductive argument, when available, is the argument of choice. This is because in deduction, given that the premises are true, we can be absolutely sure that the conclusion will be true. If we know, for instance, that John cannot get an A on the exam unless he studies hard, and we know that he did not study hard, then we can be sure that he will not get an A. Here there is no possible gap between premises and conclusion; the argument is logically tight (unfortunately for John).

It should be noted that in *deductive* logic we have an either/or sit-uation; either the premises do provide absolute support for the conclusion, in which case the argument is valid, or they do not, in which case it is invalid. In *inductive* logic, on the other hand, the "goodness" of an ar-gument is a matter of degree. We may be 99.99% sure of the conclusion, as in the case of our next airplane trip, or we may have considerably less confidence in the conclusion. There is really no point at which we can say that an inductive argument is acceptable, at which we can put confidence in the conclusion. Do we have to be 99.99% sure? 90% sure? 70% sure?

---

[1] It should be noted, however, that some philosophers, in particular Sir Karl Popper, have argued that there is no such thing as induction, and that all knowledge is based upon deductive reasoning. This controversy, while extremely interesting, would be more appro-priately discussed in a course in the philosophy of science.

Or is better than 50-50 good enough? It depends partly upon the situation. Inductive logic is not clear-cut; there are no sharp boundaries between good and bad arguments.

In this book we will be concerned exclusively with *deductive* logic, partly because it is so clear-cut, partly because it is easier and thus better suited to an introductory course in logic, and partly because there is at present no generally accepted system of inductive logic.[2] Some philosophers even claim that there is no such thing as correct induction, no such thing as a good inductive argument, and even among those who do believe in it, there is widespread disagreement on how it should be formulated.[3] This kind of controversy need not concern us, however, and we will confine our interests to the relatively undisputed areas of deductive logic.

4. Form and Validity

In deductive logic, as noted above, there are only two sorts of arguments: good, or valid, and bad, or invalid. A valid deductive argument is one in which the truth of the premises absolutely guarantees the truth of the conclusion, in which there is no *possible* way the premises could all be true but the conclusion false. Another way of putting this is to say that necessarily, if the premises are all true, then the conclusion will be true as well. An invalid argument, by contrast, is one in which it *is* possible for the premises all to be true but the conclusion false.

But this is all rather abstract, and if we take particular examples of arguments, these definitions may seem somewhat puzzling. The first argument below, for instance, is valid, and the second is invalid.

1. All horses are green.

   All green things are plastic.

   /∴ All horses are plastic.

2. If Reagan was president, then Bush was vice-president in 1986.

   Bush was vice-president in 1986.

   /∴ Reagan was president in 1986.

Now, what can it mean to say that, in the second example, it is *possible* for the premises to be true and the conclusion false, when in fact the conclusion is true? And what could it mean, in the first example, to say that *if* both the premises were true then the conclusion would be true

---

[2] Inductive logic needs to be sharply distinguished from probability theory, which is a deductive, mathematical system about which there is no disagreement, and from applied statistics, on which there is at least practical agreement.

[3] Although there is less disagreement about deductive logic than there is about inductive logic, there are still some disputes. Not all logicians accept all of the results of the "standard" logic we will be discussing in this book; in particular, there are many questions about the interpretation of "if-then." Alternatives to the standard approach might be studied in a second course in logic.

as well, when in fact both premises and conclusion are false? Furthermore, how can we ever be sure, in a valid argument, that there could be no possible way for premises to be true and conclusion false? How could we possibly prove such a thing?

The answer to all these questions lies in the concept of *form*, or structure. To say that an argument could not possibly have true premises with a false conclusion is simply to say that it has a certain kind of *form*, a form which admits no instances of that kind. To say that it *could* have true premises with a false conclusion, on the other hand, is to say that the argument has a *form* for which there are (other) possible instances which do have true premises with a false conclusion. And as we shall see, it is relatively easy (at least for the kind of logic we will be doing in the first half of the course) to determine whether a form has these invalidating instances. It is the *form* of an argument, then, which determines its logical validity, and symbolic logic turns out to be primarily the study of the abstract forms and structures used in argumentation.[4]

What, then, is logical form, and how does this concept help us determine questions about validity? Perhaps the simplest approach is to start with the notion of two arguments having the *same* form, and from there we can abstract the concept of form itself. If we consider the following two arguments, we can see that although they have completely different subject matter, they do have something in common: they have exactly the same structure, or pattern.

3. Either Reagan was elected in 1980 or Carter was elected in 1980.

Carter was not elected in 1980.

/∴ Reagan was elected in 1980.

4. John is either at the movies or at home.

John is not at home.

/∴ John is at the movies.

The pattern of both arguments is "Either p or q, not q, therefore p," where the lowercase letters p and q stand for simple sentences, such as "Reagan was elected in 1980" and "John is at home." These arguments have been analyzed according to *sentential logic*, which you will be studying in Units 2 through 9. In sentential logic, complete simple sentences, such as "John is at home," are taken as unbroken units, and are not analyzed into their component parts. The form of a compound sentence or argument

---

[4] It should be noted that there are some arguments whose validity depends upon the meaning of certain key words. An example would be: John is a bachelor /∴ John has no wife. Here it is impossible for premise to be true and conclusion false because being a bachelor *means* being unmarried, which means (for a man) having no wife. We might call this *definitional* validity, validity which depends upon meanings, as opposed to what we will call *logical* validity, which depends upon the form of an argument. It is only *logical* validity which we will be discussing in this book.

is then determined by how these simple sentences are combined with certain logical words, such as "and," "or," and "not."

In *predicate logic*, presented in Units 10 through 20, we will introduce three other logical words, "all," "some," and "equals," and another kind of variable, which represents individual objects rather than sentences. The analysis of form in predicate logic is somewhat more complex than in sentential logic, but the idea is the same: the form of a sentence is the way in which certain specified logical words are combined with the other elements of the sentence.

It should be clear that the following two arguments in predicate logic have the same form, though the form here contains the predicate logic words "all" and "some." Note that the tense is irrelevant.

5. All U.S. presidents have been male. Some U.S. presidents have been generals.

/∴ Some males are generals.

6. All snakes are reptiles. Some snakes are poisonous.

/∴ Some reptiles are poisonous.

Here the common form is "All *A*'s are *B*'s, some *A*'s are *C*'s. /∴ Some *B*'s are *C*'s," where *A*, *B*, and *C* represent class terms: common nouns.

In order to see how form determines validity, we must be very clear about the distinction between a *form* and its *instances*. The form is the general pattern, or structure, which abstracts from all specific subject matter, whereas the instances are the particular meaningful examples which exhibit that form. In examples 3 and 4, as noted, the *form*, or *pattern*, or *structure* of the two arguments is "Either *p* or *q*, not *q*, therefore *p*," and each of the two particular arguments is said to be an *instance* of that form, which simply means that it exhibits that form. In sentential logic, an instance is obtained from a form by substituting meaningful sentences (consistently) for the *p*'s and *q*'s, while in predicate logic instances are obtained by substituting class terms for the capital letters *A*, *B*, and *C*.

Having made the distinction between form and instance, we can now be a little more precise in our definition of validity. We will need to distinguish between the validity of an argument (instance) and the validity of a form, and the former definition will be dependent upon the latter. An *argument (particular instance) will be said to be valid if and only if it is an instance of, or has, a valid form. An argument form will be valid if and only if there are no instances of that form in which all the premises are true and the conclusion is false. A form will be invalid just in case there is an instance of that form with true premises and a false conclusion.*

The following two arguments, from sentential logic, which have the same form as argument 2, are invalid.

7. If coal is black, then snow is white.

Snow is white.
_____

/∴ Coal is black.

8. If the author of this book is a monkey, then the author of this book is a mammal.

The author of this book is a mammal.
_____

/∴ The author of this book is a monkey.

Again, both arguments have the same form; in this case, "If p then q, q, therefore p," where p and q stand for the simple sentences, such as "Snow is white." But in this case we can see that there *is* an invalidating instance of the form; the second argument actually *does* have all the premises true but the conclusion false. (You will have to take the author's word for this.) *Such an invalidating instance of a form, an instance with all true premises but a false conclusion, will be called a counterexample to that form*, and *any form which has a counterexample will be invalid.*

Argument 9 below, from predicate logic, is also invalid. We can see that it is invalid by extracting its form, and then finding class terms for A, B, and C that give us true premises with a false conclusion, as in argument 10. The common (invalid) form is "All A's are B's, some B's are C's, so some A's are C's." (The similarity of this form to the valid predicate form of examples 5 and 6 shows how careful we must be in using logic. A seemingly minor change in form can mean the difference between validity and invalidity. Note also that it is not obvious at first glance whether an argument is valid or invalid; argument 9 initially looks quite plausible.)

9. All cats are fur-bearing mammals.
Some fur-bearing mammals are black.
_____

/∴ Some cats are black.

10. All U.S. presidents have been males.
Some males have been on the moon.
_____

/∴ Some U.S. presidents have been on the moon.

An argument form will be *valid*, as noted, if and only if it has *no* counterexample; no instance with true premises and a false conclusion. We will not be able to show that argument forms are valid until we have the means of systematically examining all possible instances; you will learn how to do this in Unit 5, and a more detailed explanation of validity will have to be deferred until we reach that unit.

5. Truth and Validity

The single most important moral of the above story is that *the validity of a particular argument (the instance) is dependent upon its form*; again, a particular argument will be valid if and only if its form is valid. Since form has nothing to do with subject matter, it follows that what an argument says, its content, is irrelevant to its validity. Silly-sounding arguments may be valid, and weighty-sounding arguments may be invalid. Furthermore, the truth or falsity of the premises and conclusion are also irrelevant to the validity of an argument, with one exception: a valid argument may never have true premises and a false conclusion. (This would, by definition, be a counterexample, and thus would make the argument form, and thus the argument, invalid.) All other combinations are possible. Some examples of valid and invalid arguments, with various combinations, are given below. (Arguments may, of course, have more or fewer than two premises; the ones below all have two for the sake of uniform comparison. Note that some are from sentential logic and some are from predicate logic.)

VALID ARGUMENTS

1. All cats are green. (F)
   All green things are immortal. (F)

   /∴ All cats are immortal. (F)

2. All cats are reptiles. (F)
   All reptiles have fur. (F)

   /∴ All cats have fur. (T)

3. Either Carter or Bush was
   vice-president in 1986. (T)
   Bush was not vice-president
   in 1986. (F)

   /∴ Carter was vice-president
   in 1986. (F)

4. If dogs don't bark,
   then cats meow. (T)
   Cats do not meow. (F)

   /∴ Dogs bark. (T)

INVALID ARGUMENTS

5. Some Republicans are
   wealthy. (T)
   Some Democrats are not
   wealthy (T)

   /∴ Some Democrats are not
   Republicans. (T)

6. If that tree with
   acorns is a maple,
   then it's deciduous. (T)
   That tree with
   acorns is deciduous. (T)

   /∴ That tree with
   acorns is a maple. (F)

7. No Democrats are wealthy
   people. (F)
   No wealthy people have
   visited Mars. (T)

   /∴ No Democrats have
   visited Mars. (T)

8. Either dogs bark or
   cats bark. (T)
   Cats bark. (F)

   /∴ Dogs bark. (T)

It may seem a bit strange to talk about valid arguments with false premises and false conclusions, but in fact we very often want to see what follows—what can be deduced from—false premises. A scientist may try to figure out, for instance, what would happen *if* a ten-megaton atomic bomb were dropped on New York City, though we certainly hope one won't be. More optimistically, an economist may try to deduce the economic consequences of a breakthrough in solar energy technology, though unfortunately none has yet occurred. Sometimes we simply do not know whether the premises are true or not, as when we are testing some unknown chemical compound or a new engineering design. The use of logic allows us to figure out ahead of time what *would* happen *if* certain hypotheses or premises were true, and this is obviously essential in all kinds of situations. Think how expensive it would be if, instead of sitting down at a computer and simply *deducing* the flight characteristics of certain designs, an aeronautical engineer had to actually build an example of each new design he or she thought up and test it out!

Of course, some valid arguments do have true premises, and there is a special term which is used for these cases. Such arguments, with true premises (and therefore also with true conclusions), are said to be *sound*. A sound argument is "one up" on a valid argument, since in addition to being valid it also has true premises and a true conclusion. In general, however, logicians are not concerned with the truth of the premises of particular arguments; their job is to determine *validity*. This, again, is not a matter of the truth or falsity of premises and conclusion, but of the *connection* between them. Logic is concerned solely with whether the conclusion *follows from* the premises, and this, as we have seen, is a matter of the *form* rather than of the truth, falsity, or content of an argument.

6. The Nature of Symbolic Logic

Detailed systems of logic have existed at least since the time of Aristotle (384 to 322 B.C.), but symbolic logic is largely an invention of the twentieth century. The advantages of using symbols in logic are the same as in mathematics; symbols are easier to manipulate, they provide an economical shorthand, and they allow us to see at a glance the overall structure of a sentence. By using symbols we are able to deal with much more complicated arguments, and thus take logic much further than we otherwise could. In fact, since the development of symbolic logic there has been an explosion of knowledge in this area, and fascinating developments which would not have been possible without it. (In some of the extra credit sections you will be introduced to some of these more advanced topics.)

In this book we use two sets of symbols, corresponding to the various kinds of form we will be investigating. For the first half of the book, which covers sentential logic, we use an extremely easy symbolic system, consisting only of (1) single letters to stand for simple sentences, such as "Jane is blond," (2) special symbols for the five logical words, "and," "or," "not," "if-then," and "if-and-only-if," and (3) grouping symbols such as parentheses. We will be concerned only with how simple sentences are compounded into more complex sentences by using the five logical words, and we will not attempt an analysis of the simple sentences. In the second half of the book, which covers predicate logic, we go a bit deeper into sentence structure, and analyze the simple sentences into their component parts, such as subject and predicate. For this we need a few additional symbols, but the symbolism will remain surprisingly simple. Even students who have "hang-ups" about symbols need not worry about symbolic logic; the symbols are quickly grasped, and they are simply a way of making logic much easier than it would otherwise be.

There are some limitations to the use of symbols. Since we have only five symbols to represent ways of compounding sentences, and English is an extremely subtle and complex language, something is bound to be lost in translation. We are going to have to squeeze all the richness of English into just a few symbols, and this will mean, unfortunately, violating the full sense of the sentence on many occasions. What logic lacks in subtlety, however, it more than makes up for in clarity, and you will probably appreciate the ease of manipulating the logical system. Just don't look for poetry, or you will be disappointed.

7. The Scope of Symbolic Logic

a. Levels of logical structure.   As noted earlier, there are various levels of logic, depending on how deeply we analyze the structure of the English sentences. At the simplest level, *sentential* or *propositional logic*, we analyze only how complete sentences are compounded with others by means of logical words such as "and," "or," and "not," which are called *sentential operators*. We do not attempt an internal analysis of simple sentences in terms of their grammatical parts such as subjects and predicates. In this book, sentential logic is discussed in Units 2 through 9.

In *one-variable predicate logic*, presented in Units 10 through 16, our logical analysis goes deeper, into the internal structure of the simple sentences, and instead of single letters to represent complete sentences, we have various symbols to represent the various parts of a sentence. We have *individual terms* to represent single individual objects, *predicate letters* to represent predicates (roughly, what is said about an individual object), and two new logical words, the very important *quantifiers* "all" and "some." With these new symbols we can represent the internal structure of sentences such as "John is mortal" and "All men are mortal."

In one-variable predicate logic, the predicates used make sense only when applied to single individuals; examples of such predicates are "is blond," "is mortal," and "is over six feet tall." There are many predicates, however, which are used to state a relationship between two or more individuals. Examples of such predicates are "is taller than" and "lives next to," which are two-place predicates used to state a relation between two things, and "between," which is a three-place predicate used to state a relation between three things. In *relational predicate logic*, the third level we discuss, we simply add these relational predicates to the already existing machinery of one-variable predicate logic. Thus, relational predicate logic is simply an extension of one-variable predicate logic. Relational predicate logic is the topic of Units 17 and 18.

Finally, the fourth and last level we discuss is *relational predicate logic with identity*, and here we just add one more element: the very important logical relation of identity. The only new symbol needed is the familiar symbol for identity, the "equals" sign '='. Relational logic with identity is covered in Units 19 and 20.

b. Four kinds of logical investigation.

For each of the four levels of logic mentioned above there are four kinds of inquiry which must be undertaken. We must first ask what is the *formal structure* of the logical language; such an investigation might be called the *logical grammar* of the system. Here we must make very clear exactly which symbols are to be used, and how they fit together properly into meaningful formulas. The grammar of sentential logic is discussed in Unit 2, and the grammar of one-variable predicate logic, relational predicate logic, and predicate logic with identity is explored, somewhat diffusely, in Units 10 through 12, 17, and 19, respectively.

Once we know the structure of the symbolic system, we must see how it is reflected in ordinary English sentences and arguments. We might call this the *application* of the logical system, and it includes being able to put the English sentences into symbolic form, as well as being able to read off from the symbolic form the appropriate English sentence. For this you will always be provided with a "dictionary" which links simple symbolic elements with their English counterparts. In Unit 4 we undertake a careful analysis of the application of sentential logic to English, and in Units 10 through 14 and 17 through 20 we cover various aspects of the application of predicate logic to English.

The third kind of inquiry we must undertake for any branch of logic is its *semantics*. Here, as the title indicates, we explain exactly what our logical words such as "and," "not," and "all" mean. These meanings are given by stating precisely the conditions under which sentences containing them will be true or false; we give the logical meaning of "and," for instance, by saying that a conjunction "*p* and *q*" is to be true if and only if both the components *p* and *q* are true. There are a great many other logical concepts which can be explained in semantic terms, that is, in terms of their truth conditions. We have already seen that *validity* is defined in terms of the possible truth combinations for premises and conclusion: an argument form is valid if and only if there is no possible instance of that form which has all the premises true with the conclusion false. Other semantic concepts are *consistency, equivalence,* and *contingency.* We examine the semantics for sentential logic in considerable detail in Units 3, 5, and 6, and discuss the semantics for predicate logic in Unit 16.

Finally, the fourth part of a study of any branch of logic is its *proof methods.* Here we set out formal rules for deriving, that is, proving, certain symbolic formulas from others. It is interesting that this procedure is theoretically independent of any semantics—we can learn to do proofs of formulas without even knowing what they mean. The proof methods for sentential logic are developed in Units 7, 8, and 9; predicate logic proof methods are discussed in Units 15, 18, and 20. In Unit 9 we discuss the relationship between proof methods and semantics for the various branches of logic. The material covered in Units 1 through 20—the grammar, applications, semantics, and proof methods of the four branches of logic—forms the solid core of symbolic logic. As noted earlier, there are many extensions of and alternatives to this basic logic, which you might study in a more advanced course.

DEFINITIONS

1. An **argument** is a set of sentences consisting of one or more premises, which contain the evidence, and a conclusion, which should follow from the premises.
2. A **deductive argument** is an argument in which the premises provide an absolute guarantee of the conclusion.
3. An **inductive argument** is an argument in which the premises provide some degree of support for the conclusion.
4. An **argument** (particular instance) *is valid* if and only if it is an instance of, or has, a valid form.
5. An **argument form is valid** if and only if there are no instances of that form in which all premises are true and the conclusion is false.
6. A **counterexample to an argument form** is an instance of that form (a particular example) in which all the premises are true and the conclusion is false.
7. A **sound argument** is a valid deductive argument in which all the premises are true.

STUDY QUESTIONS

1. What is the advantage of thinking logically?
2. What are the two ways in which the study of logic can improve your reasoning ability?

3. What theoretical aspect of logic can you expect to learn in this course?
4. Give a brief statement of what logic is about.
5. What is a valid deductive argument?
6. What is the difference between deductive and inductive arguments?
7. Give an example of two different arguments with the same form.
8. How is the form of an argument related to its validity or invalidity?
9. Can a valid argument have false premises with a true conclusion? What other combinations are possible?
10. Give an example of your own of an *in*valid argument with true premises and a true conclusion. Give an example of your own of a valid argument with false premises and a false conclusion.
11. What are the advantages of using symbols in logic?
12. What is one disadvantage of using symbols in logic?
13. What are the four branches, or levels of logic, and what are the fundamental differences between them?
14. What are the four areas of investigation for a branch of logic? Describe each briefly.

EXERCISES*

Decide whether each of the arguments below is valid or invalid. If invalid, give a counterexample. After you have completed the assignment (and *only* after), check the answer section to see which of these are valid. If you made mistakes, this should demonstrate the need for a systematic study of logic. Our intuitions are often wrong!

*1. Arguments in sentential logic:

a. Either Reagan or Bush was president in 1986. Bush was not president. Therefore, Reagan was president.

b. Not both Reagan and Bush were president in 1986. Bush was not president. Therefore, Reagan was president.

c. Not both Carter and Mondale were president in 1986. Carter was not president. Therefore, Mondale was not president.

d. If Whiz is a male cat, then Whiz will not have kittens. Whiz will not have kittens. Therefore, Whiz is a male cat.

e. If Whiz is a male cat, then Whiz will not have kittens. Whiz will have kittens. Therefore, Whiz is not a male cat.

f. If Whiz is a male cat, then Whiz will not have kittens. Whiz is not a male cat. Therefore, Whiz will have kittens.

g. If Whiz is either a cat or a dog, then Whiz is a fur-bearing mammal. Therefore, if Whiz is a cat, then Whiz is a fur-bearing mammal.

* The answers to all exercises preceded by an asterisk appear in the back of the book.

h. If I neither diet nor exercise, I will gain weight. Therefore, if I do not diet, I will gain weight.

i. If I exercise, I will neither gain weight nor lose muscle tone. Therefore, if I exercise, I will not lose muscle tone.

j. If I don't exercise, I will gain weight and lose muscle tone. Therefore, if I don't gain weight, I have exercised.

*2. Arguments in predicate logic:

a. Some Republicans are wealthy. Some wealthy people are not Democrats. Therefore, some Republicans are not Democrats.

b. All U.S. presidents have been men. No man has experienced childbirth. Therefore, no U.S. president has experienced childbirth.

c. All U.S. presidents have been men. No U.S. president has experienced childbirth. Therefore, no man has experienced childbirth.

d. Some cats are not black. Therefore, some cats are black.

e. No cats are dogs. No dogs are horses. Therefore, no cats are horses.

f. Not all corporate executives are men. All corporate executives are wealthy people. Therefore, some wealthy people are not corporate executives.

g. Not all corporate executives are men. All corporate executives are wealthy people. Therefore, some wealthy people are not men.

h. All frogs are reptiles. Some reptiles are poisonous. Therefore, some frogs are poisonous.

i. Some preachers are wealthy people. All wealthy people will get to heaven. Therefore, some preachers will get to heaven.

j. Some wealthy people will get to heaven. No one who gets to heaven has committed a mortal sin. Anyone who commits murder has committed a mortal sin. Therefore, some murderers are not wealthy.

COUNTEREXAMPLES:

▷ Some men are mammals.
Some mammals a 4-legged animals.
Therefore, some men are 4-legged animals.

↳ ▷ No men are wives.
No wives are husbands.
∴ No men are husbands.

UNIT 2

The Structure of
Sentential Logic

A. INTRODUCTION

As we saw in the first unit, deductive logic is about correct, or valid, inference, and validity is a matter of form, or structure. In this unit you will learn the basic structure, or grammar, of one branch of symbolic logic, what is generally called *sentential*, or *propositional*, logic. The names derive from the fact that in this part of logic our most basic or elementary unit is the complete sentence, such as "Jane is blond" or "Swans are graceful." What you will be studying is the way in which these complete sentences are compounded into more complex sentences, such as "If interest rates rise and the price of lumber continues to escalate, then the housing industry is in trouble." There is a very large class of arguments whose validity depends solely upon this kind of compound structure, and the first half of the book is devoted to these relatively simple structures.

It is important to note that although there are such studies as the logic of questions and the logic of commands, we will be confining ourselves to the logic of *declarative* sentences, those which are definitely either true or false, such as "Salamanders are mammals" (which is, of course, false). Whenever the word "sentence" is used hereafter, it should be understood that we are referring to declarative sentences. More will be said about this in Unit 5, when we begin truth tables.

Any expression which is used to build up a compound sentence out of simpler sentences will be called a *sentential operator*, because it "operates" on sentences to produce more complex forms. Although there are potentially an infinite number of ways of forming compound sentences in English—that is, an infinite number of sentential operators—we will be using only five. These five operators, for which we will have special symbols, are "and," "or," "not," "if-then," and "if-and-only-if." Some examples of sentences using these five operators, with the operators italicized, are "*Either* I study hard for this exam *or* I won't get an A," "*If* math majors can add *then* John is *not* a math major," and "Inflation can be curbed *if and only if* labor cuts back on its demands *and* corporations quit being so greedy."

In this unit, then, you will learn about the basic structure and symbolism of sentential logic. Our discussion includes the difference between simple and compound sentences, the definition of "sentential operator," the five sentential operators we will be using, and how compound sentences are constructed from their components by the use of our five operators. What you will need to learn is stated somewhat more explicitly in the "Objectives" section below.

B. UNIT 2 OBJECTIVES

1. Be able to distinguish between simple and compound sentences, and be able to identify the simple components of compound sentences.
2. Learn the definitions of "sentential operator," "compound sentence," and "simple sentence," and be able to give examples of operators other than the five we will be using.
3. Learn our five sentential operators and the symbols for them.
4. Learn how compound formulas are constructed out of more elementary parts.

C. UNIT 2 TOPICS

1. Simple and Compound Sentences

It is absolutely essential in logic to be able to analyze the structure of sentences and arguments, since validity is a matter of form, or structure. Of course, there are various *levels* of structural analysis, but for the first half of the book, we are going to do only the most elementary sort of analysis. We are going to be concerned only with how complete sentences are compounded with others by words such as "and," and "or," into more complex forms. We will not analyze sentences into their "inner" elements such as subjects and predicates (this will come later, in predicate logic), but will take them as unbroken wholes, our smallest units.

The first step in analyzing sentential structure is to be able to distinguish between simple sentences and compound sentences, and to be able to identify the simple components of compound sentences. We begin by defining "compound sentence" and then define a simple sentence as one that is not compound.

A declarative sentence will be considered to be compound *if it contains another complete declarative sentence as an independent clause.* (An independent clause is simply a clause which can stand alone as a separate, meaningful sentence. A dependent clause, by contrast, is not a meaningful sentence if it stands alone.) The following sentence contains an independent clause, and so is compound: "John believes that Mary loves David." The clause "Mary loves David" is a complete sentence in its own right, which appears as a part of the larger sentence.[1] The sentence "The person who ate the cake has a guilty conscience," on the other hand, is not compound because it has no independent clauses. The phrase "who ate the cake" is not a separately meaningful declarative sentence. *Any independent clause (simple or compound) which appears as a part of a larger sentence will be called a* component *of the compound sentence.* Some examples of compound sentences are given below, with their simple components italicized. "*John likes Mary* and *Mary likes John.*" "*Nixon resigned* after *the incriminating tapes were made public.*" "Either *you tell me the truth* or *we're through.*" It is not true that *Mary won first place in the race.*" "*It is raining* because *the clouds were seeded.*"

A *simple sentence*, as noted earlier, is one which is not compound, one which does not contain another complete sentence as an independent clause. Some examples of simple sentences are: "John is going to New York," "Mary is a good student," "Manx cats are friendly," and "Dolphins are highly intelligent." It is important to realize that rather complicated-looking sentences may still be simple in our sense of the word, containing no other sentences as parts. The subject may be modified in various ways, and the predicate may say something rather intricate, requiring lengthy phrases, but as long as no other complete sentence appears as an independent clause, the sentence is still logically simple. The following is an example of a rather lengthy but simple sentence: "The odd-looking person standing to the right of the woman with the weird hat with the flowers

---

[1] Note that "compound" does *not* mean simply having parts. All sentences have parts; the question is what *kind* of parts a sentence has. A sentence is compound if it has *another complete sentence* as a (proper) part.

and cherries on it is the one who infuriated the chairman of the board by complaining in public about the high prices of the company's inferior products." Not all simple sentences are short, and as we shall see, not all short sentences are simple. The criterion of a simple sentence is decidedly not its length.

With some sentences, such as "Mary likes Bob and Bob likes Jane," it is relatively easy to see that the sentences are compound and to pick out their components. It is not always such an easy matter, however, to determine whether a sentence is simple or compound. "John and Mary like fish," for example, does not literally contain another sentence as a part. ("Mary like fish" perhaps comes closest, but this is not a grammatical sentence.) In order to clarify the concept of a compound sentence we need to be a little more precise about, and extend our notion of, what it means for one sentence to *contain* another as a component. In English, if the predicates of the components of a compound sentence are the same, as in "John went to New York and Mary went to New York" (where the predicates "went to New York" are identical), we tend to compress, or condense, the compound by using a compound subject rather than repeating the predicate for each subject. Thus, in place of the sentence above, we would probably use the more graceful form "John and Mary went to New York," with the compound subject "John and Mary." If the *subject* of the independent clauses is the same, we may condense the compound sentence by using a compound predicate. Thus, instead of saying "John went to France and John went to Italy and John went to Spain and John went to Siberia," we would usually say, using a compound predicate, "John went to France, Italy, Spain, and Siberia."

A compound subject is one which includes more than one individual or group, such as "John and Mary," "Cats and dogs," "Bill or John," etc. Examples of sentences with compound subjects (with the subjects italicized) are: "*Dogs and cats* make good pets," "*Joggers and tennis players* have to be in good physical condition," "*Johnson or Agnew* resigned in disgrace." To say precisely what a compound predicate is, we need to know a little more about the structure of our logic, but for now we can say roughly that it is one which says more than one thing, or which makes a compound claim, about the subject. Examples of compound predicates are "is lucky or intelligent" and "loves puppies and kittens." Sentences with compound predicates (with the predicates italicized) are: "Mary will be *a good student or a good tennis player*" and "John *will take first place in the race but will be sick for days afterwards*." Of course, sentences may contain both compound subjects and compound predicates. An example of such a sentence would be "*The Democratic presidential and vice-presidential candidates* will *either be defeated or will win by a narrow margin*." Most sentences with compound subjects or predicates can be considered to be compound rather than simple. (The exceptions will be noted below.)

Sentences with compound subjects and/or predicates will be considered to be compound if they can be paraphrased into sentences which are explicitly compound, which literally contain other sentences as components. Thus, our previous sentence "John and Mary like fish" is compound because it can be paraphrased into the longer version "John likes fish and Mary likes fish," which explicitly contains the two clauses "John likes fish" and "Mary likes fish." We may say that *one sentence logically contains another if it literally contains the other as a part, or if it can be paraphrased into an explicitly compound sentence which contains the other as a part.* We may now define a *compound sentence* as *one which logically contains another sentence as a component.* The sentence "Kennedy and Mondale are Democrats and liberals," for instance, is compound and logically contains the following as components: "Kennedy is a Democrat," "Mondale is a Democrat," "Kennedy is a liberal," and "Mondale is a liberal."

As noted, however, not every sentence with a compound subject can be considered to be compound, because not all such sentences can be paraphrased into explicitly compound sentences. "John and Mary are friends," for instance, has a compound subject (by most grammatical reckoning), but it cannot be correctly paraphrased as the conjunction "John is a friend and Mary is a friend." It does not state facts about John and Mary separately, but rather states a relationship between them. In order to state this relationship, "John" and "Mary" must both appear in the same clause, and no paraphrase which separates them can say the same thing. In some cases you may not be able to tell whether the sentence is genuinely compound or is just stating a relationship between two individuals. The ambiguous sentence "John and Mary are married," for instance, might mean simply that they both have spouses, in which case it would be a compound sentence, or it might mean that they are married to each other, in which case it would be stating a relationship between them and would not be a compound sentence. This demonstrates that the art of paraphrase is not exact. In many cases, you will simply have to use your own best judgment in determining whether a sentence with a compound subject can be paraphrased as a compound sentence or is simply stating a relation.

Although it may seem a little strange at first, it will be important to remember that all *negated* sentences are to be considered compound. "John is not happy," for instance, has as its simple component "John is happy," since it can be paraphrased into the longer expression "It is not the case that John is happy." When we come to symbolizing, in Unit 4, only positive sentences will be considered simple; negated sentences, as well as those with compound subjects and predicates (except for the relational sentences mentioned above), will be considered to be compound.

The following sentences are all simple because they logically contain no independent clauses as separately meaningful components:

John is happy.

Dogs like bones.

Children fight a lot.

John likes bananas with cheese.

The man standing by the door is a doctor.

The third president of the United States ate raw beef smothered with raw onions.

The following sentences are all compound. Their simple components are listed in parentheses immediately following the sentence.

John and Mary like cats. (John likes cats; Mary likes cats)

John likes cats and snakes. (John likes cats; John likes snakes)

Harvey thinks that the earth is flat. (The earth is flat)

It is possible that John is cheating on his girlfriend. (John is cheating on his girlfriend)

It is not snowing. (It is snowing)

If there are flying saucers, then fish live in trees. (There are flying saucers; Fish live in trees)

John, Mary, and Harvey like lobster and Big Macs. (John likes lobster; Mary likes lobster; Harvey likes lobster; John likes Big Macs; Mary likes Big Macs; Harvey likes Big Macs)

Exercises at the end of the unit will give you practice in recognizing simple and compound sentences and picking out the components of the compound sentences.

2. Sentential Operators

Sentential logic, as noted earlier, is concerned only with the way in which simple sentences are combined by means of sentential operators into more complex sentences. We now need to explain more precisely what a sentential operator is. We have indicated that it is an expression used to build up more complex sentences from simpler ones, but this is somewhat vague. An operator should be written with blanks to indicate where the sentences should be placed to form a proper compound, since, if they go in the wrong place and combine in the wrong way with the operator, the result will be nonsense. Thus, we will formally define "sentential operator" in the following way: *a sentential operator is an expression containing blanks such that when the blanks are filled with complete sentences, the result is a sentence.* The conjunction operator, then, would be _____ and _____, and a sentence using this operator would be (with the component sentences italicized) *"The college is bank-rupt* and *all the faculty are laid off."*

As we have noted in the introduction to this unit, there are an in-definite number of sentential operators in English, though we will be using only five in our development of sentential logic. A few of these operators are listed below, with blanks, to give you some idea of the variety.

John believes that ___
John knows that ___
John hopes that ___
John heard that ___
(Clearly, we can get an indefinite number of other, similar operators from these simply by substituting a different proper name. There are also many other phrases which will yield similar operators.)

It is possible that ___
It is necessary that ___
It is likely that ___
It is not true that ___
(We could get an infinite number of operators here simply by substituting different numerical values, such as "it is 56.3% probable that.")

Either ___ or ___
Neither ___ nor ___
___ and ___
If ___, then ___
___ if and only if ___
___ unless ___
___ after ___
___ only if ___
___ because ___

Note that the first eight of these operators contain only one blank, for a single sentence, whereas the last nine have two blanks, and so combine two sentences. The former are called "one-place" operators, and the latter, as you might expect, are called "two-place" operators. It is possible to have three-place operators, four-place operators, etc., but we have little use for these in English. An example of a three-place operator would be "Neither ___ nor ___ nor ___."

Fortunately, since our time is finite, we will not be studying all the possible operators in the English language. In fact, for the type of symbolic logic which is usually covered in an introductory course (classical two-valued, truth functional logic) there are only five operators which are particularly important: "and," "or," "if-then," "if and only if," and "not." Some of the other operators are studied in more specialized areas of logic. *Modal logic*, for instance, investigates the concepts of possibility and ne-cessity, while *epistemic logic* examines concepts such as knowing, believ-ing, thinking, and related operators. The five operators we use are listed below, with the special symbols that will represent them.

(___ · ___)	(the dot)	will stand for "___ and ___"
(___ ∨ ___)	(the wedge)	will stand for "___ or ___"
(___ ⊃ ___)	(the horseshoe)	will stand for "if ___then ___"
(___ ≡ ___)	(the triple bar)	will stand for "___ if and only if ___"
~ ___	(the tilde)	will stand for "not ___"

3. The Structure and Symbolism of Sentential Logic

Sentential operators, again, are used to build up compound sentences from simpler components. Since we will be using only five operators, we will have only five basic kinds of compound sentences. These five kinds of sentences, with some special terms used to describe and discuss them, are as follows, where p and q represent arbitrary sentences. The "and" sentence, which will be symbolized as $(p \cdot q)$, is called a *conjunction*, and its two major components are called *conjuncts*. The "or" sentence is called a *disjunction*, and will be symbolized as $(p \vee q)$; its major components are called *disjuncts*. The "if-then" sentence is called a *conditional*, and will be symbolized as $(p \supset q)$. The part to the left of the horseshoe is called the *antecedent* (what comes before), and the part to the right of the horseshoe is called the *consequent* (what comes after). In addition, we have *negations*, symbolized as $\sim p$, and *biconditionals*, symbolized as $(p \equiv q)$. There are no special names for the components of negations and biconditionals.

We can now see what the *formulas*, or symbolic expressions, of sentential logic will look like. In Unit 4 we consider the relationship between the English sentences and the symbolic formulas, but for now we will simply try to understand the structure of the formulas.

Our simplest formulas, the ones which will serve as building blocks for all the other formulas of sentential logic, will be *single capital letters* which will represent simple (noncompound) English sentences. We will generally use the first letter of the sentence being symbolized, or at least some letter which reminds us of the meaning of the sentence. We might use R, for instance, to symbolize "Reagan is president," and B to symbolize "Bush is vice-president."

We will build up more complex formulas in the following ways: given any two formulas (not necessarily simple), we may combine them by placing one of our two-place operators between them and enclosing the result in parentheses. If, for example, we use B and R as above, N to represent "Nixon is president," and A to represent "Agnew is vice-president," we can then form the following elementary compounds:

$(B \cdot R)$ Bush is vice-president and Reagan is president.

$(R \vee N)$ Either Reagan is president or Nixon is president.

$(B \equiv R)$ Bush is vice-president if and only if Reagan is president.
$(A \supset N)$ If Agnew is vice-president, then Nixon is president.

We can *negate* a formula simply by placing a tilde in front of it, *without* using extra parentheses. (If we are negating a single letter, we clearly do not need parentheses, and if we are negating a compound, it should already be enclosed in parentheses.) $\sim A$, for instance, will say that Agnew is not vice-president, and $\sim (R \cdot N)$ will say that not both Reagan and Nixon are president.

We may build formulas of any degree of complexity using only our five operators. We could disjoin $(N \cdot A)$ with $(B \cdot R)$, for instance, to get $((N \cdot A) \vee (B \cdot R))$, which would be read "Either Nixon is president and Agnew is vice-president, or Bush is vice-president and Reagan is president." $((B \cdot R) \supset \sim N)$ would be read "If Bush is vice-president and Reagan is president, then Nixon is not president." The formula $(A \equiv \sim R)$ would be read "Agnew is vice-president if and only if Reagan is not president."

To take another example, where the capital letters are again abbreviations for simple (but here undesignated) English sentences, we could first construct $(A \cdot B)$, then $(C \vee D)$, and join them with a conditional to get $((A \cdot B) \supset (C \vee D))$. We could then put E and F together in a biconditional, and conjoin the result to the negation of G, which would yield $((E \equiv F) \cdot \sim G)$. We might then disjoin these two formulas to get $(((A \cdot B) \supset (C \vee D)) \vee ((E \equiv F) \cdot \sim G))$. Obviously we could go on like this indefinitely, but this should be enough for illustrative purposes. There is no upper limit, theoretically, to the length of formulas you can construct in this way, though there are certainly some practical limits, such as the availability of paper, ink, time, and energy!

The possibility of these multiply compound formulas makes the use of parentheses absolutely essential in order to avoid ambiguity. Without parentheses, for example, we would not be able to tell whether the formula $A \cdot B \supset C$ was a conjunction or a conditional, and this makes an enormous difference in what it says. As a conjunction, which would be properly symbolized as $(A \cdot (B \supset C))$, it would say "A is definitely true, and if B is true, then C is true." As a conditional, properly symbolized as $((A \cdot B) \supset C)$, it says only "If A and B are both true, then C will be true," but it asserts nothing about whether A is actually true, unlike the first reading. The situation here is just like what we have in arithmetic; without parentheses the expression $3 - 2 + 5$ is ambiguous, and has no definite value. It could mean either $(3 - 2) + 5$, which is 6, or $3 - (2 + 5)$, which is -4, quite a difference! The necessity of using parentheses should be clear in arithmetic; it is just as important in logic.

There are a couple of useful terms which should be introduced here, since they will make it easier to discuss certain things later on. The first is the term *subformula*, which is, roughly, a meaningful formula which occurs as a part of another formula. (Any formula is a subformula of itself.) As an example, E, F, G, H, $\sim E$, $\sim G$, $(\sim E \vee H)$, $(\sim G \cdot F)$, and $((\sim E \vee H) \equiv (\sim G \cdot F))$ are all subformulas of $((\sim E \vee H) \equiv (\sim G \cdot F))$. A subformula is really just a component of a symbolized sentence.

Another useful, and very important, term is *major operator*. The major operator of a formula is the one that determines the overall form of the sentence, and is the operator introduced *last* in the process of constructing the formula from its more elementary components. Thus the major operator of $((A \cdot B) \supset (C \vee D))$ is the horseshoe, which makes that formula a *conditional* even though a conjunction and a disjunction occur as subformulas (respectively, as the antecedent and consequent of the conditional). The major operator of $(((A \cdot B) \supset (C \vee D)) \vee ((E \equiv F) \cdot \sim G))$ is the second wedge, and so the formula is a *disjunction*, even though it contains all the other operators. Often when we speak of "the form" of a formula or sentence, we will be referring simply to its major operator; we may say, for instance, that the form of $((A \cdot B) \vee (C \equiv D))$ is a disjunction.

Since the form of a sentence plays such a crucial role in logic, it is important that you be able to recognize the major operator of any given formula. With simple sentences you will be able to see this at a glance; with more complicated formulas, such as

$$(((A \cdot B) \vee ((C \supset D) \vee F)) \supset ((H \cdot E) \vee C))$$

it is a little more difficult. One way to do this is to pair parentheses, starting with the smallest formulas and working your way up to the largest. The last operator you cover in this process will be the major operator. If we take the formula above again, we can pair parentheses as follows:

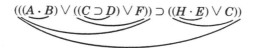

This makes it clear that the major operator, the last one to be joined up, is the second horseshoe. With a little practice this will become second nature to you; practice is supplied in Exercise 2, at the end of the unit. The use of brackets or braces in place of some of the parentheses will make formulas a bit easier to read, and we will use this notation where appropriate in the rest of the book.

DEFINITIONS

1. A sentence is **compound** if it logically contains another complete sentence as a proper part.
2. A sentence is **simple** if and only if it is not compound.

3. A **component** of a compound sentence is an independent clause (complete sentence) that occurs as a proper part of the compound sentence.
4. One sentence **logically contains** another if it either literally contains the other as a component, or can be paraphrased into an explicitly compound sentence that contains the other as a component.
5. A **sentential operator** is an expression containing blanks, such that when the blanks are filled with complete sentences, the result is a sentence.

STUDY QUESTIONS

1. Give an example of a simple sentence with more than ten words.
2. Give an example of a sentence with a simple subject and a compound predicate.
3. Give an example of a sentence with a compound subject and a simple predicate.
4. Give an example of a sentence with a compound subject and predicate.
5. What are the five operators we will be using (in English) and what are the symbols for them? What are the other symbols we will be needing in sentential logic?
6. Why is this called sentential logic? What are we mainly interested in in sentential logic?
7. What are five operators other than the five we will be using?
8. Make up a four-place operator.
9. What are the names of the five compound formulas we will be using, and what are their components called?
10. What is a subformula?
11. What is the major operator of a formula?

EXERCISES

1. State whether the following sentences are simple or compound and, for those which are compound, state all the simple components.

*a. Mary hit a home run and a triple.
 b. John enjoys a baseball game if he can have popcorn.
*c. Dogs with fleas make poor housepets.
 d. Dogs don't like bumblebees.
*e. John will get cancer or have a heart attack if he doesn't stop eating fatty beef.
 f. John's wife tries to keep him from eating junk food.
*g. No one can survive for long on junk food.

h.   Mary told John that he was getting fat.

*i.   Whatever John does is all right with Mary.

j.   Neither John nor Mary likes gooseberries.

*k.   John and David are enemies.

l.   Life on earth is doomed if the pollutants are not eliminated.

*m.   Human beings will die out or be mutated if there is an atomic war.

n.   John lay down and had a nap.

*o.   Either Mike or John will have to clean up the kitchen after dinner.

p.   John and Mary are close friends.

*q.   John will probably lose weight if Mary quits teasing him.

r.   John likes to play games for money with people who are a bit dim.

*s.   ~~Nobody dislikes John.~~

t.   John will have to either sink or swim if his father stops supporting him.

2.   Identify the major operator in the following formulas. (You may have to specify "the second horseshoe," "the third dot," etc., in very complex formulas.)

*a. $((A \lor B) \cdot C)$

b. $((A \cdot B) \lor \sim (B \cdot C))$

*c. $(\sim A \equiv \sim (B \lor C))$

d. $(A \lor (B \lor C))$

*e. $(((A \lor \sim B) \supset (C \equiv D)) \supset \sim E)$

f. $((A \cdot \sim B) \supset \sim (B \supset (C \cdot D)))$

*g. $((((F \cdot \sim G) \supset C) \equiv D) \lor (B \supset C))$

h. $(\sim \sim B \supset \sim \sim C)$

*i. $(((B \supset C) \lor (D \lor F)) \cdot E)$

j. $((((A \lor B) \cdot C) \lor D) \cdot (E \lor (F \cdot D)))$

*k. $(((A \cdot B) \cdot C) \lor ((D \cdot E) \cdot F))$

l. $((((((A \supset B) \supset C) \supset D) \supset E) \supset F) \supset G)$

*m. $(A \supset (B \supset (C \supset (D \supset (E \supset (F \supset G))))))$

n. $(((A \equiv B) \equiv (C \equiv D)) \equiv ((F \equiv G) \equiv H))$

*o. $((((A \lor B) \cdot (C \lor D)) \equiv ((A \cdot B) \lor (C \cdot D))) \supset (((G \lor H) \supset (P \lor Q)) \cdot R))$

UNIT 3

Computing Truth Values

A. INTRODUCTION

In Unit 1 it was emphasized that an argument form is valid if and only if it has no instances with true premises and a false conclusion. If we are ever to be able to determine whether or not a form is valid, then, we obviously must be able to tell whether the premises and conclusions of the instances are true or false, which means we must be able to determine the truth value of the *compound* sentence once we are given the truth values of its component parts. For reasons which need not concern us now, we will not have to worry about determining the truth values of the simple sentences; this will be done for us.

In Unit 2 we saw that there were an infinite number of possible sentence operators, out of which we have chosen just five. We will now see that one very good reason for choosing these five, aside from the fact that they are very common, is that they have a very special property, which sets them off from a great many other ways of combining sentences in the English language, and which will be of great importance for our logical purposes. That is, these five operators are all *truth-functional*, which means that the truth or falsity of the compound sentences which they form can always be determined just by knowing the truth or falsity of their component parts. Another way of putting this is to say that for these five operators, *the truth value of the compound sentence is completely determined by, is a function of, the truth values of the component sentences.*

Of course, in order to determine the truth-value of a compound sentence given the truth-values of its components, you will have to know the *rules of computation* for each of our five sentential operators. These rules will be given by means of schematic *truth tables* for the operators, which will indicate for each possible combination of truth-values for the components of the formula, what truth-value must be assigned to the compound. Once you learn the truth tables for each operator, you will be able, given the truth-values of the elementary components, to compute the truth-value of compound formulas by working your way up from smaller subformulas to larger ones.

In this unit, then, you will be learning the meaning of the term "truth-functional" (and what it means for an operator *not* to be truth-functional), the truth tables for the five operators, and how to compute the truth value of a compound formula given the truth values of the components. What you will need to know is stated more explicitly in the "Objectives" section below.

B. UNIT 3 OBJECTIVES

1. Memorize the truth tables for the five sentential operators, and be able to state informally the computation rules for each operator.
2. Be able to compute the truth value of compound sentences of any degree of complexity, given the truth values of the simple sentences they contain.
3. Learn the definition of "truth-functional."
4. Be able to show that an operator is *not* truth-functional.

C. UNIT 3 TOPICS

1. Truth Tables for the Operators

As noted above, a very important property of the operators we will be using is that they are all *truth-functional*, that is, *the truth value of the compound which they form can be determined solely by the truth values of the components*. Thus, there will be a *rule* telling us exactly what the value of the compound must be for each combination of values of the components. The rule for conjunction, for instance, is that a conjunction will be true only if *both* conjuncts are true, and thus, if one or both of the conjuncts is false, the entire conjunction must be counted false. Such rules, as we shall see, can be given more formally by means of little *truth tables* for each of the operators, which will list systematically all possible combinations of truth values for the components and the result of the computation for each of these possibilities.

One of the very important presuppositions of this procedure is that *a sentence must be either true or false.* We have noted, in the Introduction to Unit 2, that in elementary logic we will be dealing only with *declarative* sentences; we now go a little further to say that (1) our sentences must have a truth value (they cannot be indeterminate) and (2) the truth value must be either *true* or *false* (we will have no such value as "nearly true"). What we have, then, is a *two-valued logic*, which simply means that whatever sentences we use in our logical operations must have one of the truth values true or false.

It should be mentioned that there are logical systems (many-valued logics) which investigate the logical properties of sentences which have three possible values (for instance, true, false, and undetermined), four possible values (for instance, necessarily true, contingently true, contingently false, and necessarily false), or even more. But these are specialized disciplines, and standard logic assumes, as noted, that the sentences it uses are simply true or simply false.

Given that we have a two-valued logic, that each sentence is either true or false, we can draw up a list of all possible combinations of truth-values for the simple components of a compound formula.[1] In the formula $((A \supset B) \cdot (A \lor B))$, for example, we know that A must be true or false, and that B may be either true or false when A is true, and may also be true or false when A is false. (Note that another very important assumption we will be making is that the truth values of the simple sentences are *independent* of each other.) We may indicate these four possibilities in a little schematic table; each of the four possibilities will be referred to as a *row* in the truth table.

	A	B
1.	T	T
2.	T	F
3.	F	T
4.	F	F

The fact that we can systematically list all these combinations is one of the things which makes it possible to give rules for the computation of our compound formulas. We can say, for each of the combinations, what

---

[1] It is also possible in a three-valued, four-valued, or any finite-valued logic to list all possible combinations. However, it is more cumbersome for the higher-valued logics, since the number of possibilities gets very large very fast. In a four-valued logic, for example, there are 16 possibilities (instead of 4) for 2 different sentence letters, and 64 different combinations (instead of 8) for 3.

the value of the compound must be and, since these are the only possible combinations given that we have a two-valued logic, we will have stated a complete rule. The rules of computation for the five operators are given below.

a. "and."   The rule for computing the truth value of conjunctions is just what you would expect given the meaning of "and." When we say "*p* and *q*" we mean to assert that *both p* and *q* are true, so the first row is the only row in which the conjunction will be considered true. Otherwise, where one or the other (or both) of the conjuncts is false, the conjunction as a whole turns out to be false. We can give this rule for computing the value of conjunctions by means of the following *truth table for "and"*:

p	q	$(p \cdot q)$
T	T	T
T	F	F
F	T	F
F	F	F

Read horizontally, from left to right, this table tells us (once again) that if p and q are both true, $(p \cdot q)$ is true; if p is true and q is false, $(p \cdot q)$ is false; if p is false and q is true, $(p \cdot q)$ is false; and if p and q are both false, $(p \cdot q)$ is false.

Note that we use the variables p and q here. This indicates that the conjuncts may be *any* formulas, simple or complex. One instance of $(p \cdot q)$ is the more complex formula $(((A \vee B) \supset C) \cdot \sim (D \vee E))$; this will be true only if $((A \vee B) \supset C)$ is true and also $\sim (D \vee E)$ is true.

b. "or."   The truth table for disjunction will be very nearly what we would expect from the meaning of the English "or"; the only question is what happens in the top row, where p and q are both true. It is easy to say what must happen in the last three rows; if one disjunct is true and one is false, the disjunction as a whole will be true, and if both are false the disjunction will be false.

What about the top row, however? What happens if both p and q are true? Here we need to distinguish between *inclusive* disjunction and *exclusive* disjunction. Inclusive disjunction allows the possibility that both disjuncts are true, while exclusive disjunction rules this out and declares that only one or the other, *but not both*, of the disjuncts are true. Thus in inclusive disjunction the top row in the truth table comes out true, while in exclusive disjunction it comes out false. These are two different operators, and we must make a decision as to which one we will use here.

Both senses of "or" occur in English. In many cases it is clear that the inclusive sense is intended, since we count the sentence true when both of the disjuncts are true. One example would be "Ronald Reagan was either a movie actor or a politician." Even though he was both, we would still count the sentence true. Or if we see Mary at a university function we might say "Mary is either a student or the wife of a student," and of course she might very well be both, in which case we would still count the disjunction as true. Instances of exclusive disjunction would be "Either coffee or tea is included in the price of the meal" (where it is clear that you would pay extra if you wanted both), and "John is married to either Josephine or Carolyn."

It is the *inclusive* sense of disjunction, in which the top row of the truth table comes out true, that we will be using in this book. Its computation rule is given by means of the following *truth table for "or"*:

p	q	$(p \vee q)$	
T	T	T	We can summarize this table by saying that an inclusive "or" is to be counted false in only one case: where both of the disjuncts are false. If one or the other, or both, of the disjuncts is true, the disjunction will be considered true.
T	F	T	
F	T	T	
F	F	F	

The *exclusive* sense of "or" is also a truth-functional operator, and its truth table is the same as that for the inclusive sense, except that the top row comes out false instead of true. It is sometimes symbolized with a line above the wedge: $\overline{\vee}$. However, we will not be including this as one of our operators because we can always say the same thing by combining some of our other operators. If we want to say, for instance, that either John or Bob will be promoted, but not both, we can simply *conjoin* the "not both" phrase to the inclusive disjunction. This would give us the following symbolization: $((J \vee B) \cdot \sim (J \cdot B))$. In general, the exclusive sense of disjunction, where p and q are the two disjuncts, can always be symbolized as $((p \vee q) \cdot \sim (p \cdot q))$, and so a separate operator is not needed. In this book, then, we will be using (exclusively) the *inclusive* sense of "or," and this means that whenever you are computing the value of a disjunction, the top row of the truth table, where both of the disjuncts are true, will always turn out to be true.

c. "if and only if."   A biconditional statement says that one thing happens if and only if (in just the same circumstances in which) another takes place. An instance of this would be "Mary will be elected student body president if and only if Fred is elected treasurer," which could be symbolized as $(M \equiv F)$. (Note that the *bi*conditional, as the name indicates, is a two-way conditional; the sentence above asserts both that if Mary is elected student body president, then Fred will be elected treasurer, *and* that if Fred is elected, then Mary will be elected.) Now, what should be the truth table for this operator? Suppose both Fred and Mary are elected; then both M and F are true, which would correspond to the top row in our truth table. Surely we would then consider the biconditional $(M \equiv F)$ to be true. What if they are both defeated? This would correspond to the last row of the table, in which both M and F are false. Since the biconditional says that one will be elected if and only if the other is elected, or roughly, that they will stand or fall together, it will turn out to be *true* when both components are false. (In this case they have fallen together.) Thus the top and bottom rows of the truth table, where the truth values of the components are the same, will be considered true. What about the middle two rows, in which the truth values of the components are different? These would correspond to the cases in which one of the candidates was elected and the other defeated. But in these cases we would surely say that the biconditional, which asserted that one would be elected *if and only if* the other was elected, was false, and this is how we complete the truth table. The *truth table for the biconditional*, then, will be as follows:

p	q	$(p \equiv q)$
T	T	T
T	F	F
F	T	F
F	F	T

We can summarize this truth table by saying that if the truth values of the components are the *same*, as in the first and fourth rows, the biconditional will be *true*, and if they are different, as in the second and third rows, the biconditional will be false.

d. "not."   The truth table for negation will contain only two rows instead of four, because the negation sign is placed in front of a single formula (which may, of course, be complex as well as simple) instead of joining two formulas together. A single formula can be only true or false, so we need consider only two cases: we must say what happens to $\sim p$ when p is true and what happens to $\sim p$ when p is false. The truth table for negation is just what you would expect: if p is true then $\sim p$ is false, and if p is false, $\sim p$ turns out to be true. This is schematized in the following *truth table for negation*:

p	$\sim p$
T	F
F	T

This can be summarized by saying that the negation of p will have the truth value opposite that of p.

A negation, in other words, simply *reverses* the truth value of the formula being negated. This is why *double negations*, such as $\sim \sim p$, cancel each other out. The inner negation reverses the truth value once, and the outer negation reverses it again, right back to where it started. (This is why your elementary school teachers always warned you not to say things like "I ain't got no bad habits." The two negations cancel each other out, so that the sentence literally, though probably not colloquially, means "I do have some bad habits.")

e. "if-then."   We have left the conditional until last because its truth table is the least intuitive and the most difficult to justify of any of our

eliminate

five operators. We can begin by noting, however, that if the antecedent of a conditional is true, and its consequent is false, the conditional will always be false. If we had predicted in 1976, for instance, "If Reagan is elected there will be a stock market crash in 1986," that prediction would be seen as false, since Reagan *was* elected (the antecedent is true) but there was not a stock market crash in 1986 (the consequent is false). The conditional $(p \supset q)$ will be false whenever p is true and q is false. What happens if p and q are both true? Suppose we had said in 1984, "If Reagan is elected, the stock market will climb." Since Reagan was elected and the stock market did climb, we would count this sentence as true. This example corresponds to the top row in the truth table, and the previous example corresponds to the second row.

What happens when p is false, however? Suppose we let F stand for the false sentence "Ford was elected in 1976," M stand for "There was an invasion from Mars in 1978," and C stand for "Chrysler Corporation got a huge government loan guarantee." What would be the truth values of $(F \supset C)$ ("If Ford was elected, then Chrysler got a huge loan guarantee"), which would correspond to the third row of our truth table, and $(F \supset M)$ ("If Ford was elected there was a Martian invasion"), which would correspond to the fourth row? How *do* we decide cases like this, in which we can never observe the antecedent condition (in this case, Ford being elected in 1976)? There seems to be no way of judging. Classical logicians, whose lead we will be following in this book, dispose of the matter rather neatly simply by declaring that *any conditional with a false antecedent will be counted true.* If the antecedent is true, as we have seen earlier, then the conditional will be true or false according to the truth value of the consequent. The *truth table for the conditional*, then, will look like this:

p	q	$(p \supset q)$
T	T	T
T	F	F
F	T	T
F	F	T

This table can be summarized by saying that the *only* time a conditional is to be considered false is when the antecedent is true and the consequent is false. Whenever the antecedent is false the conditional is true, and whenever the consequent is true the conditional is true.

This table has some very odd consequences. For instance, since the conditional will be true whenever the antecedent is false, we will have to say, if the "if-then" is taken as the horseshoe, that all of the following are true:

If cats speak French, they produce brilliant mathematics.

If Ford was president in 1980, the Soviets dismantled all their atomic weapons.

If a woman landed on the moon in 1963, she discovered highly intelligent moon men and produced one of their offspring, who is now masquerading as president of the USSR.

If no one landed on the moon in 1969, then California experienced a severe earthquake in 1980 and has dropped into the Pacific Ocean.

If there are intelligent beings on Mars, then π is greater than 3.9.

It also follows from the truth table that whenever the consequent of a conditional is true (the first and third rows) the conditional as a whole is true, whether or not there is any connection between antecedent and consequent. This means that all the following sentences must also be counted as true:

If my cat sleeps a lot, then Mt. St. Helens erupted violently in May 1980.

If you are taking a logic course, then men landed on the moon in 1969.

If Jupiter is a planet, then the Freer Gallery in Washington, D.C. has some excellent Oriental art.

What is perhaps most disconcerting about this truth table is that conditionals which seem to contradict each other both have to be considered true, provided the antecedents are false. For instance, all of the following are true:

If an atomic bomb was dropped on New York City in May 1986, then millions of people were killed.

If an atomic bomb was dropped on New York City in May 1986, then no one was hurt and most people benefited greatly from the strong dose of radiation.

If cats speak all Western European languages, then they speak French.

If cats speak all Western European languages, then they do not speak French.

At this point you may very well wonder what gives logicians the right to make up such rules as "whenever the antecedent of a conditional is false, the conditional as a whole is true," especially when they lead to such odd results. Isn't it obviously *false* rather than true that if a bomb was dropped on New York, then no one was hurt and everyone benefited? A satisfactory reply to this sort of perplexity would take us far afield into the philosophy of logic, so just a few observations will have to suffice here.

The most important thing to keep in mind is that the logical operator, the horseshoe, is not the same as the ordinary English "if-then." The logical operator, which is sometimes called the *material conditional*, is a kind of "minimal" if-then, which captures only the common logical content of the various uses of the English "if-then." This common logical content, which is reflected in the truth table for the horseshoe, is simply that *it is not the case that the antecedent is true and the consequent false*. The horseshoe is a very weak operator which says nothing more than is indicated in its truth table. Thus the appearance of paradox in the previous examples vanishes; when we keep in mind that the "if-then" is intended to be simply the material conditional, we *may* properly claim both that if a bomb was dropped then millions were killed, *and* that if a bomb was dropped then no one was hurt and everyone benefited. Both conditionals are true simply because in neither case do we have a true antecedent with a false consequent, since in both cases the antecedent is false. One way to think of it might be the following: since the only time the conditional is false is when the antecedent is true and the consequent is false, *if we have a false antecedent, then we have not said anything false*. But if a sentence is not false, then we must count it as true, since this is a two-valued logic. Thus, a sentence with a false antecedent must be true.

What is crucial in a system of logic is that it never permits us to make incorrect inferences, that is, to go from true premises to a false conclusion. According to most logicians, classical truth-functional logic, with its rather peculiar "if-then" operator, does pass this test. The horseshoe is adequate for many of our logical purposes, and as long as it does not lead us into logical temptation, most logicians are willing to put up with its quirks.

One possibility which may have occurred to you is that we might simply leave the third and fourth rows blank in the truth table for the horseshoe. There is one very good reason, however, for filling in those blanks by making a decision one way or the other. We want to be able to determine the truth values of compound sentences *in all cases*, so that we can always tell whether an argument is valid or not. If we left blanks, many of our sentences would have unknown truth values, and so we could never determine whether the argument forms containing them had instances with true premises and a false conclusion. We want a *complete decision procedure* for sentential logic, and we can have this only if all our truth tables are complete. We want all our operators, then, to have rules determining in every case what the truth value of the compound will be.

Given that we want complete truth tables for our operators, it turns out that the rule we have given for the horseshoe is the only acceptable candidate. It is fairly clear that the top two rows must be T and F, respectively. For the bottom two rows there are only four possibilities: they may both be true, as in the truth table we have; they may both be false; the third row could be true with the fourth false; or the fourth row could be true with the third false. We can list these possibilities, including the top two rows as well and outlining the bottom two rows, as follows:

		OUR VERSION		OTHER POSSIBLE VERSIONS				
p	q	1. $(p \supset q)$	2.	$(p \supset q)$	3.	$(p \supset q)$	4.	$(p \supset q)$
T	T	T		T		T		T
T	F	F		F		F		F
F	T	T		F		T		F
F	F	T		F		F		T

Notice that we could not use the second version for the horseshoe because that is the truth table for conjunction, and we certainly want "if p then q" to mean something different from—be true in different circumstances than—"p and q." Similarly, we could not use the fourth version because that is the truth table for the biconditional. Finally, if we compare the third version with the list of possibilities on the far left, we see that it simply represents q, and we certainly want "if p then q" to mean something different from just q. Thus, if we are going to have a truth-functional logic in which all operators have complete rules for computation, the only possible truth table for the horseshoe is the one we have given.

One final comment on the use of the horseshoe. There are alternative logics which use a stronger if-then operator. Many logicians claim that these stronger operators are closer to our English "if-then" and are thus better candidates for an accurate system of logic. We will make no judgment on this point, except to say that it may well be true, and that these stronger systems of logic deserve close study. *Not*, however, in an elementary textbook. The logical systems with stronger if-then operators are considerably more complex than our truth-functional version, and even if you eventually decide that one of them is more nearly correct, you will probably not be able to understand it unless you have first thoroughly mastered the simpler system. If nothing else, then, we could justify the use of the horseshoe in elementary logic on the grounds that it is the most easily understood version of "if-then" and is thus the most suitable for an introduction to logic.

2. Computing Truth Values

By now you should have memorized the truth tables for the five operators, and should be able to state informally their computation rules (for instance, that a disjunction is false only if both disjuncts are false, and is otherwise true). You are now in a position to see how it is possible to compute the truth value of any compound formula given the truth values of the simple component sentences, and that will be the topic of this section. As in the exercises at the end of the unit, we will here assume that A, B, and C are true, while X, Y, and Z are false. We will also adopt the convention of occasionally dropping the outside parentheses on our formulas if this makes them easier to read, since no ambiguity results as long as we do not use such a formula as a part of another. In longer formulas, there will be enough parentheses as it is, without including the outermost pair.

We will begin our computations with a fairly simple example. How, for instance, would we compute the truth value of $((\sim A \lor \sim B) \supset \sim C)$? Since we are counting A, B, and C as true, $\sim A$, $\sim B$, and $\sim C$ will all be false. The disjunction $(\sim A \lor \sim B)$ will then be false, since both the disjuncts are false. Both antecedent and consequent, then, are false, and by consulting the truth table for "$\supset$," we see that $F \supset F$ turns out to be true. The truth value for the whole, then, is *true*. We could represent the computation for the formula above in the following way:

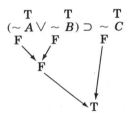

It will help to adopt the following conventions: We will place the truth values of the *simple sentences* immediately *above the sentence letters*. We will place the truth values of the various *subformulas* (formulas which occur as a part of a larger formula) immediately *below the major operator for that subformula*. The arrows indicate how the truth values of the subformulas "feed in" to the computation of the value of the next largest formula. Our computation procedures will be first to determine the truth values of the smallest subformulas, then use these to compute the values of the next-largest subformulas, and so on until we reach the value of the sentence as a whole. Another example will illustrate this procedure for a slightly more complicated case: $((A \lor B) \supset Z) \lor ((B \lor Z) \supset \sim A)$.

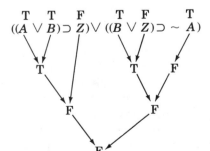

Since A and B are true, $(A \lor B)$ is true and since Z is false, $((A \lor B) \supset Z)$ is false.

Since B is true, $(B \lor Z)$ is true, and since A is true, $\sim A$ is false. This makes $((B \lor Z) \supset \sim A)$ turn out false.

Since both of the disjuncts are false, the formula as a whole comes out false.

Negations can be confusing, and it is essential that you understand what formula is affected by, that is, what is the *scope* of, the negation operator. The tilde will operate on, or negate (and thus reverse the truth value of), *the first complete formula following it*; this will be indicated by the use of parentheses. If there are no parentheses immediately following the tilde, then it negates only the sentence letter immediately following. (In ($\sim A \vee B$), for instance, we would be negating only A. The sentence would be read "Either not A or else B," and would be true since B is true. If the tilde is followed by a parenthesis, as in $\sim (A \vee B)$, then it negates the formula contained between the left parenthesis immediately following it and the right parenthesis which is paired with it. In $\sim (A \vee B)$, for instance, the tilde negates the whole formula ($A \vee B$), so the sentence would be read "It is not the case that either A or B," or in other words, "Neither A nor B," and would be false since ($A \vee B$) is true. In the more complex formula $\sim [(\sim A \equiv (B \vee C)) \cdot \sim (X \supset (Z \vee Y))]$ we have three tildes, all with different scopes. Reading from left to right, the first tilde negates the entire formula (and hence will be the *last* operator to be computed), the second negates only the A, and the third negates the conditional ($X \supset (Z \vee Y)$). The computation would be as follows:

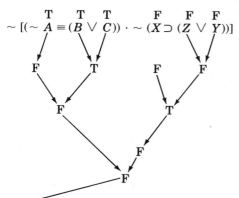

$\sim A$ will be false, and $B \vee C$ will be true; this makes $(\sim A \equiv (B \vee C))$ *false*.

X is false, and $(Z \vee Y)$ is false, so that $(X \supset (Z \vee Y))$ is *true*. Thus $\sim (X \supset (Z \vee Y))$ is false.

The conjunction, then, is false.

The major operator of this formula is the outer negation, which negates the conjunction. Since the conjunction is false, its negation, the formula as a whole, turns out to be *true*.

If you do not understand the results of the computations above, go back and review the truth tables for the operators. You will need to know them very thoroughly, so that you can compute the results for the subformulas with a minimum of effort. Remember that the procedure must always be to *work from the inside out*, to start with the smallest subformulas and work your way up step by step to the larger ones. Exercise 1, at the end of the unit, will give you practice in this sort of computation.

There are some shortcuts you can use if you know your truth table rules well. For instance, since it only takes one true disjunct to make a disjunction true, if you find that one side is true, you need make no other calculations but can conclude without further ado that the entire disjunction is true. Similarly, if you find one conjunct false, you may

conclude that the conjunction as a whole is false. And if either antecedent is false or consequent is true in a conditional, you may conclude
that the conditional itself is true. Given that A and B are true,
and X, Y, and Z are false, for instance, we know almost immediately
that $(\sim (A \vee B) \supset (\sim Z \supset \sim Y)) \vee ((Z \cdot A) \vee B)$ is true, since B is true
and this makes the second disjunct true. Similarly, we know that
$\sim A \supset [((Z \vee Y) \cdot (A \vee \sim B)) \equiv ((A \vee \sim Z) \vee (X \vee B))]$ is true as
soon as we see that $\sim A$ is false, since a false antecedent always yields
a true conditional. We need not bother at all with the very complex
consequent! A summary of these shortcut rules is as follows:

> If one disjunct is true, the entire disjunction is true.
>
> If one conjunct is false, the entire conjunction is false.
>
> If the antecedent is false, the conditional is true.
>
> If the consequent is true, the conditional is true.

Knowing these shortcuts will considerably facilitate your work when it
comes to truth tables, in Units 5 and 6. Exercise 2 at the end of the unit
will give you practice in using these shortcuts.

3. Truth-Functional Operators

The concept of truth-functionality was discussed briefly in the Introduction to this unit. Now that you know the truth tables for the five
operators and are able to do the computations, you are in a position to
understand this concept in a little more depth. As we noted in the introduction, what it means for an operator to be *truth-functional* is that *the
truth value of the compound which it forms is completely determined by
the truth values of the component parts*. There are other ways in which
this could be stated. We might say, for instance, that there are rules that
tell us what the value of the compound must be given the value of the
components, or we could say that the truth value of the compound is a
function solely of the truth values of the components. However we put it,
the important thing to remember is that it is the *truth values only*, and
not, for instance, the meanings of or the relations between the sentences,
which determine the value of the compound. *All* we need to know is the
truth values of the components in order to determine the value of the
compound. As we shall see in the next section, there are many operators,
perhaps most, for which this is not the case.

In the system of logic we will be using, all of our operators are truth-
functional, which means that we have a truth-functional logic. *A system
of logic is truth-functional* if and only if *each of its operators is truth-
functional*. There are many systems of logic in which some operators are
truth-functional and others are not, for instance, systems of modal logic,

which explore the concepts of possibility and necessity. Such systems are not considered to be truth-functional even though some of the operators have this property; to be truth-functional *every* operator in the logical system must have a complete rule for determining the truth value of the compounds.

4. Non-Truth-Functional Operators

You will have a better grasp of what it means for an operator to be truth-functional if you understand, by contrast, what it means for an operator *not* to have this property. A non-truth-functional operator is by definition, of course, simply an operator for which we *cannot* determine the truth value of a compound, given the truth values of the components. Some examples of non-truth-functional operators are "*x* believes that ____" (where *x* is any individual), "it is possible that ____," "it is necessary that ____," "____ because ____," and "____ after ____." In fact, most of the operators in general use in English are non-truth-functional. With none of these operators is it possible, given *just* the truth values of the components, to determine the truth value of the whole. Something else is needed as well, some outside information; in the case of "after," for instance, we would need the times at which the stated actions took place.

A typical example of a non-truth-functional operator involves the concept of belief. We cannot determine the truth value of "John believes that ____" *just* by knowing the truth value of the sentence that goes into the blank. That a sentence is true does not guarantee that John (or any of us, unfortunately) *believes* it; that is, it does not determine the value of the compound sentence. Nor does the fact that a sentence is false guarantee that John, or we, *won't* believe it. We all believe all sorts of false things (though we may not believe that we do.) There is, in short, no *rule* which determines the truth value of sentences of the form "*x* believes that ____", given only the values of the component sentences. Of course, we may determine the truth values of belief sentences by other means; we might, for instance, simply *ask* people what they believe. What we cannot do is to determine whether they believe something simply on the basis of whether or not it is true.

Another example, and one which raises some very interesting philosophical questions, is "because." It is particularly important not to confuse this strong non-truth-functional operator with our weak truth-functional "if-then." With "because," unlike our "if-then," the truth value of the components does *not* determine the truth-value of the compound. That two sentences are true, for instance, does not guarantee that the corresponding "because" statement is true (though it *would* guarantee the truth of the material conditional). It is true, for instance, that it rained in Morgantown, West Virginia in October 1980, and it is also true that Reagan was elected president in November 1980. But it would be absurd, or worse, to suppose that Reagan was elected *because* it rained in Morgantown in October.[2]

We need to be more precise, however, in what it means to say that an operator such as "because" is not truth-functional, and in particular we need to show how we can *demonstrate* that it is not. To show that an operator is not truth-functional, we need to be able to show that the truth values of the components do *not* determine the truth value of the compound. But how do we do this? We could begin by noting that to say an operator *is* truth-functional—to say that the value of the compound is determined *solely* by the values of the components—is to say, among other things, that given the *same* truth values for two different sets of components, you will always get the same end result. Identical input implies identical output. "Nixon was president in 1980," for instance, is false, and so is "Toyotas are made in Lichtenstein." If we negate the two sentences, using a truth-functional operator, we get the same results in both cases; both the negated sentences "Nixon was not president in 1980" and "Toyotas are not made in Lichtenstein," are true. This suggests that if we could come up with pairs of *component* sentences with *identical* truth values, and show that the results of *compounding* them were *different*, this would be a demonstration of *non*-truth-functionality, since it would show that identical truth value input does *not* imply the same truth value output, hence that the value of the compound is not a function solely of the truth values of the components. This is exactly what we do to show that an operator is not truth-functional. We will use this method to demonstrate the non-truth-functionality of "because."

It is true that much of the state of Washington was covered with ash on May 20, 1980. It is also true that Mt. St. Helens erupted in the middle of May 1980. And it is obviously true that the state was covered with ash *because* the volcano erupted. In this case, then, we have "T because T," resulting in T. Another pair of true sentences is "Toyotas are made in Japan" and "Lincoln was president during the Civil War." It is safe to say, however, that the sentence "Toyotas are made in Japan *because* Lincoln was president during the Civil War" is false (and even safer to say that "Lincoln was president during the Civil War because Toyotas are made in Japan" is false). In this case we have "T because T" coming

---

[2] In less silly cases, however, people do sometimes make the mistake of thinking that because two statements are true, especially if one event happened after another, that one thing caused another. The error is so common, in fact, that it has been given a special name: *post hoc ergo propter hoc* (after the thing, therefore because of the thing). This kind of error is probably the source of all sorts of superstitions, such as the belief in the efficacy of the rain dance. No doubt in some cases the dance was done and rain followed; thus it was believed, fallaciously, that the rain occurred *because* of the rain dance. The question of when (or even if) causal statements are justified is one of the most interesting questions in the philosophy of science.

out false. What these four sentences show is that the same truth value input does *not* yield the same truth value output, since in one case two true components yield a true compound, while in the other case two true components yield a false compound. This shows that the truth value of the output is *not* a function solely of the truth values of the input, so "because" is not a truth functional operator. (Try, just for the fun of it, to answer the following question: could you ever *yourself* show that "I believe that ____" is not truth functional? Why or why not?)

In summary, in our system of logic we will be using five operators, all of which are truth-functional. The advantage of a truth-functional sentential logic is that it makes it possible to determine the validity of any given argument form, since we can always find out for any of its instances whether or not it has true premises with a false conclusion. There are a great many operators in English, however, which are not truth-functional, including causal operators, temporal operators, possibility and necessity operators, and operators using terms such as "believes" and "hopes." To show that an operator *is* truth-functional, we need to come up with a *rule* for computing the truth value of the compound given the truth values of the components. This we have done for all our operators by using the little truth tables. To show that an operator is *not* truth-functional, we need to come up with examples which show that the same truth values for the components may result in different truth values for the compound.

DEFINITIONS

1. An **operator is truth-functional** if and only if the truth value of the compound which it forms is completely determined by the truth values of the component parts.
2. A **system of logic is truth-functional** if and only if each operator of that system is truth-functional.

STUDY QUESTIONS

1. Write down the truth tables for the five operators we will be using.
2. State informally the rules for computing the values of our five operators.
3. What does it mean for a logic to be two-valued?
4. What is the advantage of using a two-valued (or at least a finite-valued) logic?
5. Give an example of a truth-functional operator other than the ones we will be using, and write down the truth table for it. (You may make one up if you wish.)

6.   What does it mean for an operator *not* to be truth-functional? Give two or three examples other than the one below.

7.   Show that "____ after ____" is not truth-functional.

8.   What are the advantages and disadvantages of using material implication, the horseshoe, as our "if-then" operator?

EXERCISES

1.   Compute the truth values of the following, given that A, B, and C are true, and X, Y, and Z false.

*a. $\sim A \lor \sim B$

 b. $\sim A \supset \sim C$

*c. $(A \cdot X) \lor (B \cdot Y)$

 d. $(X \lor A) \supset \sim B$

*e. $\sim B \supset \sim (X \lor Y)$

 f. $\sim (A \lor B) \supset (X \lor Y)$

*g. $(A \equiv B) \equiv (Z \equiv X)$

 h. $(A \equiv X) \equiv (B \equiv Y)$

*i. $(X \supset (Y \supset Z)) \supset ((Z \supset X) \supset Y)$

 j. $(\sim X \cdot \sim Y) \equiv (A \supset \sim (X \cdot Y))$

*k. $\sim ((A \lor B) \supset C) \supset ((X \lor Y) \supset Z)$

 l. $\sim ((A \lor B) \lor C) \lor (C \supset \sim (X \lor \sim A))$

*m. $\sim (A \supset (\sim A \supset (\sim B \lor X)))$

 n. $\sim (((A \supset \sim B) \supset \sim C) \supset \sim X)$

*o. $\sim (\sim A \lor \sim (B \lor \sim (C \lor \sim X)))$

 p. $\sim \sim (A \cdot \sim (B \supset \sim (C \supset \sim (X \lor Y))))$

*q. $((X \cdot A) \lor \sim (X \cdot B)) \supset ((A \lor X) \cdot Y)$

 r. $((X \cdot Y) \lor (A \cdot \sim B)) \supset ((X \lor A) \cdot (Y \lor \sim B))$

*s. $\sim (\sim (A \lor \sim B) \lor \sim (\sim A \lor X)) \supset \sim (X \lor A)$

 t. $((((A \supset B) \supset X) \supset Z) \supset Y) \supset (((X \supset Z) \supset A) \supset X)$

2.   Compute the truth values of the following, using the values given above, but *without* being given the values for G, H, or I.

*a. $(G \lor H) \cdot \sim A$

 b. $(B \lor G) \cdot (A \lor H)$

*c. $\sim \sim A \lor (X \cdot (G \supset \sim A))$

 d. $(A \lor B) \equiv ((G \lor H) \supset A)$

*e. $\sim (A \lor B) \supset \sim (G \lor H)$

 f. $(X \lor Y) \supset ((G \lor A) \supset \sim (X \lor Z))$

*g. $\sim (X \lor \sim Y) \supset (G \equiv (\sim H \cdot I))$

 h. $(G \equiv (H \cdot A)) \supset ((\sim H \lor I) \supset (X \supset Z))$

*i. $\sim (A \lor (G \equiv ((H \cdot I) \lor (X \cdot Y)))) \supset (A \supset X)$

 j. $\sim ((G \lor (H \equiv \sim A)) \cdot \sim A)$

3. Given the values above, can you figure out the values of the following? Why or why not?

*a. $(A \cdot G) \lor (B \cdot H)$

 b. $\sim (X \lor G)$

*c. $H \supset (G \supset H)$

 d. $(A \lor G) \cdot (B \lor H)$

*e. $\sim (A \lor G)$

 f. $H \supset (G \cdot \sim G)$

*g. $(H \equiv G) \supset (A \equiv B)$

 h. $(H \equiv A) \supset (G \equiv B)$

*i. $\sim (A \cdot G)$

 j. $\sim (X \cdot G)$

UNIT 4
Symbolizing English Sentences

A. INTRODUCTION

Learning the techniques of formal symbolic logic would not have much point (except as an intellectual exercise, or perhaps a computer game) if you did not learn as well how to apply them in particular cases, and the first step in application is learning how to *symbolize* English sentences, how to put them into their logical form. Since an argument is valid just in case its form is valid, we need to be able to determine the logical form of arguments, and this means giving the logical structure of, symbolizing, the English sentences that make them up.

Since in this unit we will be talking about sentential logic, the only sort of structure we need to consider is the way in which compound sentences are constructed out of their simple components by means of our five sentential operators. The first thing you will need to do is to refresh your memory on the relationship between simple and compound sentences, and be able to pick out the simple components of the compounds. You will then need to learn how certain English operators are symbolized by means of our logical operators (the sentential operator "but," for instance, will always be translated into a conjunction). Finally, you will learn how to symbolize multiply compound sentences, those which contain several operators.

Since we have only the five operators, and since English is an extremely rich and complex language, with any number of ways of forming compound sentences, there will necessarily be a fair bit of "squeezing" involved to get the English to fit our simple logical language. Something, sometimes a good bit, is bound to be lost in translation. In some cases, in fact, there is just no way to symbolize an English compound by means of one of our five logical operators, particularly in those cases in which the English operator is not truth-functional. But there are a great many cases where the fit between English and the symbolic language is very close, and a great many arguments whose validity depends upon the logical structure reflected in the logical language. It is these cases with which we will be concerned, and it will be your task in this unit to learn to recognize, and symbolize, the logical structure of the English sentences.

B. UNIT 4 OBJECTIVES

1. Be able to pick out the simple components of compound sentences.
2. Be able to distinguish between non-truth-functional compounds, which cannot be symbolized with our operators, and truth-functional compounds, which can be.
3. Learn the English words and phrases which are associated with, and can be symbolized by, our five sentential operators.
4. Be able to symbolize multiply compound sentences.

C. UNIT 4 TOPICS

1. Simple Sentences

Remember from Unit 2 that a *compound* sentence is one which logically contains another, that is, one in which another sentence logically occurs as an independent clause. Also keep in mind that most sentences with compound subjects and/or predicates must be considered to be compound, since they are equivalent to "expanded" sentences in which the simpler sentences are literally contained. A *simple* sentence is just one that is not compound, one that does not logically contain other sentences as components. We will be symbolizing simple sentences by means of single capital letters.

In order to start symbolizing, then, you must first pick out the simple sentences which go into the compound. Once you have identified them you may want to set them off in some way, perhaps by underlining or putting them in parentheses, in order to make clearer the structure of the English sentence. In the following compound sentence, for instance, the simple components are in italics. *"John has a TV set*, and either *Mary has a stereo* or *David has an FM radio* but *Fred will have the best sound system of all*, if *he loves music."* Having set off the components, we can get a better picture of how the operators combine these sentences into the compound.

Once you have identified the simple sentences, you must pick out capital letters to stand for them. When you can, it is helpful to choose letters which remind you in some way of the meaning of the sentence. Thus for the above sentence, we might use the letters J, M, D, F, and L, respectively, to stand for the simple components. It is important to *use different letters for different simple sentences*, and of course you must *use the same letter for repeated occurrences of the same simple sentence*. Thus we would symbolize "Either John has a TV set or David has an FM radio, but John doesn't have a TV set" as $((J \lor D) \cdot \sim J)$.

It is sometimes not possible, of course, just to underline the simple sentences in a compound as we have suggested. Compound sentences which are "compacted" by the use of compound subjects and/or predicates do not lend themselves to this treatment, since the components do not occur separately. In the sentence "Tom, Mary, and John are all going to the party," for instance, the simple components are "Tom is going to the party," "Mary is going to the party," and "John is going to the party," but none of these occurs separately in the compound sentence. What you need to do in such cases is to work out the "expanded" version of the sentence, and then identify and choose capital letters for the simple components which are literally contained in the "expanded" version. The expanded version of the above sentence would be *"Tom is going to the party* and *Mary is going to the party* and *John is going to the party"* (where the simple components have been italicized). We might abbreviate the simple components, in order, as T, M, and J, and we could then symbolize the compound sentence as $((T \cdot M) \cdot J)$. It will be extremely important to remember that *compacted sentences, sentences with compound subjects and/or predicates, must be considered to be compound sentences and must be symbolized accordingly*.

It will always be important in symbolizing to be very explicit about which capital letter stands for which simple sentence. We will adopt the convention in this book of using a triple bar between a capital letter and an English sentence to indicate that the capital letter is the abbreviation for, the symbolization of, that English sentence.[1] We might have written

---

[1] Thus the triple bar will do "double duty": as a symbol for "if and only if," and as a symbol indicating abbreviations. There should be no confusion between the two, however, since it will always be clear from the context which sense is being used.

above, for instance, "$T \equiv$ Tom is going to the party; $M \equiv$ Mary is going to the party; $J \equiv$ John is going to the party." Such an explicit statement of abbreviations can be considered a kind of "dictionary," which makes possible a translation between the English sentence and its symbolic counterpart. When this "dictionary" is given to you in the exercises, as it almost always will be, you should use exactly those abbreviations, since otherwise it will be impossible, when checking your answers, to tell whether or not you have done the problems correctly.

It is also important to remember that the capital letters should always be used to abbreviate simple, that is, *unnegated* sentences. You would *never* use N, for instance, to stand for "Nixon is not president." You must use N to stand for "Nixon *is* president" (the positive version), and then symbolize that Nixon is *not* president by using the negation sign. The correct symbolization for "Nixon is not president" would thus be $\sim N$. As we saw in Unit 2, negated sentences must be considered to be compound, since they contain simple sentences as components; thus they may not be symbolized by means of a single capital letter.

One final point, which may seem theoretical now but which will be important later on, is that the capital letters are to be considered as *abbreviations* for the simple sentences, shorter ways of writing them. Thus their status is that of ordinary, meaningful English sentences, with particular truth values. The capital letter is *not* to be considered a variable, which stands for no sentence in particular, but is to be thought of as a *constant*, a definite, meaningful entity. Again, the capital letter should be considered as simply an easier, shorter way of writing down the English sentence.

2. Truth-Functional and Non-Truth-Functional Compounds

As mentioned in the introduction to this unit, not every English compound sentence can be symbolized by means of our five sentential operators. In particular, we will not be able to represent the structure of sentences whose major operators are not truth-functional, such as "John went to the party because Mary was going to be there". To represent this by means of a truth-functional operator, such as the horseshoe, would be to claim erroneously that the truth value of the compound could be determined by the truth value of the components. But as we saw in Unit 3, the truth value of compound sentences using "because" cannot be so computed.

What do we do, then, with operators such as "because," or "possibly"? We may distinguish between *truth-functional* and *non-truth-functional* compounds; the former have a truth-functional major operator and the latter have a non-truth-functional major operator. The question is how to symbolize non-truth-functional compounds, sentences that have a major operator such as "because" or "after." Since we cannot represent their structure by means of truth-functional operators, and these truth-functional operators are all we have in classical logic, we simply cannot represent their structure at all, so we must represent them as noncompound, simple sentences. Thus we will symbolize them by means of a single capital letter. We might symbolize "Mary went to the party because she did not feel like studying," for instance, just as M.

Of course, it may happen that we have two non-truth-functional compounds joined together with a truth-functional operator, such as "John went to the party because Mary was going to be there, and Mary went to the party because Fred was going to be there." In this case, if we let J stand for "John went to the party because Mary was going to be there" and M stand for "Mary went to the party because Fred was going to be there," we could symbolize the sentence as $(J \cdot M)$.

There are a great many operators in English which are truth-functional, and which can be symbolized by our logical operators, aside from the simple ones we have mentioned so far ("and," "or," "if-then," "if and only if" and "not"). Some of these other English operators are "but," "still," "unless," "only if," "neither-nor." These are only a small sample of the operators you will be learning to symbolize in the next section.

3. Symbolizing English Operators

a. Conjunction. The truth-functional operator "and," as its truth table indicates, has the logical force of stating that both conjuncts are true. Thus, any operator in English that has the same logical force can be symbolized as a conjunction. If we say, for instance, "John likes TV but Mary hates it," we are making two separate claims: that John likes TV and that Mary hates TV, and the sentence could be symbolized as $(L \cdot H)$. There is, of course, in the ordinary meaning of "but" more than *just* conjunction; otherwise we would have no use for a separate word. But what more there is in the term is a matter of suggestion, or style, or rhetoric, rather than a matter of logical content. The *logical* content of "but" is simply that both of the component sentences are true, and this is why we can symbolize it as a conjunction.

There are many operators in English that have the same logical force as "and," and which can therefore be symbolized with the dot. It would be impossible (unless perhaps one were Webster) to come up with a *complete* list of such words and phrases, but some of the most common are the following: "however," "nevertheless," "still" (used as an operator rather than an adverb), "but still," "although," "even though," "also," "and also," "not only-but also," "while" (used as an operator), "despite the fact that," "moreover," and so on. Some examples of sentences using these operators, which would be symbolized with the dot, are given below, along with their symbolizations.

John loves Mary; however, she barely tolerates him. $(J \cdot T)$
It's raining; nevertheless, we will go on a picnic. $(R \cdot P)$
John is a bit flaky; still, I like him. $(F \cdot L)$
I know it's a beautiful day, but still, I want to stay home and read. $(B \cdot S)$
John got the job, although he didn't wear a tie to the interview. $(J \cdot \sim T)$
I won't give up, even though it is not certain that I will succeed. $(\sim G \cdot \sim S)$
John is a sweety; also, he's rich. $(S \cdot R)$
Mary likes classical music and also rock. $(C \cdot R)$
Mary is not only a musician, but also a first-rate scientist. $(M \cdot S)$
'But' is truth-functional, while 'because' is not. $(B \cdot \sim C)$
John will marry Anne, despite the fact that she can't cook. $(J \cdot \sim C)$
John doesn't have a job; moreover, he can't cook. $(\sim J \cdot \sim C)$

This should give you some idea of the wide variety of English sentences that can be translated as conjunctions. One thing you must be careful about, however, is that there are some uses of these operators which are not truth-functional; sometimes, in fact, the words are not even used as operators. "Still," for instance, is often used as an adverb, as in "It is still raining." Sentences like this, however, should give you little trouble, since it is clear that the function of the word in this case is not that of an operator. What is a little trickier is the use in non-truth-functional compounds of operators which are often, or usually, truth-functional. In the sentence "John waited in the car while Mary cashed a check," for instance, the "while" indicates not only that both things were done, but also that they were done *simultaneously*. It would not be correct to symbolize these sentences with the dot, because that would imply that *so long as both sentences were true* the compound would be true, and this is simply not the case. In order for the sentence to be true, not only must both conjuncts be true, but the two events mentioned must have occurred at the same time. Thus, it would be wrong to represent it with the dot. If you keep these special cases in mind, however, as exceptions to the general rule, you should have no trouble symbolizing conjunctions.

b. Disjunction. In general, you will have little difficulty recognizing when to symbolize a sentence by means of the '$\lor$'. Almost any sentence which contains an "or," either alone, paired with an "either," or in the form "or else," can be symbolized as a disjunction. Examples would be "Reagan or Bush was president in 1986" $(R \lor B)$, "Either you get out of here or I'll call the police" $(G \lor P)$, and "It's hot in here, or else I'm getting sick" ($H \lor S$). There may be other phrases where disjunction is indicated as well, but you should have little trouble spotting them. We will usually be interpreting "or" sentences in the *inclusive* sense, but, as noted in Unit 3, there is an *exclusive* sense of "or" as well, which means "one or the other *but not both*." An example of this might be the following: "John wants to be a lawyer or a physician, but not both." We could symbolize this, as indicated in Unit 3, by conjoining to the inclusive disjunction ($L \lor P$) a clause which explicitly says "not both," which in this case would be $\sim (L \cdot P)$. We could then symbolize the entire sentence as ($L \lor P$) $\cdot \sim (L \cdot P)$. Unless there is a phrase explicitly indicating that an exclusive disjunction is intended, however, you should symbolize the "or" as an inclusive disjunction.

c. Negation. "And" and "or" are quite straightfoward, and will probably provide few challenges. "Not," however, can be tricky, and we need to discuss some special problems that come up with negated simple sentences, and some that arise with negated compounds. As noted in Unit 2, negations of simple sentences, such as "John is not a good student," must be symbolized by means of a tilde prefixed to a simple sentence letter. We may *not* use sentence letters alone to symbolize negated sentences, since negated sentences are compounds, and our capital letters are supposed to stand only for simple sentences (or non-truth-functional compounds.) In the sentence above we could let J stand for "John is a good student," and then symbolize the compound "John is not a good student" as $\sim J$.

When negations occur in a sentence, however, it is not always crystal clear whether it is the whole sentence which is being negated, so that we must consider it a compound, or whether it is just a particular predicate that is being negated, such as "unusual." If we say "John is unlucky," for instance, this should probably *not* be interpreted as $\sim L$, where L means "John is lucky." $\sim L$, would then say "John is not lucky," which might simply mean that he doesn't win sweepstakes; "John is unlucky," however, probably means something stronger, for instance, that every time he walks under a ladder a bucket of paint falls on his head. There is some element of judgment involved in determining whether a sentence should be considered a negation or a simple sentence with a negative predicate. Since in most cases you will be given the letters which are supposed to represent simple sentences, you will not have to worry about this after Exercise 1, but you will need to be aware of the problem when you do your own symbolizations.

Negations, of course, may operate on compound sentences as well as simple ones; a sentence of any degree of complexity may be negated. Negated compound sentences are the source of a lot of logical grief, and you will need to be very careful with them. There are two very common phrases which are particularly bothersome, and which students generally have trouble with at first (sometimes also at last, unfortunately). These phrases are "not both" and "neither." The *not both* is a *negated conjunction*, while the *neither* is a *negated disjunction*. If we say, for instance, that *not both* Carter and Reagan were elected in 1980, we are negating a conjunction, and the proper symbolization would be to place a tilde in front of a conjunction: $\sim (C \cdot R)$. On the other hand, if we say that neither (not either) Nixon nor Ford is president, we are negating a disjunction, and the appropriate symbolization would be to place the tilde in front of a disjunction: $\sim (N \vee F)$.

Part of the confusion comes from the fact that there are equivalent ways of saying the same thing: two different, and equally correct ways of symbolizing "not both" and "neither." To say that *not both Carter and Reagan were elected in 1980*, for instance, is to say that *one or the other of them was defeated*, that is, *either Carter was not elected or Reagan was not elected*. This could be symbolized as $(\sim C \vee \sim R)$, and is just as correct as $\sim (C \cdot R)$. To say that *neither Nixon nor Ford is president* is to say that *both of them are absent from the current presidential scene*, that is, *Nixon is not president and also Ford is not president*. This could be symbolized as $(\sim N \cdot \sim F)$ as well as $\sim (N \vee F)$. In general, *the negation of a conjunction is equivalent to a disjunction with the disjuncts negated, and a negated disjunction is equivalent to a conjunction with both conjuncts negated*. These equivalences are so important in logic that they have been enshrined as rules, and are generally referred to as "DeMorgan's Laws." These rules tell us that the two forms $\sim (p \cdot q)$ and $(\sim p \vee \sim q)$ are interchangeable, as are the forms $\sim (p \vee q)$ and $(\sim p \cdot \sim q)$. The most straightforward symbolization for "not both" is probably $\sim (p \cdot q)$, and for "neither" is probably $\sim (p \vee q)$, but either of their equivalents will do as well.

What may be confusing, and what you will have to watch very carefully, is that $\sim (p \cdot q)$ is *not* equivalent to $(\sim p \cdot \sim q)$, and $\sim (p \vee q)$ is not equivalent to $(\sim p \vee \sim q)$. A couple of examples should make this clear. If your instructor solemnly informs you that you will *not* get *both* an A and a B in his introduction to basket-weaving course, $\sim (A \cdot B)$, you will probably not see any cause for alarm. On the other hand, if he tells you you will *not* get an A *and* will *not* get a B, $(\sim A \cdot \sim B)$, you may very well be unhappy, since this implies you will get a C or worse. The two statements are not at all the same; one is utterly trivial, and the other says something significant. Or suppose an instructor tells you, in a very challenging course, that you will get *neither* a D nor an F, $\sim (D \vee F)$. This may be a relief, since it means you will get at least a C. On the other hand, if she tells you that you will either not get a D or not get an F $(\sim D \vee \sim F)$, she is only telling you what you could have learned yourself from reading the college catalogue section on grades. It is certainly no cause for reassurance, since it still leaves open the possibility that you *will* get one or the other (just not both)!

There are other compounds, of course, which may be negated. Someone may say, for instance, "It is not true that John will quit school if he fails logic." Here what is being *denied* is that he will quit school *if* he fails logic; that is, we are negating $(F \supset Q)$, so the proper symbolization would be $\sim (F \supset Q)$. Or we might say, "It is not true that Mary will come to the party if and only if John does not." This would be symbolized as a negated biconditional, $\sim (M \equiv \sim J)$. We will say more about negated compounds when we begin to symbolize more complex sentences in Section 4.

d. Conditional. In symbolization the conditional is the most challenging of our five operators, and requires the most discussion. This is partly because we have to be careful *not* to use it where it is unwarranted (as in causal statements) and partly because there are so many kinds of sentences for which it *is* appropriate. In the paragraphs below, we will talk about the many English operators, such as "unless" and "only if," that can be symbolized by means of the horseshoe.

Sentences of the form "If p, then q" are the most obvious candidates for the use of the horseshoe. The sentence "If it doesn't rain, then we can have a picnic," for instance, can be symbolized as $(\sim R \supset P)$. In some cases the "then" may be omitted, as in "If John was not at the party, he is sick," which could be symbolized as $(\sim P \supset S)$. It is the "if," rather than the "then" which is the clue to the conditional. In general, sentences of the form "if p, then q" or "if p, q" will be symbolized as $(p \supset q)$, where *the clause immediately following the "if" serves as the antecedent of the conditional.*

It is extremely important to realize that the first clause in the *English* sentence is *not* necessarily the antecedent of the conditional. In many cases in English we reverse the order of antecedent and consequent for stylistic reasons. We might very well put the sentences above as "We can have a picnic if it doesn't rain" and "John is sick if he was not at the party." The *logical force* of the sentences, however, *what is being claimed*, is exactly the same. In the sentence about John, for instance, we are asserting that, given one thing, John's not being at the party, we are entitled to state the other, that John is sick. Since the symbolization is supposed to represent the *logical structure* of the sentence rather than the typographical order, and since the logical structure is the same in both cases, we would symbolize both sentences about John as $(\sim P \supset S)$. Again, *it will be the clause immediately following the "if" which becomes the antecedent of the horseshoe statement, no matter where the "if" occurs in the English sentence.*

There are many other phrases in English which mean approximately the same as "if," such as "provided," "given that," "supposing that," "in the event that," and so on. These sentences will also be symbolized as material conditionals, where the clause following the "provided," the "supposing that," and so on, is the antecedent. "John will get good grades provided he studies," for instance, could be symbolized as $(S \supset G)$. Some examples of basic "if-then" sentences with their symbolizations, are the following; abbreviations for the simple sentences are given in parentheses.

> There will be dirty rain if the volcanic ash cloud passes over. $(V \supset D)$
> $(D \equiv$ There is dirty rain; $V \equiv$ The volcanic ash cloud passes over)[2]
> The cat will not scratch the furniture, provided you train her. $(T \supset \sim S)$
> $(S \equiv$ The cat scratches the furniture; $T \equiv$ You train her)
> If John quits school, he will regret it. $(Q \supset R)$ $(Q \equiv$ John quits school;
> $R \equiv$ He regrets it)
> Supposing that interest rates remain high, the housing industry will not recover any time soon. $(I \supset \sim R)$ $(I \equiv$ Interest rates remain high; $R \equiv$ The housing industry recovers some time soon)
> Rescue teams will be launched in the event that the volcano explodes. $(V \supset R)$ $(V \equiv$ The volcano explodes; $R \equiv$ Rescue teams are launched)
> Provided that Mary goes with him, John will move to Alaska. $(G \supset A)$
> $(G \equiv$ Mary goes with him; $A \equiv$ John moves to Alaska)

It is extremely important to distinguish between "if" and "only if." "q if p," as we have just seen, is symbolized as $(p \supset q)$. "q only if p," however, says something quite different, just the converse, in fact, and will be symbolized as $(q \supset p)$. The difference is particularly clear if we consider the example of a lottery. The sentence "You will win the state lottery *only if* you have a ticket" is certainly true, since that is how lotteries are played, but "You will win the state lottery *if* you have a ticket" is almost certainly false, so the two sentences cannot mean the same. To say you will win *if* you have a ticket would be symbolized as $(T \supset W)$ (where $T \equiv$ You have a ticket and $W \equiv$ You win). To say you will win *only if* you have a ticket, however, is to say that the only way you can win is by having a ticket, so that *if* you win, then you must have had a ticket. Thus we would symbolize the sentence "You will win the state lottery only if you have a ticket" as $(W \supset T)$. Another example is the sentence "It will snow only if it is cloudy." This says that the *only* time it will snow is when it is cloudy, so that *if* it snows, then we can infer that it is cloudy. The sentence is thus symbolized as $(S \supset C)$. Note that this does *not* imply that

---

[2] It should be noted at this point that tense is generally irrelevant in symbolizing. Thus if a clause like "It will rain" occurs in a conditional, it is perfectly appropriate to use it interchangeably with the tenseless "It rains." Simple sentences will usually be given in this tenseless form.

if it is cloudy then it will snow; it might be 80°F outside! Thus it would *not* be correct to use $(C \supset S)$.

There is another correct way to symbolize "only if" sentences. In the second example, for instance, we might say that *without* clouds it cannot snow, so that if it is *not* cloudy, then it will *not* snow. This would be symbolized as $(\sim C \supset \sim S)$, and is equivalent to the symbolization $(S \supset C)$ given originally. In the first example, to say that you will win *only if* you have a ticket is to say that *without* a ticket you *won't* win. In other words, if you don't have a ticket, then you won't win, which may be symbolized as $(\sim T \supset \sim W)$. This is equivalent to the original symbolization $(W \supset T)$. The following patterns summarize the symbolizations for "only if" and show the difference between this operator and the simple "if."

> "*q* if *p*" is symbolized as $(p \supset q)$
> "*q* only if *p*" is symbolized as $(q \supset p)$
> or as $(\sim p \supset \sim q)$

"Only if" is the phrase that students generally have the most trouble with. What you should do is to memorize the standard patterns given in the paragraph above, and also keep in mind examples that are particularly clear, such as the lottery example. If you remember that "You win only if you have a ticket" is symbolized as $(W \supset T)$, then it will be easier to do analogous sentences such as "The dog bites only if he is provoked" $(B \supset P)$ or "John will get an A only if he quits partying" $(A \supset Q)$.

To complicate things just a bit further, the order of clauses in "only if" sentences may be reversed, just as in "if" sentences. "Only if the dog is provoked will it bite" and "Only if he quits partying will John get an A" are symbolized as $(B \supset P)$ and $(A \supset Q)$, exactly as in the examples above.

It is important to remember that the "only if" asserts *only* a one-way conditional; this will be stressed again when we come to the biconditional in the next section. It does sometimes happen in ordinary conversation that the converse conditional (the conditional with the antecedent and consequent reversed) is taken for granted, even though not explicitly stated, but it should *not* be included as part of a formal symbolization. The sentence "John gets paid only if he does a good day's work," for instance, would be symbolized as $(P \supset G)$. (Roughly, "If he got paid, he must have done a good day's work.") It would *not* be symbolized as $(G \supset P)$, which would say *if* he does a good day's work he will be paid, or as $(G \equiv P)$, which would say he gets paid if and only if he does a good day's work. It may well be that if he does a good day's work he will be paid, and normally we would probably take this for granted, but *the sentence does not explicitly state that*, and so it should not be included as a part of the symbolization. You should symbolize only what the sentence explicitly says, and not something you take for granted or read into the sentence.

Although there are undoubtedly many other words and phrases in English which are appropriately symbolized by the horseshoe, we will discuss only one more, which is very common and whose symbolization is quite simple: "unless." In the sentence "John's dog will not come home unless it is called," for instance, the "unless" can be read almost literally as an "if not," and the sentence could be reworded as "John's dog will not come home if it is not called," or equivalently, "If John's dog is not called, it will not come home." In this form, it is easy to see that the correct symbolization would be $(\sim C \supset \sim H)$. "p unless q," in general, may be paraphrased as "p if not q," and symbolized as $(\sim q \supset p)$.

There is also another way to symbolize the "unless," which is neatly illustrated by the following example: if we say "Mary will be fired unless she shapes up," we could symbolize this, using the above formulation, as $(\sim S \supset F)$ (If she does not shape up, Mary will be fired). This also means, however, as Mary no doubt realizes, that *either* she shapes up *or* she gets fired; in other words, "shape up or ship out," which could be symbolized as $(S \vee F)$ (Either Mary shapes up or she will be fired). The "unless," then, may also be symbolized simply as a *disjunction*.

Some examples of English sentences using these various operators, which can be symbolized by means of the horseshoe, are given below, along with their symbolizations. Abbreviations for the simple sentences are given in parentheses.

> Interest rates will drop only if there is a recession. $(D \supset R)$ or $(\sim R \supset \sim D)$ ($D \equiv$ Interest rates drop; $R \equiv$ There is a recession)
>
> Interest rates will not drop unless there is a recession. $(\sim R \supset \sim D)$ or $(R \vee \sim D)$[3]
>
> Unless it stops raining soon, we can't go on a picnic. $(\sim S \supset \sim P)$ or $(S \vee \sim P)$ ($S \equiv$ It stops raining soon; $P \equiv$ We can go on a picnic)
>
> Only if he publicly apologizes will I withdraw my lawsuit. $(\sim A \supset \sim W)$ or $(W \supset A)$ ($A \equiv$ He publicly apologizes; $W \equiv$ I withdraw my lawsuit)
>
> John will come to the party if Mary is there. $(M \supset J)$ ($J \equiv$ John comes to the party; $M \equiv$ Mary comes to the party)
>
> Henry will come to the party only if there is good food. $(H \supset G)$ or

[3] This could also be symbolized as $(D \supset R)$; it means exactly the same as the previous sentence, which could also be symbolized as $(R \vee \sim D)$. I have not given all possible symbolizations for each of these sentences, but only those which seem to capture most directly the sense of the English. You may well be able to figure out symbolizations other than those given here.

$(\sim G \supset \sim H)$ $(H \equiv$ Henry comes to the party; $G \equiv$ There is good food at the party)

Henry will be at the party provided there is good food. $(G \supset H)$

Below is a list of common operators which can be symbolized with the horseshoe, and their schematic representations:

if p, then q	$(p \supset q)$	q only if p	$(q \supset p)$ or $(\sim p \supset \sim q)$
if p, q	$(p \supset q)$	p only if q	$(p \supset q)$ or $(\sim q \supset \sim p)$
q if p	$(p \supset q)$	p unless q	$(\sim q \supset p)$ or $(p \vee q)$
q provided p	$(p \supset q)$	not p unless q	$(\sim q \supset \sim p)$ or
provided p, then q	$(p \supset q)$		$(\sim p \vee q)$

e. Biconditional.   The biconditional, as its name implies, is a two-way conditional. To say "p if and only if q" is really to assert the conjunction, "p if q" and "p only if q." Thus one symbolization of the biconditional could be $((q \supset p) \cdot (p \supset q))$, or equivalently, $((p \supset q) \cdot (q \supset p))$. We have a special operator for this form, however, so we will symbolize "p if and only if q" with the triple bar, as $(p \equiv q)$. It would not be wrong, however, to symbolize it as a conjunction of conditionals (this form will be useful later on), and you should keep in mind that this is the *logical force* of the biconditional. There are relatively few phrases in English that will call for the use of the biconditional, and it is far more common to overuse it—to use it where it is not warranted—than to underuse it and fail to recognize its application. "If and only if," of course, will always call for the use of the biconditional; other common phrases would be "just in case," or "just in the event that." Any English sentence which can be paraphrased into the form "p will happen in exactly the same circumstances in which q will happen" can be symbolized with the biconditional.

One thing you must keep in mind is that the biconditional will *not* be used for sentences of the form "p only if q." The latter phrase, remember, is only a one-way conditional, and is properly symbolized as $(p \supset q)$. "p if *and* only if q" is the phrase which will signal the use of the biconditional, rather than just the conditional.

Some examples of English sentences with their symbolizations, which make use of the triple bar, are given below:

Interest rates will drop if and only if there is a recession. $(D \equiv R)$ $(D \equiv$ Interest rates drop; $R \equiv$ There is a recession)

John will come to the party just in the event that he finishes his paper. $(J \equiv F)$ $(J \equiv$ John comes to the party; $F \equiv$ John finishes his paper)

Mary will come to the party just in the event that she doesn't have a date. $(M \equiv \sim D)$ $(M \equiv$ Mary comes to the party; $D \equiv$ Mary has a date)

The cat will play if she is fed, and she will not play if she is not fed. $(P \equiv F)$, or $((F \supset P) \cdot (\sim F \supset \sim P))$, or $((F \supset P) \cdot (P \supset F))$ $(P \equiv$ The cat will play; $F \equiv$ The cat is fed)

4. Symbolizing Multiply Complex Sentences

In Unit 2, when we were talking about the structure of sentential logic, we noted that formulas may be of any length; they need not contain only one or two operators. We continually encounter sentences such as the following, in newspapers, magazine articles, etc.: "If inflation continues at its present pace and unemployment continues to rise, then there will be no economic recovery, unless either grain prices rise substantially or both OPEC falls apart and gasoline prices drop substantially." In this section your job will be to learn how to analyze and symbolize such sentences containing multiple operators. You already know how to pick out the simple components, and what the meaning of each of the five operators is; what is left is to learn how to fit all the parts together in the proper way.

The best way to work out the logical structure of a multiply compound sentence, once you have identified and abbreviated the simple components, is to pick out the major operator first, write that down, and then go on to symbolize the components. For instance, the sentence, "If John gets A's in both Physics and Chemistry, then, if he gets at least a B in Math, he will get into medical school" is fairly complex, but we can attack it in the following way. We can let P, C, B, and G stand for the simple components (in the order in which they appear), and we should then notice that the *overall* structure of the sentence, that is, its major operator, is an "if-then." The "if" part is that he gets A's in both Physics and Chemistry, and the "then" part is another "if-then," namely, that if he gets at least a B in Math, then he will get into medical school. So we can symbolize the sentence as $((P \cdot C) \supset (B \supset G))$.

It may help to partially symbolize the sentence first, by abbreviating the simple sentences while retaining the English words for the operators. The sentence in quotation marks in the first paragraph of this section, for instance, could be partially symbolized as:

1. If both I and U, then not R unless either P or both O and G.

The next step is to identify the major operator of the sentence; this is a matter of understanding the grammar. In the case above, for instance, it should be clear from the comma that the major operator is the "if-then." Once you have identified the major operator, you can write it down, and place its components in parentheses in the proper order. In the sentence above, the antecedent is (I and U), and the consequent is (not R unless either P or both O and G). We could thus write:

    2. (I and U) $\supset$ (Not R unless either P or both O and G)

The next step, as should be obvious, is to symbolize the components. In this case the antecedent is easy; it is simply a conjunction. So we can write as the next stage:

    3. ($I \cdot U$) $\supset$ (Not R unless either P or both O and G)

The consequent in this example is fairly complicated, and the thing to do is start with what is easiest (which is, of course, a matter of judgment). We know that "not R" is to be symbolized as $\sim R$, and we know that "both O and G" will be ($O \cdot G$). The disjunction "either P, or both O and G" should then be easy to manage: it will look like ($P \vee (O \cdot G)$). We can now write down, as a further preliminary step, the following:

    4. ($I \cdot U$) $\supset$ ($\sim R$ unless ($P \vee (O \cdot G)$))

Now the only thing left is to symbolize the "unless"! Remember that "p unless q" can be symbolized either as ($\sim q \supset p$) or as ($p \vee q$). The latter is certainly easier, so let us finally finish off the symbolization as follows:

    5. ($I \cdot U$) $\supset$ ($\sim R \vee (P \vee (O \cdot G))$)

Note that an equivalent formulation, using the first form for "unless," would be:

    6. ($I \cdot U$) $\supset$ ($\sim (P \vee (O \cdot G)) \supset \sim R$)

If you follow this general method, of identifying and symbolizing the major operator first, and only then moving to the components, you will be more likely to come up with a correct symbolization than if you just try to write everything down from the beginning.

    You may have noticed in the above symbolizations that the outermost parentheses have been omitted. Since we now have several pairs of parentheses to contend with, we will relax our definition of "formula" and will often eliminate parentheses where they are not needed to avoid ambiguity. Thus we will often omit outside parentheses, and also inner parentheses if they are used to join a sequence of conjunctions or a sequence of disjunctions. Since it does not matter how conjunctions or disjunctions are grouped, we may write ($p \cdot q \cdot r \cdot s$), or even just $p \cdot q \cdot r \cdot s$ instead of the more cumbersome (($p \cdot q$) $\cdot$ ($r \cdot s$)), and we may use ($p \vee q \vee r \vee s$) or even just $p \vee q \vee r \vee s$ in place of (($p \vee q$) $\vee$ ($r \vee s$)). Where unnecessary parentheses are eliminated, the formula is generally easier to read. Note, however, that parentheses may *not* be omitted around a sequence of *mixed* conjunctions and disjunctions; $(p \lor q \cdot r)$ would simply be ambiguous. Nor may we omit parentheses in a series of horseshoes; $((p \supset q) \supset r)$ does *not* mean the same as $(p \supset (q \supset r))$.

One thing you must be alert to is the fact that the *logical* structure of the sentence, which will be represented in the symbolization, is not always reflected in the order in which the components appear in the English sentence, especially for conditionals. Conditionals, as we have seen, are not always neatly phrased as "if A then B." They may be stated as "B if A," "B provided A," and so on. You should be safe, however, if you keep in mind that what follows the "if" (or its equivalent such as "provided"), wherever it may appear in the English sentence, *always* becomes the antecedent of the symbolic form, and what follows the "then" becomes the consequent. Thus in the sentence "John will come to the party if Mary does, provided he is not sick," if $J \equiv$ John comes to the party, $M \equiv$ Mary comes to the party, and $S \equiv$ John is sick, the logical structure is "if not S, then if M then J," and the symbolization would be $(\sim S \supset (M \supset J))$. Notice here that the order of the simple components in the symbolization is exactly the reverse of their appearance in the English sentence. The divergence will not always be this extreme, but do watch out for the *logical*, as opposed to the literary, structure of the sentence.

As we have seen in Unit 2, the use of parentheses in complex formulas (and punctuation in the corresponding English sentences) is essential if we are to avoid ambiguity (the exceptions have been noted above). Without parentheses, for instance, the formula $(P \lor O \cdot G)$, could be interpreted in two different, and nonequivalent ways: $(P \lor (O \cdot G))$ or $((P \lor O) \cdot G)$. In general, changing the placement of parentheses changes the meaning of the sentence.

The grammatical structure of the English sentence, and hence its logical form, which will be represented by the symbolization, will in many cases be indicated by the use of punctuation. Semicolons (which should be interpreted as conjunctions) and commas will indicate where the major breaks in the sentence occur. Where there is no punctuation, the structure may be indicated by strategic placement of the operators, or the way in which subjects and predicates are compounded. (Some sentences, of course, are downright ambiguous, and with these the only thing you can do is make the best guess about the intended meaning.) In the sentence "Carolyn likes wine, and John likes beer if the weather is muggy," for instance, the comma indicates that the major break in the sentence comes after "wine," so that the major operator is "and." The sentence is symbolized as $(C \cdot (M \supset J))$, where $C \equiv$ Carolyn likes wine; $J \equiv$ John likes beer; and $M \equiv$ The weather is muggy. If the sentence read "Carolyn likes wine and John likes beer, if the weather is muggy," on the other hand, the major break would come after "beer," so the major operator would be "if." The symbolization then would be $(M \supset (C \cdot J))$.

There is no punctuation in the sentence "John will be disappointed if either Carolyn or Ronald wins the election," but the structure is clear nevertheless. Since the "either," followed by its two disjuncts, comes after the "if," it is clear that the major operator is the conditional, and that the disjunction should be the antecedent. The correct symbolization is $((C \vee R) \supset D)$. In the sentence "John plays tennis and squash but Mary plays neither," there is no punctuation, but from the fact that "John plays tennis and squash" is a compound unit (say, $(T \cdot S)$), and "Mary plays neither" is also a unit (say, $\sim (E \vee Q)$), it is clear that the major operator is "but." The sentence quite clearly should be symbolized as $((T \cdot S) \cdot \sim (E \vee Q))$.

Let us take one more example and apply the above techniques: "If John is going to either study or watch TV, then if Mary plays tennis only if John plays with her, then Mary will make dinner and pout." (Let the simple sentences, in order of their appearance, be abbreviated as S, T, M, J, D, P). Our first step would be:

    1.   If either S or T, then if M only if J, then D and P

Since the first "if-then" is clearly the major operator of the sentence, we next have:

    2.   (either S or T) $\supset$ (if M only if J, then D and P)

and so to:

    3. $(S \vee T) \supset$ (if M only if J, then $(D \cdot P)$)

The major operator of the consequent is clearly "if, then," so we can move to:

    4. $(S \vee T) \supset ((M$ only if $J) \supset (D \cdot P))$

This leaves only the "M only if J," and we know that this should be symbolized as $(M \supset J)$. Our final symbolization, then, will be:

    5. $(S \vee T) \supset ((M \supset J) \supset (D \cdot P))$

In general, when symbolizing complex sentences, once you have identified and abbreviated the simple components, your next move should be to identify and symbolize the major operator. Then go on to symbolize the components, which may themselves require analysis into the major operator of the components and the components of the components. When you have symbolized all the parts, your analysis will be complete.

One final reminder: in symbolizing, especially very complex sentences, there are often many equivalent, and equally correct, ways of putting things. As already noted, "neither" may be symbolized as $\sim (p \lor q)$, or as $(\sim p \cdot \sim q)$, while "not both" is equally correct as $\sim (p \cdot q)$ or as $(\sim p \lor \sim q)$. A sentence of the form $(p \supset q)$ will always be equivalent to $(\sim q \supset \sim p)$, as you saw from "only if." And the biconditional can be symbolized either as $(p \equiv q)$, or as the conjunction of conditionals, $((p \supset q) \cdot (q \supset p))$. Keep this in mind when you are checking your answers in the back of the book; if you don't have *exactly* what is there, you are not necessarily wrong. You may be saying the same thing in another way. If you are in doubt, you should check with your instructor or teaching assistant; it can be very confusing to think that your answer is wrong when in fact it is right. Of course, there are some forms which are not equivalent. If the answer says $\sim (p \cdot q)$, for instance, and you have $(\sim p \cdot \sim q)$, you have simply made a mistake, because the latter is *not* equivalent to the former.

EXERCISES

1.   Identify and abbreviate the truth-functionally simple components of the following compound sentences, and give a complete truth-functional symbolization of the compounds. Keep in mind that only truth-functional compounds can be symbolized by means of our five operators, and that non-truth-functional compounds will have to be symbolized by using single capital letters.

*a.   Man is not descended from monkeys.

 b.   Man is descended either from small primates or great apes.

*c.   Mary is not the most athletic girl in her class, but she is the smartest.

 d.   People like to drive automobiles because it gives them a sense of freedom.

*e.   It is possible that we will run out of oil by the year 2000.

 f.   If Americans don't stop driving big cars, gasoline will have to be rationed.

*g.   Americans will stop driving big cars only if there are comfortable small cars on the market.

 h.   Detroit did not make small cars earlier because there was not enough profit in them.

*i.   John is pleased that his new Honda gets 35 miles to the gallon.

 j.   John drives his van only if he is camping or needs to haul large loads.

*k.   John does not drive his van unless he needs the space.

 l.   It is necessary that we conserve energy, because we may lose Mideast oil.

*m.   John and Mary are married, but not to each other.

n.   John was married before Mary, but after Stephen.

*o. John thinks that Mary married the wrong person, but Mary knows she married the right person.

p.   Neither John nor Bob nor Andrew wants to play pro football.

*q. John and Mary will not both bring pies to the picnic if Ted is going to bring a birthday cake.

r.   Ted will not bring a cake with candles if no one is having a birthday, unless he thinks that candles taste good for dessert.

*s. Either Ted or John will bring the dessert, but not both.

t.   Ted will bring dessert or a salad, but not both, and John will bring salad or wine, but not both.

2.   Symbolize the following, which require frequent use of the horseshoe. Use only the abbreviations provided.

J = John will make supper; W = Mary is working late; M = Mary will make supper; L = John is working late; H = John is very hungry; B = John's boss requires his working late; O = It is a holiday; D = John washes the dishes.

*a. John will make supper only if Mary is working late.

b.   Mary will make supper if John is working late.

*c. John will not make supper unless he is very hungry.

d.   John works late if and only if Mary does not.

*e. Not both John and Mary will make supper.

f.   John will not work late unless his boss requires it.

*g. John and Mary both work late only if it is not a holiday.

h.   John or Mary will work late, but not both.

*i. Neither John nor Mary makes supper if it is a holiday.

j.   Only if John works late and it is not a holiday will Mary make supper.

*k. Mary will make supper if John does not, and only if John does not.

l.   John will make supper and wash the dishes unless it is a holiday or he is working late.

3.   Symbolize the following, using the abbreviations that have been provided.

D = I diet; E = I exercise; G = I gain weight; M = I am motivated; T = I get too tired; P = I am depressed; L = I am lazy; S = I swim; R = I run; W = I drive to work.

*a. If I don't either diet or exercise, I will gain weight.

b.   I will diet only if I am motivated, and I will exercise only if I don't get too tired.

*c.   I will not both diet and exercise unless I am motivated.

 d.   I will exercise if I am motivated, provided I am not either lazy or depressed.

*e.   I won't gain weight if I either swim or run, unless I drive to work and don't diet.

 f.   I will be depressed if and only if I am lazy and neither diet nor exercise.

*g.   If I gain weight if and only if I neither diet nor exercise, then I will either swim or run and will not drive to work.

 h.   I gain weight if and only if I don't diet, provided I am not motivated and neither swim nor run.

*i.   Only if I am motivated and don't get too tired do I diet and exercise.

 j.   Unless I am depressed or lazy, I diet or exercise if I gain weight, provided I am motivated and don't get too tired.

4.   Symbolize the following, using the abbreviations that have been provided.

$G \equiv$ There will be gasoline rationing; $L \equiv$ We lose Mideast oil; $M \equiv$ We continue to consume more than we produce; $R \equiv$ We discover vast new reserves of oil; $N \equiv$ We develop new oil exploration technologies; $F \equiv$ There are financial incentives; $C \equiv$ There is conservation; $S \equiv$ There will be an oil shortage; $O \equiv$ We develop coal conversion techniques; $A \equiv$ Americans drive smaller cars; $V \equiv$ Americans take fewer vacations by automobile; $Z \equiv$ Gasoline prices zoom out of sight.

*a.   There will be gasoline rationing only if we lose Mideast oil and continue to consume more than we produce.

 b.   There will be no gasoline rationing if either we do not lose Mideast oil or we discover vast new reserves of oil.

*c.   We won't discover vast new reserves of oil unless we develop new oil exploration technologies, and we will develop new oil exploration technologies only if there are financial incentives.

 d.   There will not be both gasoline rationing and conservation, but there will be one or the other.

*e.   If there is neither gasoline rationing nor conservation, there will be an oil shortage.

 f.   There will be no oil shortage if we discover vast new reserves of oil, provided there is conservation.

*g.   Provided we develop coal conversion techniques, there will not be both a shortage of oil and gasoline rationing.

 h.   If and only if we develop coal conversion techniques will we avoid both gasoline rationing and conservation.

*i.   There will be an oil shortage and gasoline rationing if we lose Mideast oil and do not conserve, unless we discover vast new reserves.

 j.   Either there will be an oil shortage or gasoline rationing, or both, unless we conserve and do not lose Mideast oil.

*k. There will be no gasoline rationing and no oil shortage, provided we either develop coal conversion techniques or discover vast new reserves of oil.

l. There will be gasoline rationing if there is an oil shortage, but there will be an oil shortage only if there are no financial incentives and no development of new oil exploration technologies.

*m. Unless we discover vast new reserves of oil, there will be an oil shortage if there is neither gasoline rationing nor conservation.

n. There will be an oil shortage and gasoline rationing unless either there is conservation and we develop coal conversion techniques, or Americans drive smaller cars and take fewer vacations by automobile.

*o. Americans will either drive smaller cars or take fewer vacations by automobile, but they will not do both unless either there is gasoline rationing or gasoline prices zoom out of sight.

UNIT 5

Truth Tables for Testing Validity

A. INTRODUCTION

In Unit 1 it was emphasized that the validity of arguments is a matter of their form rather than of their content. In Units 2 and 3 you learned the structure of sentential logic, and the rules for computing the truth values of compound propositions. In Unit 4 you learned how to analyze the structure of English sentences and to symbolize them, to represent their form by means of our logical language. In this unit you will finally put it all together and learn how it is possible in sentential logic to test argument forms, and hence arguments, for validity.

Remember that an *argument* is valid if and only if its form is valid and an *argument form* is valid if and only if it has no counterexample—no instance in which all the premises are true but the conclusion is false. To test an argument form for validity, then, we need to examine its instances, to see whether there are among them any bad cases—any counterexamples. What you will be learning in this unit is the *truth table method*, which is simply a way of setting out all the possible instances of the form (that is, all possible combinations of truth values for the simple component sentences), and then computing the result for each of these possible instances. Once we have constructed our list of possibilities and computed the results, all we need to do is look them over to see whether there are any instances with true premises and a false conclusion. If there are, of course, the form is invalid; if not, it is valid.

What you will learn in Sections 1 and 2 is how to list all possible substitution instances for a form and compute the results; that is, you will learn how to test argument forms for validity using the truth table method. Since the full truth table method can be rather long and cumbersome (for a form with only five variables, for instance, there will be 32 possibilities which need to be inspected), Section 3 will introduce you to some shortcuts: the *partial* method and the *short* method for determining validity, which enable you to bypass some of the tiresome detail of the fully completed truth tables. Finally, in Section 4 we discuss some of the more interesting theoretical aspects of the truth table method, for example, the fact that it is purely "mechanical," and provides what is called a *decision procedure* for sentential logic.

B. UNIT 5 OBJECTIVES

1. Learn how to construct the *base columns* for a truth table, that is, how to list all possible substitution instances.
2. Learn to test argument forms for validity, using the full truth table method.
3. Learn the definition of "valid argument form" and related definitions, given in the glossary at the end of the unit.
4. Learn the "partial" and the "short" truth table methods.
5. Be able to explain what a decision procedure is, and what are the three things which make possible the decision procedure for sentential logic.

C. UNIT 5 TOPICS

1. Constructing Base Columns for Truth Tables

According to our definition, a valid argument form is one in which there is no possible substitution instance with true premises and a false conclusion. In order to test argument forms for validity, then, we must be able to determine whether or not there is such an instance. Since there are an infinite number of substitution instances for each form, we cannot possibly inspect each actual instance, each meaningful argument, of that form. What we can do is something which serves our logical purposes just as well: we can write down all possible combinations (all possible assignments) of truth values for the variables. Such a list of possible truth-value combinations will be called the *base column* for the truth table, and what you will learn in this section is how to construct these base columns.

You are already familiar with the process of listing truth value combinations for formulas with one or two variables, from your study of the truth tables for our five operators. Remember that since we have a two-valued logic, the variable q can take as substitution instances only formulas which are true or false, and p can then take a true or a false substitution instance in either of these two cases. This gives us the following four possibilities for a form with two variables.

p	q
T	T
T	F
F	T
F	F

If we have three variables, p, q, and r, r can be either true or false, q can be true or false for either value of r, and p can be either true or false for any of these four combinations for q and r. We thus have the following list of possibilities for three variables:

p	q	r
T	T	T
T	T	F
T	F	T
T	F	F
F	T	T
F	T	F
F	F	T
F	F	F

If we were to add a fourth variable, it could be true in all these eight cases for p, q, and r, and also false for each of these cases, yielding 16 possible combinations. A fifth variable would double the number again, since it could be true *or* false for each of the 16 possibilities for the four variables, and so on. *Every time you add another variable, the number of possible combinations will double.* Since we start out with two possible values for a single variable, this gives us a neat formula for the number of possible combinations for any number of variables. In general, for n different variables in a statement form or argument form, the number of possible truth combinations for the variables will be 2^n (*not* $n^2!$).

Notice that given this formula, the number of possible truth combinations gets very large very fast: for 6 variables there are 64 possibilities, for 8 variables 256 possibilities, and for 10 variables over a thousand possibilities. If you had 20 variables in an argument form, the number of possible truth combinations for the variables would be over a million! (You will not be given any truth table exercises with 20 variables.) In any case, you should keep in mind that the number of truth combinations, or as we will be saying, *rows in the truth table*, will always be some power of 2. A truth table may have 8 rows, or 32, or 16, or 64, but *never* 10, or 12, or 24, or 48.

The list of possible truth combinations for the variables will be called the *base column for the truth table*. Each possible combination will be called a *row in the truth table*. We could then say, for instance, that a form with three variables will have a base column of 8 rows, or equivalently, will have 8 rows in its truth table. A form with five variables will have 32 rows in its truth table, etc. The *rows* (as in planted fields) will be the lines going across horizontally, while the *columns* (as in Greek architecture) will be the vertical listings.

It might occur to you that if you have more than just a very few variables, it might be difficult to be sure that you have listed all the possible rows in the truth table. Even if you have the right number of rows, 16, or 32, or perhaps 64, you might have duplicated some and left others out. A simple way to make sure you cover all the possibilities is to list them *systematically*, and to follow the same procedure consistently. The procedure we will be using in this book is the following: first, write the variables *in alphabetical order*, from left to right. Then, starting on the *right*, alternate the truth values for the far right variable. Your column on the far right, then, should read, from top to bottom, T, F, T, F, . . . Moving one variable to the left, *double* the number of T's and F's you use before alternating. Your column second to the right, then, should read, from top to bottom, T, T, F, F, T, T, F, F, . . . Moving one more to the left, you should double the number of T's and F's again, so that you have a list of four T's on the top, followed by four F's directly underneath, followed by another four T's and another four F's, and so on. As you move one variable to the left, always double the number of T's and F's. When you reach the leftmost variable, if you have done things right, you will always have the top half of the column all T's, and the bottom half all F's. (Notice that this is verified by our eight-row truth table for p, q, and r.) You should practice drawing up several base columns, with different numbers of variables, so that it becomes second nature to you.

Although it may sound picky (since you *can* get the same results by listing possibilities haphazardly, as long as you get them all) *you should always draw up your base columns in this systematic manner*. There are several reasons for this unimaginative approach. For one thing, it is much quicker to draw up the tables in a systematic way than to write down the various possibilities randomly, and then have to go back to check that you did not omit or duplicate any rows. For another, if you list the possibilities systematically, you will begin to find patterns in the truth tables for the various formulas, which will both speed up the process of constructing the tables and also serve as a check in case you did something wrong. Finally, if the possibilities are out of order, the final truth table will look different, and in checking your answers it will be difficult to tell whether you have done things right.

As one more illustration of the proper method, we have drawn up the list of 32 possibilities for a form with five variables. The portion of the table in the enclosed rectangle would be the truth table for four variables. The variables for this 16-row table are listed *underneath* the table instead of on top. Notice that in both cases the results conform to our standard procedure.

p	q	r	s	t
T	T	T	T	T
T	T	T	T	F
T	T	T	F	T
T	T	T	F	F
T	T	F	T	T
T	T	F	T	F
T	T	F	F	T
T	T	F	F	F
T	F	T	T	T
T	F	T	T	F
T	F	T	F	T
T	F	T	F	F
T	F	F	T	T
T	F	F	T	F
T	F	F	F	T
T	F	F	F	F
F	T	T	T	T
F	T	T	T	F
F	T	T	F	T
F	T	T	F	F
F	T	F	T	T
F	T	F	T	F
F	T	F	F	T
F	T	F	F	F
F	F	T	T	T
F	F	T	T	F
F	F	T	F	T
F	F	T	F	F
F	F	F	T	T
F	F	F	T	F
F	F	F	F	T
F	F	F	F	F

| | p | q | r | s | |

Finally, note that the rows in the base columns represent all possible substitution instances of the argument form. Since this is a two-valued logic, any substitution instance of the form must have elementary sentences that are either true or false. But the base columns list *all* possible combinations of truth or falsity for the elementary sentences, so any substitution instance will be represented by one of the rows in the base columns. This means that we can talk about rows in the truth table instead of substitution instances; in fact, in the next sections we will use the following terms more or less interchangeably: *substitution instances, rows in the truth table, assignments of values to the variables,* and (*possible*) *combinations of truth values for the variables.*

2. The Truth Table Test for Validity

Once you have drawn up the base columns for an argument form, testing the form for validity is relatively easy: you simply compute the truth values for the premises and the conclusion of the argument form for each row in the truth table, and then check the completed table to see whether there are any rows with all the premises true and the conclusion false. If there are, then the form is invalid, and if there are not, the form is valid.

A *row in the truth table that has all the premises true with the conclusion false is called a counterexample to that argument form,* since it shows that the form is not valid. Another definition of validity, then, could be the following: *an argument form is valid if and only if it has no counterexamples.*

The procedure for testing an argument form for validity, including drawing up the base columns, can be summarized as follows:

1. List the premises and conclusion horizontally at the top of the table.
2. List all the variables that occur anywhere in the premises or conclusion at the top left, in alphabetical order.
3. Write down the list of all possible combinations of truth values for the variables. (That is, construct the base columns.)
4. Compute the truth values of premises and conclusion for each possible combination, that is, for each row in the truth table.

The result of these computations will be the *complete truth table for the argument form.*

5. Check the truth table for counterexamples.

Let us take a relatively simple example and test it for validity.

p	q	$(p \supset q)$,	$(q \supset p)$,	$(p \lor q)$	$/\therefore$ $(p \cdot q)$
T	T	T	T	T	T
T	F	F	T	T	F
F	T	T	F	T	F
F	F	T	T	F	F

Here we have three premises, $(p \supset q)$, $(q \supset p)$, and $(p \lor q)$, and the conclusion is $(p \cdot q)$. Since we have only two variables, we have four rows in the truth table. You should have no trouble understanding the truth tables for the first and third premises and for the conclusion. We arrived at the table for the second premise by noting that a conditional is false only if the antecedent (in this case q) is true, while the consequent (in this case p) is false. This happens in the third row and in that row only; so the third row for this premise is false, while all the other rows are true. This form then turns out to be *valid* because there is no row in which *all* the premises are true but the conclusion is false. In each of the three rows in which the conclusion is false, one of the premises is false as well. Thus we have no counterexample.

Let us take a slightly more complicated example, in which you must compute the value of the subformulas of the premises and conclusion before you can compute the value of the formulas themselves. What you should do in cases like this is to *write down the values of the subformulas directly underneath the major operators for those subformulas.* (This is the procedure we followed in Unit 3, when we were computing the values for single substitution instances.) When you have computed the values for the subformulas (the components), you then use those values in computing the value of the whole, applying the truth table rules which you learned in Unit 3. You may want to repeat the values for the single letters under the letters, although it is not really necessary—you can simply refer back to the base columns for the values of the variables.

p	q	$\sim$	$(p \cdot q)$	$\supset$	$\sim$	$(p \lor q),$	$\sim$	$(p \cdot \sim q)$		$/\therefore \sim$	$(q \cdot \sim p)$	
T	T	F	T	[T]	F	T	[T]	F	F	[T]	F	F
T	F	T	F	[F]	F	T	[F]	T	T	[T]	F	F
F	T	T	F	[F]	F	T	[T]	F	F	[F]	T	T
F	F	T	F	[T]	T	F	[T]	F	T	[T]	F	T
		(2)	(1)	(3)	(2)	(1)	(3)	(2)	(1)	(3)	(2)	(1)

Explanation: In the first premise, we first compute the values for $(p \cdot q)$ and $(p \lor q)$. We write down those values underneath the dot and the wedge, respectively. Then, since both formulas are negated, we write down just the *reverse* values under their negation signs. Finally, we compare the tables under $\sim (p \cdot q)$ and $\sim (p \lor q)$, and we see that in the second and third rows $\sim (p \cdot q)$ is true, while the consequent $\sim (p \lor q)$ is false. The conditional, the premise as a whole, is false in these two rows and true in the others. In the second premise and in the conclusion, we first write down the values for $\sim q$ and $\sim p$, which will be the reverse of those for q and p. We then compute the *conjunctions* $(p \cdot \sim q)$ and $(q \cdot \sim p)$, and write their values underneath the dots. Finally, we compute the values of the negations of the conjunctions, which will be exactly the opposite of those for the conjunctions. The numerals at the bottom of the table indicate the order in which the computations are done for each of the premises and the conclusion. The *highest* number indicates the *last* step in the computation process, and so indicates the truth table for the formula as a whole. We have outlined this column to make clear that it is the result of the final computation. It is this column which you will inspect to determine validity.

Finally, having completed the table, we inspect it to see whether it has any counterexamples, or invalidating instances. In this case there is only one row, the third, in which the conclusion is false. In this row, however, the first premise is also false, so again, the argument form is valid. It has no instance in which both of the premises are true but the conclusion is false.

A few practical hints on constructing truth tables: it is, of course, possible to work horizontally, starting with the top row (where the values of the variables are all true), and working your way across to the right. In general, however, it will be much easier to work vertically. That is, instead of starting with a particular *row*, at the left, start with a particular *formula*, at the top, and work out the entire *vertical column* for that formula. Then move on to the next most complex formula, and compute the entire vertical column for that one, and so on. In our second example, for instance, we know immediately the whole truth tables for our subformulas $(p \cdot q)$ and $(p \lor q)$; there is no deep thought required. Then, we know that negations reverse truth values, so the truth tables for the negations will be exactly the opposite. You will find that your truth tables go much more quickly if you work in this way rather than trying to work your way across from left to right for each separate substitution instance.

It is very important that you *list the truth tables for the subformulas directly underneath the operators for those formulas*; otherwise you will get terribly confused about which formula has which truth table. Also, it is a *very* good idea to use lined paper so that the rows are even. Another good idea is to use a numbering system, such as the one we will be using in the examples here, to keep track of the order in which your computations for each formula are done. You might also want to set off the truth tables for the final results, the premises and conclusion as a whole, from the tables of the subformulas. You can do this by underlining or circling your final results. This makes it much easier, especially in complex examples, to inspect the table once you have completed it, to see whether there are any counterexamples. If you are looking at the wrong column, you may miss a counterexample, or may think that there is one when there is not. We will, in the following examples, outline our final results, as well as number the preliminary computations.

If you are having trouble understanding the results of the computations given here, you probably have not learned your truth tables for the five operators well enough. Go back and review them; memorize the tables, and be sure you know the informal verbal rules which state what the results of the computations must be.

In an argument form with three variables, of course, there will be eight rows in the truth table, and in general these computations won't be quite so automatic as they often are with only four rows. You will have to look a little more carefully to see that you are using the right values for the components, and you must be a little more self-conscious about applying the truth table rules. Let us test one such example:

p	q	r	(q	⊃	~	(p ∨ r)),	(p · q)	∨	(p · r)	/∴	(p ⊃ r)	⊃	(p ⊃ q)
T	T	T	T	[F]	F	T	T	[T]	T		T	[T]	T
T	T	F	T	F	F	T	T	[T]	F		F	[T]	T
T	F	T	F	T	F	T	F	[T]	T		T	[F]	F ←
T	F	F	F	T	F	T	F	[F]	F		F	[T]	F
F	T	T	T	F	F	T	F	[F]	F		T	[T]	T
F	T	F	T	T	T	F	F	[F]	F		T	[T]	T
F	F	T	F	T	F	T	F	[F]	F		T	[T]	T
F	F	F	F	T	T	F	F	[F]	F		T	[T]	T
			(1)	(3)	(2)	(1)	(1)	(2)	(1)		(1)	(2)	(1)

In the first premise, we computed the value of (p ∨ r) first, then the value of ~ (p ∨ r), and we then compared this table to the table for q to get our final result, listed under the horseshoe and above the (3). In the second premise we first computed the tables for (p · q) and (p · r), and then computed the table for the disjunction of the two formulas. In the conclusion we first computed the tables for (p ⊃ r) and (p ⊃ q), and then applied the rule for the horseshoe to those results, which gives us a false in the third row, where the antecedent is true and the consequent is false.

When we inspect this table, being sure to look in the end *only* at the final results, which have been outlined, we find that there *is* an instance with true premises and a false conclusion, in the third row. *This is a counterexample, and so this argument form is invalid.* It is a good idea to indicate the invalidating row in some way, perhaps by an arrow, as we have done here.

Notice that in the example above there was only one counterexample (only one row with a false conclusion, in fact). The fact that the other seven rows are not counterexamples means nothing. *It takes only a single counterexample to show that an argument form is invalid,* no matter how many rows there are in the truth table. Even in a form with 20 variables, and thus over a million truth table rows, if there are 1,048,575 rows which pass the test, which are not counterexamples, and a single row out of the million which is a counterexample, the argument is still invalid. It takes only one bad apple to spoil the batch. This is what deductive validity is all about. We want, in a deductive argument, to be *sure* that if the premises are all true, then the conclusion will be true as well. Thus, a single counterexample is enough to guarantee invalidity.

So far we have talked only about testing argument *forms* for validity, rather than specific arguments. Remember, however, that an *argument* is valid just in case its *form* is valid, so to test a specific argument you need to determine and then test its form. This requires nothing you don't already know. To determine the form of an argument is simply to symbolize its sentences, which you learned to do in Unit 4. The symbolized argument will look exactly like its argument form, except for the use of constants, capital letters, in place of variables. Thus, you can test a symbolized argument in the same way as you test an argument form, by listing the sentence letters, making up the base columns, and computing the truth values for premises and conclusion. Again, if there is a counterexample, the argument is invalid, and if there is no counterexample, the argument is valid.

The following argument illustrates this procedure: "Neither Bob nor John will attend the reception. Mary will attend only if Bob attends and John doesn't. Therefore, Mary will not attend the reception." The symbolization for this argument is $\sim (B \vee J)$, $M \supset (B \cdot \sim J) \mathbin{/\therefore} \sim M$. The truth table for the argument, which shows that it is valid, is given below.

B	J	M	$\sim$ $(B \vee J)$	$M \supset (B \cdot \sim J)$	$/\therefore$	$\sim M$
T	T	T	**F** T	**F** F F		**F**
T	T	F	**F** T	**T** F F		**T**
T	F	T	**F** T	**T** T T		**F**
T	F	F	**F** T	**T** T T		**T**
F	T	T	**F** T	**F** F F		**F**
F	T	F	**F** T	**T** F F		**T**
F	F	T	**T** F	**F** F T		**F**
F	F	F	**T** F	**T** F T		**T**

3. Shortcut Validity Tests

As you have seen in Section 2 of this unit, for argument forms with large numbers of variables, the truth tables tend to get out of hand. For only 10 variables, there will be 1024 rows in the truth table (2^{10}) and for 20 variables there will be over one million (2^{20} or 1,048,576, to be exact). Even for 5 and 6 variables, 32 and 64 rows will be required, and you can probably think of better ways to spend your afternoons than making up 64-row base columns. Fortunately, it is possible to bypass some of this tedious detail, provided you have a thorough understanding of what the truth table test for validity is all about. We discuss two of these shortcut methods in this section: what we will call the *partial* method and what we will call the *short* method. (The appropriateness of these names will be evident.)

We test the validity of argument forms by checking to see whether they have any counterexamples, substitution instances with true premises and a false conclusion. By using the *partial* truth table method it will be possible to do this without going through all the computations, though we will have to draw up the base columns, to list all the possible instances. In the *short* method we will not even have to list the instances, but will simply try to "zero in" on whatever counterexamples there may be. We will discuss the partial method first.

In many argument forms there are only a relatively small number of ways in which the conclusion can be false. If the conclusion is $((p \cdot q) \supset r)$, for example, there will be only one row out of eight in which it is false (the second row). The form $(p \lor q) \lor (r \lor s)$ in a 16-row truth table will be false only in the last row, where all the variables are false. The rows in which the conclusion is false are, of course, the only possible candidates for counterexamples. (Obviously, if the conclusion is true in a row, that row will not be one with true premises and a false conclusion.) Thus, once we have computed the truth table for the conclusion, the only rows we need to compute for the premises are those in which the conclusion turns out to be false. This information will be enough to tell us whether the form is valid, since the rows we cover are the only ones which could *possibly* contain counterexamples. Let us look at an instance of a form for which we can use this *partial truth table method*:

p	q	r	s	$(p \lor q) \supset (r \lor s)$,	$p \supset \sim r$	/∴	$p \supset (s \lor q)$
T	T	T	T			[T]	T
T	T	T	F			[T]	T
T	T	F	T			[T]	T
T	T	F	F			[T]	T
T	F	T	T			[T]	T
T	F	T	F	T [T] T	[F] F	[F]	F
T	F	F	T			[T]	T
T	F	F	F	T [F] F	[T] T	[F]	F
F	T	T	T			[T]	
F	T	T	F			[T]	
F	T	F	T			[T]	
F	T	F	F			[T]	
F	F	T	T			[T]	
F	F	T	F			[T]	
F	F	F	T			[T]	
F	F	F	F			[T]	

In this example there are only two rows in which the conclusion is false, the sixth and the eighth. (Note that we took another shortcut as well: we didn't bother to compute the values for $(s \lor q)$ in the last eight rows, since we know that the conditional must be true in those rows because the antecedent, p, is false.) In the sixth row, where r is true, the second premise comes out false, so that row is not a counterexample. In the eighth row, where r is false, the first premise turns out false, since $(p \lor q)$ is true and $(r \lor s)$ is false. Thus that row is not a counterexample either, and since these are the only two possible cases, we can conclude that there is no counterexample and that the argument form is therefore valid. We have obviously saved ourselves a good bit of work by testing only those rows in which the conclusion is false.

It may also happen that in computing the *premises*, you find that they are only *true* in a few rows, so that those rows are the only possible candidates for counterexamples. In the following argument form, for instance, there is only one row in which the first premise turns out to be true, so that is the only possibility for an invalidating instance. In this row, however, the conclusion turns out to be true as well, so again there is no counterexample and the form is valid.

p	q	r	$(p \cdot (q \cdot r))$		$p \supset (r \lor \sim q)$		/∴	$p \supset (q \lor \sim r)$	
T	T	T	T	T	T	T		T	T
T	T	F	F	F		F			
T	F	T	F	F		F			
T	F	F	F	F		F			
F	T	T	F	T					
F	T	F	F	F					
F	F	T	F	F					
F	F	F	F	F					

Of course, it may happen that in the course of making up the truth table you run across a counterexample right away. In that case you are entitled to stop right there and declare the form invalid, since it takes only a single counterexample to demonstrate invalidity. This intriguing fact brings us to our next topic, the *short* truth table method.

The *short* truth table method is a way of making a systematic search for counterexamples without even drawing up the base columns. This is a considerable savings in time and effort, since an argument form with ten variables, as noted, has 1024 rows in its base column, and it is certainly not unusual to come across argument forms with ten variables. How can we manage a test for validity, then, without actually going through the process of constructing base columns?

The trick is to learn how to "zero in" on what would be the possible counterexamples, those cases in which we have a false conclusion with all the premises true. What we do is try to *construct* a counterexample; try to assign truth values to the variables which will result in a false conclusion with all the premises true. It is usually easier (though it is not necessary) to start with the cases in which the conclusion is false. Let us take the following example as an illustration of the way in which the short method works. Suppose we have the following argument form: $((p \lor q) \supset \sim r), ((r \cdot s) \lor t), (t \supset (w \supset s)) / \therefore (p \supset s)$. In this example there is only one way to make the conclusion false: by making p true and s false. Now we need to see whether it is possible, with a false conclusion, to make all the premises true. If we can, the argument is invalid, since we will have a counterexample; if we cannot, it is valid. Since p is true, $p \lor q$ is true, so the only way to make the first premise true is to make $\sim r$ true; thus *we must make r false.* Then, given that r is false, $(r \cdot s)$ will be false; so to make the second premise true, *t must be true.* Given that t is true, $(w \supset s)$ must be true to make the third premise true. But even though s is false, we can make $(w \supset s)$ true by making *w false.* Thus *there is a counterexample, and so the argument form is invalid.* It is invalidated by the row, or assignment of truth values, which makes p and t true, and r, s, and w false. (Note that the truth value of q is here irrelevant). And we have been able to *prove* this *without* making up the full 64-row truth table which would otherwise be required. All we needed to do was to find a case with all premises true and the conclusion false, and this we have done. It is convenient to make up a little table when you are finished, displaying the values that yield a counterexample. For the argument form above, we would have

p	q	r	s	t	w
T	T or F	F	F	T	F

You should also, in your exercises, indicate your computations rather than just writing down 'T' for the premises and 'F' for the conclusion. We could represent the results of the computations above, for instance, in the following way, much as we did in Unit 3.

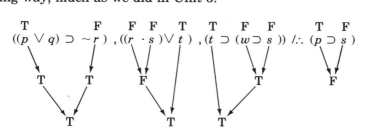

In general, in using the short truth table method to show *invalidity*, the procedure should be to try to make the conclusion false, and then, using those values, to try to assign values to the other variables so that the premises come out true. Notice that it is essential that you *be consistent in your value assignments.* You may not, for instance, make p true in the conclusion and false somewhere else. Whatever value you assign to a variable you are stuck with, for that particular instance. The reason for this should be obvious: we are looking for an instance, a row in the truth table, which results in a counterexample. If you are inconsistent in your value assignments, making a variable true in one place and false in another, then you have not got a single row in the truth table, but are mixing rows up.

The test for invalidity will not always proceed exactly as in the example above. In some cases you may want to start with the premises; if you have a pure conjunction as a premise, say, $((p \cdot q) \cdot (r \cdot s))$, you know that there is only *one case* in which it will turn out to be true—where p, q, r, and s are all true—so you might start there in trying to construct a counterexample. Also, you need not go from the conclusion to the first premise, then to the second, and so on, as we have done here. Once you have made the conclusion false (or the premise true), you should go next to a formula whose other truth values are *determined* by the values you already have, where there is no question what the values must be. And finally, the problems will not always be so easy; there will be cases, for instance, in which there is more than one way to make the conclusion false, so that you need to run through several possibilities. We will look at some of these more complicated examples later on.

The short truth table method is much easier for proving *invalidity* than for proving validity, since for *in*validity all we need to do is find a single counterexample. For validity, however, we need to show that there cannot *possibly* be a counterexample, and this is much more difficult to establish, though it can be done. What we need to do is to go through a series of steps which show that *there is no possible way*, when the conclusion is false, that the premises can all come out true. That is, we must show that one or more of the premises *must* come out false if the conclusion is false, which shows that there is no counterexample, *no* way of getting all premises true with a false conclusion.

Let us take an example. One of the valid argument forms you will be learning in the proof chapters is the *dilemma*: $(p \supset q)$, $(r \supset s)$, $(p \lor r)$ $/ \therefore$ $(q \lor s)$. Here, in order to make the conclusion false, q and s must both be false. If q is false, the only way to make the first premise true is by making p false. The only way to make the second premise true, given that s is false, is to make r false. Thus, in order for the conclusion to be false and the first two premises true, we *must* have p, q, r, and s all false; it is the only possible way we could begin to get a counterexample. But then notice that the third premise comes out false. Hence, with two true premises and a false conclusion, the third premise is false, which means that there is *no way* to have *all* premises true with the conclusion false. Hence there can be no counterexample, and so the argument form is *valid*. We must always go through this sort of extended argument in order to prove that an argument form is valid, since we have to show that it is *impossible* to get a counterexample. This is obviously more difficult than just coming up with a single counterexample, as we would for invalidity.

For validity, the format must always be to run through a short argument, which shows beyond a shadow of a doubt that there can be no counterexample. It will be helpful to number the basic steps in this demonstration. Let us take one more argument form to illustrate:

$$((p \cdot q) \supset r), (r \supset (s \cdot t)) / \therefore \sim s \supset (\sim p \vee \sim q)$$

(1) The *only* way to make the conclusion false is to make $\sim s$ true and $(\sim p \vee \sim q)$ false. This means making s *false* and $\sim p$ and $\sim q$ *both false*; hence p and q both true. (2) Since p and q are true, in order to make the first premise true *we must make r true* as well. (3) But since s is false, $(s \cdot t)$ is false, and since r is true, the second premise, $(r \supset (s \cdot t))$ comes out false. (4) But this means that it is *not possible* to make the conclusion false and both of the premises true; hence the argument form is *valid*.

We have been looking at relatively simple examples so far: those in which there is just one way to make the conclusion false, a simple way to come up with a counterexample for the invalid cases, and, in the valid cases, an easy way to reason through to the determination that there is no counterexample. But the short method is not always so easy, unfortunately. We may have cases in which there is more than one way to make the conclusion false, or more than one way to make premises true, or, heaven forbid, both. Let us take another example: $(p \supset q), (\sim q \supset \sim p)$ $/ \therefore (p \equiv q)$. Here there are two ways to make the conclusion false: we may make p true and q false, or p false and q true. Suppose we take the former assignment. This gives a false conclusion, all right, but both of the premises come out false as well! Does this mean that the argument is *valid*? Definitely not, since we have not tried all possible ways of getting a counterexample. If we take the second possibility—p false and q true—we do get a false conclusion, and also *both premises are true*; hence, *this* assignment yields a counter-example, and the argument is *invalid* after all. A good format for such cases, where there is more than one way to make the conclusion false, is to list all these ways under the conclusion, and then try each one out to see whether the premises are all true in any one of them. The example above, then, would look like this:

$(p \supset q),$	$(\sim q \supset \sim p)$	$/\therefore$	$(p \equiv q)$
F	F		T F F—No counterexample
T	T		F F T—Counterexample

If, of course, you try all possible ways of making the conclusion false, and in each of them you get at least one false premise, you may conclude that the argument is *valid*, since you will have shown that there *can* be *no* counterexample. For instance, if we have $(p \supset q)$, $(\sim p \supset \sim q)$ $/\!\!\therefore (p \equiv q)$ (which looks very much like the one above, except for the second premise), again we have two ways of making the conclusion false. Suppose *p* is true and *q* false. This makes the *first* premise false, so this is not a counterexample. Let us then make *p* false and *q* true. This makes $\sim p$ true and $\sim q$ false, which makes the *second* premise false. This does indeed show that *however* we make the conclusion false, one of the premises will come out false as well, so there is no way to get both premises true with the conclusion false. Hence the argument is *valid*. A simple format for this would be:

$p \supset q,$	$\sim p \supset \sim q$	$/\!\!\therefore$	$p \equiv q$
F	T		T F F
T	F		F F T

Since neither way of making the conclusion false yields a counterexample, the argument is *valid*.

It is extremely important to remember when doing these problems that you must test *all* possible ways of making the conclusion false before you can conclude that the argument is *valid*, because for validity you need to show there *can't possibly* be a counterexample. On the other hand, once you do find a counterexample in an *invalid* argument form, you are finished; one is all it takes.

4. Mechanical Decision Procedures

As mentioned in the introduction to this unit, the truth table method is a completely "mechanical" method, requiring no ingenuity or deep thought; the procedures could be, and have been, carried out by a machine. Another way of putting this is to say that there is an *algorithm* for the procedures described here; that is, a set of rules which, if followed faithfully, will always yield the correct answer. (As you will find in the next units, this is *not* the case with the proof method.) Such a mechanical procedure, which will *always* give an answer for *any* validity problem, is called a *decision procedure* for validity, and one of the great advantages of truth-functional logic, which offsets its rather odd "if-then" operator, is that it does have such a decision procedure.

Not all systems of logic have mechanical decision procedures; relational predicate logic, for instance, which we will be discussing in the last few units, does not lend itself to these methods. What, then, are the special features of sentential logic that make possible these mechanical tests? There are three important properties of sentential logic, which many systems lack, that make possible the mechanical decision procedures. (1) In the first place, *there are only a finite number of variables in each argument form*. (2) Second, *this is a two-valued logic*, which means that each variable can take only true or false substitution instances. Properties 1 and 2 together, as noted earlier, make it possible to list all possible combinations of truth values for the variables—all possible (kinds of) substitution instances. (3) *All our operators are truth-functional.* This means that, once we have drawn up our list of truth-value possibilities, we can compute the result for each possibility. Thus we know what the truth-values of the premises and conclusion are for each possible instance, and all we need to do is check to see whether any of those instances contain true premises with a false conclusion. Hence in theory, at least, we can always draw up a finite table and check the results, which is what is meant by saying that there is a mechanical decision procedure. In *practice*, of course, this may well not be possible—who would want to draw up a million-row truth table? But the important thing is that it is possible in principle.

Finally, lest you are tempted to underestimate the importance of these three properties, it is worth pointing out that for each of the three, there are systems of logic that lack them. There are, for instance, three-valued logics, in which the values are true, false, and indeterminate. This would still give a finite truth table, but just a much longer one. But there are also logics which are *infinite-valued*, in which a sentence can take a "truth" value ranging anywhere from 0 to 100%. Obviously there could be no truth table for that; we wouldn't even be able to list the possibilities for a single variable! When you get bored with making up truth tables remember infinite values, and be grateful.

Second, there are logics in which there are an *infinite* number of variables in an argument form. This may seem rather strange, but there are logics which may contain formulas of infinite length, and also logics in which we might have an infinite number of premises in an argument. Finally, there are logics, which you have already been made aware of, in which the operators are not truth-functional. *Modal logic*, which is the logic of "necessary" and "possibly," is not truth-functional, since the necessity and possibility operators are not truth-functional.

All these "alternative" sorts of logic have been developed in a high degree, but our two-valued, finite variable, truth-functional logic is by far the easiest to work with, especially for beginning students. Should you go on to study logic further, you may well run across some of these alternatives.

DEFINITIONS

1. A **counterexample to an argument form** is a substitution instance of that form (or a row in the truth table for the form) in which all the premises are true and the conclusion is false.

2. An **argument form is invalid** if and only if it has a counterexample, that is, a substitution instance (or row in the truth table) with all true premises and a false conclusion.
3. An **argument form is valid** if and only if it has no substitution instance (row in the truth table) in which all the premises are true and the conclusion is false.
4. An **argument is valid** if and only if its form is valid.
5. An **argument is invalid** if and only if its form is invalid.

STUDY QUESTIONS

1. How many rows will there be in the truth table for
    $[(p \lor \sim q) \lor [(r \supset (s \cdot r)) \supset (t \cdot \sim p)]] \equiv (w \supset (q \lor s))$?
2. What is the general formula for the number of rows in a truth table?
3. What is an alternative definition to the one given above for the validity or invalidity of an argument?
4. What do you have to do, using the short truth table method, to demonstrate invalidity?
5. What do you have to do, using the short method, to demonstrate validity?
6. What is a mechanical method (a decision procedure)?
7. What are the three properties of our sentential logic which make it possible to use the truth table method? Explain how they make it possible.
*8. Are the following true or false?
    a. A valid argument may have a false conclusion.
    b. An invalid argument *must* have true premises and a false conclusion.
    c. An argument with a true conclusion will always be valid.
    d. No argument with false premises can be invalid.
    e. An argument form with ten variables has over a million rows in its truth table.
    f. If a form has n different variables, it will have n^2 rows in its truth table.
    g. In the short truth table method, we are looking for a substitution instance with true premises and a true conclusion.

EXERCISES

1. Use the full truth table method to determine whether the following argument forms are valid or invalid.

*a. $p \supset q$
    p
    /∴ q

b. $p \supset q$
    $\sim p$
    /∴ $\sim q$

*c. $p \lor q$
    $\sim q$
    /∴ p

d. $p \supset q$
    $\sim q$
    /∴ $\sim p$

*e. $p \lor \sim q$
     $q \lor \sim p$
     $\overline{/\therefore \sim (p \equiv \sim q)}$

f. $p \equiv \sim q$
     $q \lor p$
     $\overline{/\therefore \sim q \lor \sim p}$

*g. $p \supset \sim q$
     $\sim (p \lor q) \lor p$
     $\sim (p \cdot q) \supset (q \lor p)$
     $\overline{/\therefore \sim (p \cdot q) \supset \sim p}$

h. $(p \supset q) \lor (q \supset r)$
     $\sim r \supset \sim (p \cdot q)$
     $\overline{/\therefore q \supset \sim p}$

*i. $p \supset (q \cdot r)$
     $\sim p \supset \sim (q \cdot r)$
     $q \cdot \sim p$
     $\overline{/\therefore \sim r}$

j. $(p \lor q) \supset (r \lor s)$
     $p \equiv \sim (r \cdot s)$
     $q \equiv \sim (p \cdot r)$
     $\overline{/\therefore (s \cdot p) \supset \sim (p \lor q)}$

2. Use the *partial* truth table method to test the following for validity.

*a. $\sim (r \lor q) \supset \sim (p \lor r)$
     $\sim (p \lor q) \supset \sim (q \cdot r)$
     $(p \lor \sim q) \supset \sim (r \cdot p)$
     $\overline{/\therefore p \supset (\sim r \lor \sim q)}$

b. $(p \cdot q) \supset \sim r$
     $(q \cdot r) \supset \sim p$
     $\sim (p \lor q) \supset \sim (q \lor r)$
     $\overline{/\therefore \sim (p \cdot (\sim q \cdot r))}$

*c. $(p \cdot q) \cdot \sim r$
     $(p \lor q) \supset \sim r$
     $\sim (p \supset q) \supset \sim (q \supset p)$
     $\overline{/\therefore \sim (p \supset r)}$

*d. $(p \lor q) \supset \sim (r \cdot s)$
     $\sim (r \cdot p) \supset \sim s$
     $\sim (p \lor q) \lor s$
     $\overline{/\therefore (s \lor p) \supset \sim (q \cdot r)}$

e. $(\sim p \lor \sim q) \equiv \sim (r \cdot s)$
     $q \equiv \sim (p \lor r)$
     $(\sim p \cdot \sim r) \supset (\sim q \lor \sim s)$
     $\overline{/\therefore \sim p \equiv \sim (q \cdot s)}$

*f. $(p \equiv r) \equiv (s \equiv \sim q)$
     $\sim ((p \lor q) \lor \sim r)$
     $((\sim p \supset q) \lor (\sim r \supset s)) \equiv \sim q$
     $((r \supset s) \cdot \sim (p \lor q)) \supset \sim (q \lor s)$
     $\sim (r \supset s) \supset (\sim (p \lor q) \supset (r \cdot \sim s))$
     $\overline{/\therefore \sim (p \lor q) \supset (r \supset s)}$

*3. Use the *short* truth table method to determine the validity or invalidity of all the argument forms above, as well as the following, which contain more variables. (Here premises and conclusion are listed horizontally.)

a. $((p \lor q) \supset r), ((r \lor s) \supset \sim t) /\therefore (p \supset \sim t)$

b. $\sim (p \cdot q) \supset \sim (r \lor s), (\sim p \equiv (t \lor w)), ((r \cdot w) \equiv z) /\therefore (z \supset \sim s)$

c. $(p \cdot q) \supset (r \supset (s \lor t)), s \equiv (p \cdot t), \sim t \equiv (q \lor \sim r) /\therefore r \supset (s \lor \sim p)$

d. $(p \lor q) \supset (r \lor s), p \equiv \sim (r \cdot t), s \equiv t /\therefore p \supset t$

e. $(p \cdot \sim q) \supset (r \equiv \sim s), \sim (s \lor z), (\sim q \supset r) \supset (t \lor w), (w \cdot \sim r) \supset t$
    $/\therefore p \supset (\sim t \supset r)$

4. Symbolize the following arguments, and use either the long or short truth table method to test their forms for validity. Do all of the arguments which sound valid to you turn out to be valid?

*a. This steak is tender only if it's fatty, and if it's fatty it's not good for me. I won't eat it unless it's good for me. Therefore, I won't eat this steak.

b. If Americans continue to eat a lot of grain-fed beef, then there is not a lot of grain which is exported. Only if a lot of grain is exported will there be enough to feed people in the underdeveloped countries. If either there is not enough grain to feed people in underdeveloped countries or their populations continue to soar, there will be mass starvation, and if there is mass starvation there will be a worldwide revolt. Thus, if Americans continue to eat grain-fed beef, there will be a worldwide revolt.

*c.   Pollution will increase if government restrictions are relaxed, and if pollution increases there will be a decline in the general health of the population. If there is a decline in health in the population, productivity will fall. The economy will remain healthy only if productivity does not fall. Therefore, if the government restrictions are relaxed, the economy will not remain healthy.

d.   If a coal liquefaction plant is built in Morgantown, there will be an increase both in employment and in pollution levels. The Morgantown economy will be strong if either there is an increase in employment or an increase in college faculty salaries. Faculty will not be healthy unless either there is an increase in their salaries or there is no increase in pollution. The Morgantown economy will be strong if and only if the college faculty are healthy. Therefore, if a coal liquefaction plant is built in Morgantown, there will be an increase in college faculty salaries.

*e.   The corn will produce well only if the weather remains hot and it neither hails nor rains hard. Only if it rains hard if and only if the weather remains hot will the beans produce well. Therefore, the corn and beans will not both produce well.

f.   The corn will survive if there is a light frost, but it won't survive if it snows. It will snow only if there is both moisture and a cold front. There is no moisture, so the corn will survive.

*g.   If I don't both diet and exercise then I gain weight. I exercise only if I am neither too tired nor too lazy. If I diet, then I get too tired unless I take vitamins. I am not taking vitamins, so I will gain weight.

h.   If I gain weight then I get depressed, and if I get depressed, then I eat too much. If I eat too much then I get lazy, and if I get lazy then I don't exercise. If I don't exercise then I gain weight. Therefore, I will gain weight.

*i.   I will get an A in logic only if I learn the truth tables, and I will learn the truth tables if and only if I stop watching the soaps. I will stop watching the soaps provided I can watch either "Dallas" or "Wheel of Fortune." I can watch "Wheel of Fortune" if I do my homework early. Therefore, I can get an A in logic if I do my homework early.

j.   If I ride my bicycle to work then I'll get caught in the rain, but if I don't ride my bicycle to work I'll be late. If I'm late my pay will be docked and I won't be able to eat lunch, and if I don't eat lunch then I'll be tired and cranky. If I get caught in the rain I'll be soaked and cranky, and if I'm cranky I'll be fired. Therefore, I'll be fired.

*k.   Gold rises if and only if the dollar falls, and the dollar falls only if the trade deficit worsens and interest rates rise. Interest rates won't rise unless the Fed tightens the money supply or the federal deficit goes up. I will make money only if gold rises. Therefore, I won't make money unless the Fed tightens the money supply.

l.   John will get a raise if and only if he works hard and doesn't insult his boss. He will insult his boss only if he is neither promoted nor complimented on his work. John will work hard only if he is complimented, and he will be promoted only if he works hard. Therefore, John will get a raise if and only if he is complimented on his work.

*m.   John will not get both a raise and a promotion unless he works hard. If he works hard then he will neglect either his family or his health. If John neglects his health then he will get exhausted, and he won't work hard if he is exhausted. If John neglects his family then he will be depressed, and he won't work hard if he is depressed. Thus, John will not get a raise.

Further Applications of the Truth Table Method

A. INTRODUCTION

Now that you know the basic process of constructing truth tables, you can apply this method in a variety of other circumstances. You can use it to show whether someone is contradicting himself, whether a statement has any real significance, or whether two statements have the same meaning. If someone claims, for instance, that he has been offered a job as a petroleum engineer which he will take provided it pays $40,000 a year, and that if it doesn't pay that much he will instead take a job as vice-president of a small coal company, but that he won't take either job, you will be able to show by the truth table method that what he says cannot be true, since he has just contradicted himself. Or, if some self-appointed economic expert "predicts" that either there will be inflation and unemployment, or no inflation and no unemployment, or inflation but no unemployment, or unemployment but no inflation, you will be able to recognize, using the truth table method, that such a claim really says nothing at all, has absolutely no content. The truth table method can also be used to show when two statements, though they seem to be making different claims, really mean the same. "The rate of inflation will be reduced provided there is a steep increase in interest rates and high unemployment" and "If there is no reduction in the inflation rate then either unemployment will not be high or there will be no steep increase in interest rates," for instance, really say exactly the same thing.

In this unit you will learn all these things and more: how to test a set of statements or statement forms for consistency, how to determine whether two statements or forms say the same thing, and whether a given statement or form is significant or is simply a tautology, giving no real information. You will need to learn to distinguish clearly the kinds of problems which can be solved using truth tables, and you will learn some interesting relationships between the various truth table concepts, such as the fact that if the premises of an argument form are inconsistent, that is, contradictory, the argument form is valid rather than invalid. The things you will be expected to master are listed below in the "Objectives" section.

B. UNIT 6 OBJECTIVES

1. Learn the definitions of "tautology," "contradiction," and "contingency," given in the glossary at the end of the unit.

2. Be able to use the truth table method to determine the logical status of single statement forms, that is, whether they are tautologous, contradictory, or contingent.

3. Learn the definitions of "logical implication" and "logical equivalence," and the relation between them.

4. Be able to use the truth table method to determine whether two or more statement forms are logically equivalent, or whether one logically implies another.

5. Learn the definition of "consistency."

6. Be able to use the truth table method to determine whether a set of statement forms is consistent.

7. Be able to state clearly the four different kinds of truth table problems which we have encountered, and the concepts which are applicable to each. (It will make no sense, for instance, to say that a single statement form is valid.)

8. Be able to state several relationships between the various truth table concepts.

C. UNIT 6 TOPICS

1. Tautologies, Contradictions, and Contingencies

So far you have been using the truth table method just to test argument forms for validity, which meant making up a joint truth table for the premises and conclusion and then inspecting the results for all the statement forms taken together. It is also possible to make up a truth table for just a single, individual statement form, and we sometimes want to do this to determine the logical status of the form. There are certain forms, for instance, which have an interesting property: they can never, under any circumstances, turn out to be false, and this is something which can be determined by the truth table. There are other forms which have just the opposite property: they can never, in any case, turn out to be true. The former sort of statement forms, which can never be false, are called *tautologies*, and the latter, which can never be true, are called, sensibly enough, *contradictions*. The intermediate cases, forms which are true in some instances and false in others, we will call *contingencies*. (In ordinary parlance a contingency is something which might or might not occur; we are using the term in a different, but closely related, sense.)

Testing single statement forms to determine their logical status, that is, whether they are tautologous, contradictory, or contingent, is a simple matter once you know how to construct truth tables. You simply take all the variables which occur in the form, construct your base columns for those variables, and then compute the result for each instance. An example should make this process clear:

p	q	$\sim$	$(p$	$\supset$	$q)$	$\equiv$	$(p$	$\cdot$	$\sim$	$q)$
T	T	F		T		$\boxed{T}$		F	F	
T	F	T		F		$\boxed{T}$		T	T	
F	T	F		T		$\boxed{T}$		F	F	
F	F	F		T		$\boxed{T}$		F	T	
		(2)		(1)		(3)		(2)	(1)	

This form turns out to be a tautology because there is no instance in which the value under the major operator is false.

We will compute the results here in the same way as we have done before, starting with the smallest components first, placing the truth tables for the subformulas underneath the operators for those formulas, and gradually working our way up to the major operator. It is, of course, the truth values under the *major operator* which determine the logical status. Notice that in this example, for instance, we have many *F*'s in various places in the truth table, but under the *major* operator the result is always *T*, so the form is a tautology. We may formally define "tautology" as follows: *a tautology is a statement form that is true (under its major operator) for every substitution instance.* Another way we could put this would be to say that for every row in the truth table, the result of the computation under the major operator is "true."

There are other statement forms which can *never* turn out to be true, no matter what the values of the component parts. Such forms, as already noted, are called *contradictions*. An example of a contradiction, with its truth table, is the following:

p	q	~	(p	≡	~	q)	≡	~	(p	≡	q)	
T	T	T		F	F		F	F		T		This is a contradiction because, as the truth table shows, there is no instance in which it turns out to be true.
T	F	F		T	T		F	F		T		F
F	T	F		T	F		F	F		T		F
F	F	T		F	T		F	F		F		T
		(3)		(2)	(1)		(4)	(2)		(1)	

A contradiction is a statement form that is false (under its major operator) for every substitution instance, or equivalently, one which turns out false for every row in the truth table. Notice that *the negation of a tautology will be a contradiction,* since a negation changes the truth values, so that if we start out with all trues—a tautology—then if we negate it, we will end up with all falses—a contradiction. Similarly, the negation of a contradiction will always be a tautology.

A contingency is a form that is neither a tautology nor a contradiction; that is, it does not have either all T's in its truth table or all F's, which means simply that it has some of each. We can define "contingency" as follows: *a contingency is a statement form that is true for some substitution instances and false for others, under its major operator.* An example of a contingency is given below, with its truth table.

p	q	((p	·	~ q)	∨	(~ p	·	q))	∨	(~ p	·	~ q)	
T	T	F	F	F	F	F		F	F	F	F		This form is contingent because it has some T's and some F's under its major operator.
T	F	T	T	T	F	F		T	F	F	T	
F	T	F	F	T	T	T		T	T	F	F	
F	F	F	T	F	T	F		T	T	T	T	
		(2)	(1)	(3)	(1)	(2)		(4)	(1)	(2)	(1)	

Notice that the negation of a contingency will be another contingency, because all the T's will change to F's and all the F's will change to T's, so there will still be some of each.

Tautologies play little role in ordinary language because they give us no information. Since they are always true under any circumstances, they don't make any definite claim about the way things actually are. If the weathermen tell you, for instance, (as they are wont to do) that it will either rain tomorrow or not rain, that is not much of a forecast. It certainly doesn't help you decide whether to go on a picnic. A tautology is an "empty" claim; it really says nothing about the world. Tautologies do play a very important role in logic, however. They are the axioms and theorems of formal logical systems, the "truths" of the system, just as "$x + y = y + x$" is a truth of most of our mathematical systems. They are useful in logic precisely because they do *not* make any claim about the empirical world, but are, we might say, true no matter what the worldly facts. They are formulas whose truth we can absolutely depend upon, and whatever its limited usefulness in ordinary language, this is a highly desirable property in logic. In doing the problems, you should pay particular attention to formulas which turn out to be tautologies (and also those which turn out to be contradictions); some of them will be important later, and, in any case, it will help to develop your logical intuitions.

If you thoroughly understand the concepts of "tautology," "contradiction," and "contingency," you should be able to answer the following sort of question: what would be the result of *disjoining* a tautology and a contradiction? Answer: it would be a tautology, since a disjunction is true provided at least one disjunct is true, and since one side is a tautology, we would have at least one true disjunct in every row. What would be the result of conjoining a tautology and a contradiction? A contingency? No. It would be a contradiction, since in every row there would be one false conjunct, which would make the conjunction false everywhere. What would be the result of conjoining two contingent forms? Here we don't know for sure; it might be contingent (if, for instance, the two forms were p and $\sim q$), or it might be a contradiction (for instance, if the two forms were p and $\sim p$). The only thing we can be sure of is that it *won't* be a tautology, since there will be at least one false row in the truth table. To take one more example, what would be the result of placing a triple bar between two contradictions? Another contradiction? No, here we will have a *tautology*, since we will have "F $\equiv$ F" in every row, which will turn out to be true. Exercises at the end of the unit will give you more practice in working out such combinations; you might even try to think up other combinations and work out their results.

2. Logical Implication and Logical Equivalence

In Unit 4, when learning to symbolize statements, you learned that there were certain forms which were interchangeable and which meant exactly the same, such as $\sim (p \lor q)$ and $(\sim p \cdot \sim q)$. We are now in a position to see *why* they mean the same and why one can be used in place of the other. Such statement forms have the very important property of being *logically equivalent* to each other, which means that they turn out to be true or false in exactly the same circumstances. In other words, they have identical truth tables under their major operators. This will be our primary definition of logical equivalence: *Two or more statement forms will be logically equivalent if and only if the truth tables under their major operators are identical.* You test statement forms for logical equivalence by constructing a *joint* truth table for them and computing the results for each of the formulas. If the truth tables have the same values under the major operator, in every single row, then the formulas are equivalent. If there is some row in which the values are different, then they are not equivalent. The following two forms, then, are equivalent, since they are each false in the second row and true in all the others.

p	q	~	(p · q)	⊃	~ p		~ q	⊃	~	(p ∨ q)
T	T	F	T	**T**	F		F	**T**	F	T
T	F	T	F	**F**	F		T	**F**	F	T
F	T	T	F	**T**	T		F	**T**	F	T
F	F	T	F	**T**	T		T	**T**	T	F
		(2)	(1)	(3)	(1)		(1)	(3)	(2)	(1)

We can also test more than two formulas at a time. In the following example we have four. We can say which are equivalent to which just by comparing their truth tables.

p	q	1. ~(p · q)		2. ~p · ~q			3. ~(p ∨ q)		4. p ⊃ ~q	
T	T	**F**	T	F	**F**	F	**F**	T	**F**	F
T	F	**T**	F	F	**F**	T	**F**	T	**T**	T
F	T	**T**	F	T	**F**	F	**F**	T	**T**	F
F	F	**T**	F	T	**T**	T	**T**	F	**T**	T
		(2)	(1)	(1)	(2)	(1)	(2)	(1)	(2)	(1)

In this instance, we can say that formulas 2 and 3 are equivalent and that formulas 1 and 4 are equivalent, but neither 2 nor 3 is equivalent to either 1 or 4.

It is also important to realize that formulas may be logically equivalent even though they have different numbers of variables! You can determine this, again, by drawing up the joint truth table and working out the results for both formulas. In this way we can see that ($\sim p \cdot \sim q$) and (($\sim p \cdot \sim q$) · r) ∨ (($\sim p \cdot \sim q$) · $\sim r$) are equivalent. The truth table follows.

p	q	r	~p ·	~q		((~p · ~q) · r) ∨ ((~p · ~q) · ~r)					
T	T	T	F	**F**	F	F	F	**F**	F	F	F
T	T	F	F	**F**	F	F	F	**F**	F	F	T
T	F	T	F	**F**	T	F	F	**F**	F	F	F
T	F	F	F	**F**	T	F	F	**F**	F	F	T
F	T	T	T	**F**	F	F	F	**F**	F	F	F
F	T	F	T	**F**	F	F	F	**F**	F	F	T
F	F	T	T	**T**	T	T	T	**T**	T	F	F
F	F	F	T	**T**	T	T	F	**T**	T	T	T
			(1)	(2)	(1)	(1)	(2)	(3)	(1)	(2)	(1)

There is an interesting connection between the concepts of logical equivalence and tautology: if two formulas which are logically equivalent are joined into a biconditional, the result will be a tautology; and if the two formulas are not equivalent, the result of joining them will not be a tautology. The reason for this is that formulas which are logically equivalent, as we have defined it, have identical truth tables; this means their truth values are the same for every row in the truth table. Thus, given that a biconditional is true if and only if its components have the same truth values, the biconditional with logically equivalent components must be true for every row. Hence, the biconditional will be a tautology. On the other hand, if the formulas are not equivalent, there will be some row in which their values differ; in that row the biconditional will turn out to be false, and so will not be a tautology. Thus we can give the following alternative definition of logical eqivalence: *Two statement forms are logically equivalent if and only if the result of joining them with the biconditional is a tautology.* This definition is, in fact, used by some textbook authors, and it should be evident that the two definitions amount to the same thing.

A third definition of logical equivalence can be obtained from a closely related concept, that of *logical implication*, which is a relationship between two statement forms. *One statement form logically implies a second if and only if there is no row in their joint truth table in which the first is true and the second false.* If we make up the truth tables for $\sim (p \lor q)$ and $\sim p$, we see that $\sim (p \lor q)$ logically implies $\sim p$, because there is no row in the truth table in which $\sim (p \lor q)$ is true while $\sim p$ is false.

p	q	$\sim$	$(p \lor q)$	$\sim$	p
T	T	F	T		F
T	F	F	T		F
F	T	F	T		T
F	F	T	F		T

Logical implication, unlike logical equivalence, is not a symmetric relation: that one form logically implies another does not necessarily mean that the second will imply the first (although it may). In the example above, for instance, although $\sim (p \lor q)$ does logically imply $\sim p$, the converse relation does not hold; $\sim p$ does not logically imply $\sim (p \lor q)$ because in the third row $\sim p$ is true while $\sim (p \lor q)$ is false.

When two formulas do both logically imply each other, they must be logically equivalent. To say that the first implies the second is to say there is no row where the first is true and the second false. To say the second implies the first is to say there is no row where the second is true while the first is false. If we have this two-way logical implication, then there can be no row in which one is true and the other is false, that is, no row in which the truth values are different. Thus their truth tables must be identical, so they are logically equivalent. Our third definition of logical equivalence, then, is the following: *Two statement forms are logically equivalent if and only if they logically imply each other.*

Since logically equivalent formulas, by definition, have identical truth tables, there will be no row in the truth table, no substitution instance, in which one is true and the other false. This is a very important feature, because it means that if two statements have equivalent forms, then we can correctly *infer* one from the other, because we can be sure that we will never be going from a true statement to a false one. If two statements have equivalent forms, then they must have exactly the same truth values. This feature of equivalent forms will be very important when we come to inference rules in the next few units. There is one very large and important set of rules which will be based upon the fact that formulas are equivalent, and thus have exactly the same truth values for every instance. Many of the exercises on equivalence at the end of this unit contain forms which will later be used as rules, so it would be a good idea to familiarize yourself with them now.

The concepts of logical implication and logical equivalence are among the most important in logic. If you haven't done so already, you should be sure you know the definitions that have been given in this section and can state the relationships between them.

3. Consistency

Consistency may be the hobgoblin of little minds, but it is also a very important property of sets of formulas. If a set of premises is *in*consistent, for instance, it makes the argument form virtually worthless, since absolutely anything will follow from it, including both the conclusion and the negation of the conclusion. If you say something inconsistent you might just as well have remained silent, because what you said could not possibly be true. Thus we all, or at least most of us, want to be consistent. But what does this mean? It will be a little easier to understand consistency if we talk first about inconsistency, so we will give the following definition: *a set of formulas is inconsistent if there is no row in their joint truth table in which they all come out true at once.* The following set of formulas, for instance, is inconsistent:

p q	~q ⊃ ~p	(p · ~q) ∨ (~p · ~q)	~p ⊃ q
T T	F [T] F	F F [F] F F	F [T]
T F	T [F] F	T T [T] F T	F [T]
F T	F [T] T	F F [F] F F	T [T]
F F	T [T] T	F T [T] T T	T [F]
	(1)(2)(1)	(2)(1)(3)(1)(2)(1)	(1)(2)

In the first row of this truth table the middle formula is false; in the second row the left-hand formula is false; in the third row the middle formula is again false; and in the last row the right-hand formula is false.

Thus there is no row in which all the formulas come out true at once, and so the set is inconsistent.

There is obviously a very close relationship between inconsistency and contradiction; the difference is that a contradiction is a *single* formula, whereas consistency is a property of *sets* of formulas. We can say, however, that *a set of formulas is inconsistent if and only if the conjunction of all the formulas is a contradiction.* If we conjoined all the formulas above, for instance, we would have a contradiction, since a conjunction is false if at least one of the conjuncts is false. Here there would be a false conjunct in every row, so the conjunction as a whole would be a contradiction.

Given the definition above, it should be obvious that *a set of formulas is consistent if and only if there is at least one row in their joint truth table in which they all come out true.* The following set of formulas is consistent:

p	q	p ⊃ (p · q)		~p ⊃ ~(p ∨ q)		~ (p · q) ≡ (~p · ~q)		
T	T	[T]	T	F [T] F	T	F	T	[T] F F F
T	F	[F]	F	F [T] F	T	T	F	[F] F F T
F	T	[T]	F	T [F] F	T	T	F	[F] T F F
F	F	[T]	F	T [T] T	F	T	F	[T] T T T
		(2)	(1)	(1) (3)(2)	(1)	(2)	(1)	(3) (1) (2)(1)

These formulas all come out true in both the top row and the bottom row; thus, there is at least one row where they all come out true, so they form a consistent set.

It is important not to confuse the concept of tautology with that of consistency. A *single* statement form is a tautology, remember, if it comes out true in every row. This means that the *vertical column* underneath the major operator consists entirely of "trues." With a *consistent* set of formulas, however, we are saying that there is a *horizontal row* in which each of several formulas comes out true.

4. Statements and Statement Forms; Applying Truth Table Concepts

The definitions we have given in this unit have been about statement *forms.* How, then, can we apply them in particular cases? How can we say about an actual set of sentences, such as the one in the introduction to this unit about jobs, that it is consistent or inconsistent? And how can we say that two *statements,* as opposed to their forms, are equivalent? The answer should be fairly obvious; we will do here what we did for validity, and characterize the *instances* in terms of their forms. We can say, then, that a set of statements is consistent or inconsistent just in case its corresponding set of forms is consistent or inconsistent, in the sense just defined in Section 3; and we can say that two statements are equivalent just in case their corresponding forms are logically equivalent. We can test statements and sets of statements in exactly the same way as in Unit 5: just symbolize them and test the formulas. We could symbolize the inconsistent set of statements about the jobs, for instance, as $J \cdot (F \supset T)$, $(\sim F \supset V)$, and $\sim (V \vee T)$. We can then test this set of formulas with a 16-row truth table. We find that there is no row in which all the formulas are true at once, so the set of statement forms, and hence the set of statements, is inconsistent. Similarly, we could symbolize the two statements about inflation as $(I \cdot U) \supset R$ and $\sim R \supset (\sim U \vee \sim I)$. The truth table for these formulas will reveal that the forms, and thus the statements themselves, are logically equivalent, since the final results are the same in every row. In doing Exercises 6, 7, and 8 at the end of the unit, use this method of just symbolizing and testing the symbolized statements.

5. Four Kinds of Truth Table Problems and the Relations Between Them

In the preceding two units you have learned four or five different applications of the truth table method—different kinds of problems which can be solved using this method. Some are used on a single form, some on pairs of forms, and some on sets of forms. It is extremely important that you keep these sorts of problems straight, and are clear about the various concepts, so let us summarize:

The concept of validity is applied to argument forms, that is, *sets of statement forms consisting of premises and a conclusion*. When you test an argument form for validity you are checking to see whether it has any counterexamples.

The concepts of tautology, contradiction, and contingency are applied to single statement forms, and concern the kind of truth tables the forms have. A form with all T's in its final table, for instance, is a tautology.

The concepts of logical implication and logical equivalence apply to pairs of statement forms. We say that *one* form logically implies *another*, or is logically equivalent to another. (We can also apply the concept of equivalence to more than two formulas.)

Finally, the concept of consistency applies to any set of formulas.

You should never say, then, that a single statement form is *valid*, since validity is a property of argument forms (or arguments) only; it doesn't even make sense to apply it to single statement forms. Nor would it make sense to say that an argument form is a tautology, since argument forms have several formulas in them. There are, however, some very interesting relationships between these concepts, and in order to thoroughly understand the concepts you need to understand these relationships. We will first discuss the way in which validity is related to the other truth table concepts.

As mentioned earlier, a very interesting fact about argument forms is that *if the premises are inconsistent the argument form is valid* (not invalid)! A closely related property is that *if the conclusion of an argument form is a tautology, then the argument form is also valid*. Why should this be so? Well, in both cases the reason is that *there cannot possibly be a counterexample*. In the first case, if the premises are inconsistent, this means by definition that there can be no row in which they are all true at once. But if there can be no row in which they are all true, there can certainly be no row in which they are all true *and* the conclusion is false; thus there is no counterexample, and so the argument form is valid. The same sort of reasoning, turned around a little, goes for the second case as well. If the conclusion is a tautology, there is no row in which it is false, but if there is no row in which the conclusion is false, then there is no row in which the conclusion is false and the premises are all true, so, again, there is no counterexample, and thus the argument form is valid.

There is a slightly more complicated relationship between the concepts of *logical implication* and *validity*: *if the conjunction of the premises logically implies the conclusion, then the argument form is valid*. The reason for this is found simply in the definition of validity. If the conjunction of the premises logically implies the conclusion, then there is no row in which the conjunction of the premises is true, but the conclusion false. Since the *conjunction* of the premises can only be true if *all* the premises are true in that row, this means that there is no row in which all the premises are true but the conclusion false. Another way of putting this, which involves the notion of tautology, is to say that if the conditional formed by putting the conjunction of premises as the antecedent and the conclusion as a consequent is a tautology, then the argument form is valid. The reason for this, again, is that the statement above implies that there is no instance with all the premises true and the conclusion false.

We have already seen that two forms are logically equivalent if and only if the biconditional they form is a tautology. Similarly, formula 1 logically *implies* formula 2 if and only if the conditional with formula 1 as antecedent and formula 2 as consequent is a tautology. This is because in both cases there is no row in the truth table in which the first formula is true and the second false.

Keep in mind that one of your definitions of logical equivalence is in terms of a dual logical implication. Other interesting facts about logical equivalence are the following: *any two contradictions are logically equivalent*, and *any two tautologies are logically equivalent*. This is evident, since all contradictions have identical truth tables (all false), and the same holds for tautologies (all true). You may be able to think of even more ways in which these concepts are related. If so, and if you understand these, you can be reasonably sure you have mastered the material of these last two units.

DEFINITIONS

1. A **tautology** is a single statement form that is true for every substitution instance, that is, which comes out true under the major operator for every row in the truth table.
2. A **contradiction** is a single statement form that is false for every substitution instance, that is, which comes out false under the major operator for every row in the truth table.
3. A **contingency** is a single statement form that is false for some substitution instances and true for others, that is, which has both T's and F's in its truth table under the major operator.
4. Two (or more) statement forms are **logically equivalent** if and only if their truth tables are identical under their major operators.
5. One statement form **logically implies** another if and only if there is no row in their joint truth table in which the first comes out true and the second comes out false.
6. Two statement forms are **logically equivalent** if and only if they logically imply each other.
7. Two statement forms are **logically equivalent** if and only if the result of joining them with a biconditional is a tautology.
8. A set of statement forms is **inconsistent** if and only if there is *no row* in their joint truth table in which they all come out true at once.
9. A set of statement forms is **consistent** if and only if *there is* a row in their joint truth table in which they all come out true at once.

STUDY QUESTIONS

1. Why is an argument form with inconsistent premises valid rather than invalid?
2. What is the relationship between validity and logical implication?
3. Why do all three definitions of "logical equivalence" amount to the same thing?
4. Can one say that a statement form is valid? Why or why not?
5. What are the four sorts of truth table problems you have learned?
6. What are some of the relationships between the various truth table concepts you have learned in the last two units?

EXERCISES

1. Answer the following questions.

*a. What is the negation of a tautology? Why? Contradiction.

 b. What is the negation of a contingent form? Why? Conti.

*c.   What is the negation of a contradiction? Why?   *taut*

d.   What is the disjunction of a contingent form and a contradiction? Why? *cont.*

*e.   What is the disjunction of a contingent form and a tautology? Why? *tauto.*

f.   What is the conjunction of a contingent form and a tautology? Why? *cont.*

*g.   What is the biconditional of two contradictions? Why?   *Taut.*

h.   What is a conditional with a contradiction for an antecedent and a contingent form for a consequent? Why?   *Taut.*

*i.   What is a conditional with a tautology as an antecedent and a contingent form as a consequent? Why?   *Cont.*

j.   What is a conditional with a tautology as an antecedent and a contradiction as a consequent? Why?   *contra.*

*k.   What is the biconditional of two contingent forms? Why? *Depends*

l.   What is the disjunction of two contingent forms?   *Contingent Depends.*

2.   Use the truth table method to decide whether the following statement forms are tautologies, contradictions, or contingencies.

*a. $p \supset (q \supset p)$

b. $(p \lor q) \cdot (\sim p \lor \sim q)$

*c. $\sim (p \supset (p \lor q))$

d. $(p \equiv \sim q) \lor (\sim p \equiv \sim q)$

*e. $((p \supset q) \cdot (q \supset p)) \supset (p \lor q)$

f. $(p \equiv q) \equiv (p \equiv \sim q)$

*g. $(p \cdot (q \lor r)) \equiv ((p \cdot q) \lor (p \cdot r))$

h. $((p \lor r) \supset q) \cdot \sim (\sim q \supset \sim r)$

*i. $((p \lor q) \lor \sim r) \supset (p \lor (q \lor \sim r))$

j. $((p \lor q) \cdot \sim r) \equiv (p \lor (q \cdot \sim r))$

*k. $(p \cdot (q \lor \sim r)) \equiv (((p \cdot s) \cdot (q \lor \sim r)) \lor ((p \cdot \sim s) \cdot (q \lor \sim r)))$

*3.   Use the truth table method to decide whether the following pairs of statement forms are logically eqivalent.

a. $\sim (p \cdot q)$ and $(\sim p \lor \sim q)$

b. $p \equiv \sim q$ and $(p \supset q) \cdot (p \supset \sim q)$

c. $\sim p \equiv q$ and $\sim (p \equiv q)$

d. $(p \cdot q) \supset q$ and $p \supset (p \cdot q)$

e. $(p \lor q) \supset p$ and $\sim q \supset \sim (p \cdot q)$

f. $\sim (p \cdot q) \supset \sim (p \lor q)$ and $\sim (\sim p \equiv q)$

g. $(p \lor q) \cdot (p \lor r)$ and $(p \cdot q) \lor (p \cdot r)$

h. $p \supset (q \supset r)$ and $(p \cdot q) \supset r$

i. $(p \cdot \sim q) \supset r$ and $\sim r \supset (\sim q \supset \sim p)$

j. $\sim (p \lor (q \cdot r))$ and $\sim ((p \lor q) \cdot (p \lor r))$

4.   Use the truth table method to decide for the following pairs of formulas whether 1 logically implies 2, or 2 logically implies 1, or both, or neither.

*a.   (1) $\sim (\sim p \lor \sim q)$ (2) $(p \lor q)$

 b.   (1) $\sim p \supset \sim (p \lor q)$ (2) $(p \lor q) \supset q$

*c.   (1) $\sim (p \supset q) \cdot q$ (2) $(p \cdot q) \cdot (p \equiv \sim q)$

 d.   (1) $(p \cdot q) \lor \sim r$ (2) $p \cdot (q \lor \sim r)$

*e.   (1) $\sim (p \lor q) \lor \sim r$ (2) $\sim (r \cdot p) \cdot \sim (r \cdot q)$

*5.   Use the truth table method to decide whether the following sets of statement forms are consistent.

a. $(p \supset q)$, $(\sim p \lor \sim q)$, $\sim (q \supset p)$

b. $(p \equiv \sim q)$, $\sim (p \supset \sim q)$, $(p \supset (p \lor q))$

c. $(\sim p \supset \sim q)$, $(\sim q \supset \sim p)$, $\sim (p \lor q)$, $\sim (p \cdot q)$

d. $((p \lor q) \supset p)$, $(\sim (p \lor q) \supset \sim p)$, $(p \equiv \sim q)$

e. $(\sim (p \lor q) \cdot \sim (p \lor r))$, $(\sim (p \cdot q) \lor \sim (p \cdot r))$, $(\sim p \supset \sim (q \cdot r))$

6.   Symbolize the following and then test their forms to determine whether they are tautologous, contradictory, or contingent.

*a.   If it doesn't rain we will go on a picnic, but if it does rain, we won't.

 b.   Inflation will be stopped only if interest rates fall, but interest rates will neither rise nor fall, and inflation will be stopped.

*c.   Inflation will be stopped if either interest rates fall or do not fall.

 d.   If interest rates neither rise nor fall, then there will be less unemployment if interest rates fall.

*e.   Either John and Mary will both go to the party, or Fred will go to the party while either John doesn't go or Mary doesn't go, or Fred will not go and not both John and Mary will go.

 f.   John likes Mary or Beth, but he doesn't like both Mary and Alice and he doesn't like both Beth and Alice.

*g.   John will get a job if and only if he runs out of money, but if he doesn't have a job then he runs out of money.

 h.   John will get a job if and only if he will run out of money if and only if he doesn't get a job.

*i.   If I am elected I will not raise taxes, but if I am elected I will support education, and I can support education only if I raise taxes.

 j.   John will get an A in logic if and only if he does not get an A in physics, but he won't get an A in either one.

7.   Symbolize the following pairs of statements and test their forms for logical implication and equivalence.

*a.   (1) Unemployment will rise if interest rates either fall or do not fall.
     (2) Unemployment will rise.

b. (1) There will be war if the country is invaded.
   (2) As long as the country is not invaded there will be no war.

*c. (1) Mary will go to the party if and only if George does not go.
   (2) Either George or Mary will go to the party.

d. (1) If there is an increase in production and a decrease in interest rates, then there will not be an increase in unemployment.
   (2) If there is an increase in unemployment, then if there is a decrease in interest rates, there is no increase in production.

*e. (1) If John was elected, then taxes were raised only if education was supported.
   (2) If education was not supported, then if taxes were raised, John was not elected.

f. (1) If taxes were raised if and only if education was supported, then John was not elected.
   (2) If John was not elected, then either taxes were raised but education was not supported or education was supported but taxes were not raised.

*g. (1) If John was elected only if education was supported, then education was not supported.
   (2) John was not elected.

h. (1) John was elected if and only if Bob was elected, and Mary was elected if and only if Bob was not elected.
   (2) John was elected if and only if Mary was not elected.

8. Symbolize the following sets of statements and test them for consistency:

*a. (1) There will be war only if the arms race continues.
   (2) There will be war only if the arms race does not continue.
   (3) The arms race will continue.
   (4) There will be no war.

(b) (1) Postal rates will increase only if the number of postal workers is not reduced and their salaries increase.
   (2) Postal salaries will increase if and only if the number of postal workers is reduced.
   (3) If either salaries increase or the number of postal workers is not reduced, then postal rates will increase.

*c. (1) If John was elected, then education was supported but taxes were not raised.
   (2) John was elected if and only if Mary was not elected.
   (3) If Mary was elected, then taxes were not raised but education was supported.

d. (1) If the human population continues to explode, then the planet will become polluted and most animal species will become extinct.
   (2) If the planet is polluted, then the human population will not continue to explode.
   (3) If the human population does not continue to explode, then it is not true that most animal species will become extinct.

The Proof Method: Eight
Basic Inference Rules

A. INTRODUCTION

In this unit you will begin to learn an entirely different method for demonstrating the validity of deductive arguments: the method of proofs. With this procedure, instead of testing the premises and conclusion simultaneously, "in one fell swoop," to see whether there is a counterexample, you will proceed step by step through a series of relatively simple intermediate inferences until you arrive at the conclusion. This procedure is much closer to the way we reason in everyday life; you probably quite often work out step by step the implications of certain hypotheses, but it is a good bet that you never made up a truth table before you picked up this book. The proof method is historically prior as well; 2300 years ago Aristotle developed the first system of proofs, and Euclid, of course, developed the proof method for geometry at about the same time. The truth table method, by contrast (though not the general method of refutation by counterexample), is an invention of the twentieth century. The two methods are obviously independent of each other: you were able to learn the truth table method without recourse to proofs, and the method of proofs was clearly possible without the truth table method. It is interesting that there are two such different methods for doing the same thing—

testing validity. As we shall see, however, there is a very important connection between them: the truth table method, in a way, serves to validate the rules we will be using in constructing proofs. Fortunately, the two methods yield the same results, at least in logic. (In mathematics, as it turns out, the results of the two methods are somewhat different—one of the most exciting discoveries of the twentieth century.)

In using the proof method, proceding step by step from premises to conclusion, you will not, of course, be permitted to make whatever inferences you wish. (You will not, for instance, to be allowed to infer $\sim B$ from $(A \supset B)$ and $\sim A$.) You will be given a certain set of *inference rules*, and will be permitted to use those and only those rules in your proofs. The reason for this restriction will be explained in Section 3.1. In this unit you will be given eight very basic inference rules, consisting of a premise or set of premises and a conclusion. In the next two units you will be given additional rules of a slightly different form.

It cannot be emphasized enough that it is absolutely essential that you learn these rules thoroughly. You must memorize them, or you will simply not be able to do the proofs. You should be able to reel off the rules as easily as you can (I hope) reel off your multiplication tables. *You will not be able to master this unit, or the next few units, unless you first learn the rules.* To assist you in this process, the rules will always be carefully explained in the text.

You will probably find the proof method more interesting and challenging than the use of truth tables, and will very possibly find it quite frustrating initially as well. This is because, unlike the truth table method, there are no set procedures to follow. I cannot give you a prescription for constructing proofs which is guaranteed to work; *there is no algorithm, no mechanical procedure, for the proof method.* You will have to use your imagination and ingenuity; this is the source of both the challenge and the frustration. On the other hand, when you do solve a problem, you will feel that you have really accomplished something. You may consider these problems as puzzles, like chess problems or crossword puzzles; your goal will be to get to the conclusion from the premises by using only the rules of inference given you.

Although there are no set procedures for reaching your goal (the conclusion), you will learn certain "rules of thumb," certain "tricks of the trade," and methods, such as "working backwards," which will help you plan your strategy. The more exercises you do, of course, the more adept you will become, and by the end of the unit you should be fairly proficient in planning strategies and figuring out how to reach your goal. *But again, it must be emphasized that unless you thoroughly learn the rules you will be able to go nowhere.* You cannot hope even to begin to construct proofs without knowing the means of construction! For convenient reference, the rules are all given at the end of the unit, and are also given on the inside front cover of the book.

In order to fully understand the application of rules of inference it will be important to master the distinction between a statement *form* and the particular *statement*, the substitution instance. Section 1 is thus devoted to this topic, before we proceed to the discussion of rules and proofs in the succeeding sections.

B. UNIT 7 OBJECTIVES

1. Learn the difference between forms and substitution instances, and learn the definitions of these and related terms in the glossary at the end of the unit.
2. Be able to determine whether a statement is or is not a substitution instance of a given form.
3. Be able to write out the symbolic forms of the inference rules and to understand their verbal explanations.
4. Be able to identify correct applications (substitution instances) of the rules, and be able to spot incorrect applications.
5. Given a derivation, be able to supply the justification for each step.
6. Be able to state and apply the following definitions: (a) justified step, (b) derivation, (c) proof.
7. Be able to construct derivations.
8. Given a set of premises and a conclusion, be able to construct a short proof (up to five intermediate steps).
9. Learn to plot proof strategies, such as the method of "working backwards."
10. Given a set of premises and a conclusion, be able to construct longer proofs (up to 25 or 30 intermediate steps).

C. UNIT 7 TOPICS

1. Form and Substitution Instance

In Unit 1 we introduced the distinction between "form" and "substitution instance," and we have used these terms already in defining validity and in discussing truth tables. So far, however, we have been relying on a rather informal understanding of these concepts; in Unit 1, for example, the form was said to be the general *pattern*, or *structure*, of the sentence or argument, with no particular meaning or truth value, while the *instance* was the *particular*, *meaningful* sentence or argument. The time has now come to make these concepts more precise.

The basic distinction is that between *constant* and *variable*. A *constant* is simply a term that has a *definite, particular value*, while a *variable*, as the name indicates, may stand for any value. In algebra, for example, quantities such as 2, 6, 101½, (8 + 7), even π are *constants*, since their value always remains the same. (Lowercase letters such as a and b are sometimes used to represent arbitrary constants.) By contrast, the letters x, y, and z are used as *variables*, which means that they may take *any* value, or stand for *any* number. Constants may be *substituted for* variables; that is, we may replace a variable with a constant. The result of such a substitution may be called a *substitution instance*, or just *instance*, of the variable or formula which contains the variable. In the formula $x + y = y + x$, for instance, we may substitute 3 for x (consistently) and 5 for y, to get the substitution instance $3 + 5 = 5 + 3$.

In sentential logic we are using sentential variables, of course, rather than numerical variables, and sentential constants rather than particular numbers, but the idea is the same. A sentential variable, or *statement variable*, as we will call it, is simply a *letter* (we use lowercase letters from the middle of the alphabet, p, q, r, s, . . .) that in itself has no meaning or truth value, but which is used to stand generally for, and which may take as a substitution instance, any particular statement, simple or complex. Our sentential constants, on the other hand, which we will call *statement constants*, are simple capital letters that stand for definite, particular, meaningful instances. They are just the letters, such as A and N, that we have been using as abbreviations for truth-functionally simple English sentences. Sentential constants, unlike variables, have definite truth values.

Statement variables and statement constants, then, are the *simple* components of our formulas. *Statement forms* and *statements*, on the other hand, may be either simple or complex, and the difference between them is that the *forms* are built up out of variables, while the *statements* are built up out of the capital letters, or constants. More precisely, *a statement form is a formula (simple or compound) that has as its smallest units statement variables*, while a *statement is a formula (simple or compound) that has as its smallest units statement constants*. Some examples of statement forms would be p, $(p \lor q)$, and $((p \cdot q) \supset \sim (r \lor s))$. Examples of statements would be A, $(A \lor B)$, and $(\sim (C \cdot D) \cdot (A \supset \sim B))$. Again, *forms have variables* and *statements have constants* as their smallest units.

The following table classifies and illustrates the various sentential elements which we have described above.

	FORMS	INSTANCES
Simple	Statement variables $p, q, r, \cdots$	Statement constants $A, B, N, \cdots$
Simple or Complex	Statement forms $p, (q \lor r), ((p \supset q) \cdot \sim r)$	Statements $N, (A \supset \sim N), ((A \lor B) \equiv \sim C)$

As noted, statements will be the *substitution instances of statement forms. We will get a substitution instance of a form by (uniformly) substituting some statement, simple or complex, for each variable in the form.* Thus, from the form $((p \lor q) \supset r)$, we could obtain the following: (1) $((A \lor B) \supset \sim C)$, where A, B, and $\sim C$ have been substituted, respectively, for p, q, and r, and (2) $((B \cdot C) \lor (C \supset \sim D)) \supset \sim (A \lor (B \cdot \sim C))$, where $(B \cdot C)$ has been substituted for p, $(C \supset \sim D)$ has been substituted for q, and $\sim (A \lor (B \cdot \sim C))$ has been substituted for r. There are, of course, an unlimited number of substitution instances for each form. It is extremely important to keep in mind that each simple variable, p, q, r, and so on, may take as substitution instances, *complex* formulas as well as simple ones. In general, in fact, the substitution instance will be more complex than the form itself, since in most cases we will substitute complex statements for at least some of the variables. Some instances of the form $(p \supset q)$, are (1) $(A \supset B)$, (2) $((A \lor B) \supset (G \lor H))$, and (3) $((F \equiv G) \supset ((A \lor B) \lor (C \cdot D)))$. Any statement, in fact, whose major operator is a conditional would be a substitution instance of $(p \supset q)$, since we just put in the antecedent for p and the consequent for q. By contrast, $((A \supset B) \lor C)$ is *not a substitution instance of* $(p \supset q)$ since, although it does contain a conditional, the conditional is not the major operator, and there is *no way* we can substitute formulas for p and q to come out with the statement. The best we could do would be to substitute A for p, and $B) \lor C)$ for q, but this would not do, since $B) \lor C)$ is not a formula. $((A \supset B) \lor C)$ would, of course, be a substitution instance of $(p \lor q)$.

Note that according to the definition we *may* substitute the *same* statement for different variables, so that $(A \lor A)$ *is* a substitution instance of $(p \lor q)$. But we must *not* put in *different* statements for repeated occurrences of the same variable. We must be *consistent* in our substitution; that is, whatever we put in for p at one place, we must put in for p at all other places. Thus $((A \lor B) \supset C)$ is *not* a substitution instance of $((p \lor p) \supset q)$, but $((A \lor A) \supset B)$ would be a substitution instance, as well as $((A \lor A) \supset A)$.

Note that the requirement of consistent substitution, and the possibility of putting in the same statement for different variables, is exactly analogous to the use of variables in mathematics. If we have the formula $x + y = y + x$, a correct substitution instance would be $3 + 4 = 4 + 3$, but $2 + 3 = 4 + 5$ would *not* be an instance, since whatever we put in for x at one place we must also put in for any other occurrence of x. On the other hand, $3 + 3 = 3 + 3$ is a perfectly good instance, since we *can* substitute 3 for both x and y if we wish.

Some examples are given below to help you see what kinds of statements do or do not count as substitution instances of forms. To help you figure out in each case why a formula *is* an instance, the statements that were substituted for each variable to get the substitution instance have been outlined. For those formulas that are not instances, you should try to see why it is not possible to make substitutions for the variables which would yield the formulas as instances. Exercise 1, at the end of the unit, will also contribute to your understanding of these very important concepts of form and substitution instance, and it is important that you do these exercises before you go on to the sections on proofs.

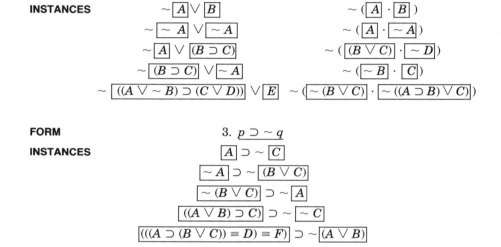

FORM

1. $\sim p \vee q$
2. $\sim (p \cdot q)$

INSTANCES

$\sim \boxed{A} \vee \boxed{B}$

$\sim \boxed{\sim A} \vee \boxed{\sim A}$

$\sim \boxed{A} \vee \boxed{(B \supset C)}$

$\sim \boxed{(B \supset C)} \vee \boxed{\sim A}$

$\sim \boxed{((A \vee \sim B) \supset (C \vee D))} \vee \boxed{E}$

$\sim (\boxed{A} \cdot \boxed{B})$

$\sim (\boxed{A} \cdot \boxed{\sim A})$

$\sim (\boxed{(B \vee C)} \cdot \boxed{\sim D})$

$\sim (\boxed{\sim B} \cdot \boxed{C})$

$\sim (\boxed{\sim (B \vee C)} \cdot \boxed{\sim ((A \supset B) \vee C)})$

FORM

3. $p \supset \sim q$

INSTANCES

$\boxed{A} \supset \sim \boxed{C}$

$\boxed{\sim A} \supset \sim \boxed{(B \vee C)}$

$\boxed{\sim (B \vee C)} \supset \sim \boxed{A}$

$\boxed{((A \vee B) \supset C)} \supset \sim \boxed{\sim C}$

$\boxed{(((A \supset (B \vee C)) \equiv D) \equiv F)} \supset \sim \boxed{(A \vee B)}$

FORM	1. $\sim p \vee q$	2. $\sim (p \cdot q)$	3. $p \supset \sim q$
NONINSTANCES	$A \vee \sim B$	$\sim A \cdot B$	$\sim A \supset B$
	$\sim (A \vee B)$	$\sim A \cdot \sim B$	$(A \supset \sim B) \vee C$
	$\sim (A \vee B) \supset C$	$\sim (A \cdot B) \vee C$	$\sim (A \supset \sim B)$
	$(\sim A \vee B) \vee (C \vee D)$	$\sim \sim (A \cdot B)$	$A \supset (B \supset \sim C)$

2. The Proof Process

In the proof method for demonstrating validity we start with the premises, and then deduce from the premises a series of intermediate steps that finally results in the desired conclusion. What we do is to construct a *chain* of reasoning in which each new step follows from the previous steps. Suppose, for instance, that we have the following argument: John is weak and has a stomachache. If he is weak and has a stomachache, and has either a rash or a fever, then he either has food poisoning or mononucleosis. He would have food poisoning only if he ate bean sandwiches at Joe's Bar and Grill last night. But he didn't have bean sandwiches at Joe's last night; instead he had lobster at Maxmillian's. He doesn't have a rash, but he does have a fever. Therefore, he must have mononucleosis. We can symbolize this argument as follows, numbering the premises:

1. $(W \cdot S)$
2. $((W \cdot S) \cdot (R \vee F)) \supset (P \vee M)$
3. $(P \supset B)$
4. $\sim B \cdot L$
5. $\sim R \cdot F \; /\therefore M$

Given these five premises, we could deduce that John has mono in the following way (we will number and symbolize every step):

6.	F	John does have a fever (from step 5)
7.	$(R \vee F)$	John has a rash or a fever (from step 6)
8.	$(W \cdot S) \cdot (R \vee F)$	John is weak and has a stomachache, and has either a rash or a fever (by combining steps 1 and 7)
9.	$(P \vee M)$	John either has food poisoning or mono (from steps 2 and 8)
10.	$\sim B$	John didn't eat bean sandwiches last night (from step 4)
11.	$\sim P$	John does not have food poisoning (from steps 3 and 10)
12.	M	John has mono (from steps 9 and 11)

Some of these steps, such as step 6, may seem obvious and trivial, and you may wonder why we even bother to list them. It is essential to a proof, however, that every single move be recorded; the whole point of a *proof* (as opposed to an intuitive leap to the conclusion) is to be absolutely *certain* that the conclusion follows, and the only way we can obtain this certainty is by being very precise and very thorough. This means taking nothing for granted, but actually writing down every inference we make, no matter how trivial. Too often, people (including logicians) take things for granted and then discover that they have omitted something, or that there is some premise they need which they don't have, so that the argument is not valid after all. In order to avoid this kind of mistake, it is necessary that you write down every step. Notice also that we have provided a justification for every step by indicating what previous steps it was derived from; this is an important part of the proof process. It should be obvious by now why we number the steps: since each later step is justified in terms of previous steps, we need a simple way of referring back to the previous steps; numbering them is the easiest way to do this.

As mentioned in the Introduction to this unit, not any old inference will do; in constructing proofs, you cannot deduce just anything that strikes your fancy. In a proof, every step must *logically follow from* previous steps; each individual inference must be *valid*. The way we guarantee this is to require that each step in a proof follow from the previous steps *according to certain specified rules of inference*. These inference rules are patterns, or forms (they will be given in terms of variables) which are "tried and true," certifiably correct, guaranteed never to result in true premises but a false conclusion. They are all truth-functionally valid argument forms, a fact you may want to verify for yourself by using the truth table method on them. If you use just these rules (and use them correctly) you can be sure that your arguments will be valid. No one is claiming, of course, that these are the *only* possible correct rules; there are, in fact, an infinite number of rules which we might use. The rules you will be given in the next three units, however, are very basic, very common, and are completely sufficient for sentential logic. There is no valid sentential argument which cannot be proven using just these rules. The first eight of these rules, the basic inference rules, will be introduced in the next section.

3. Eight Basic Inference Rules

a. Rules for "if-then," "and," and "or."   An inference rule is a basic pattern of reasoning, according to which a conclusion of a certain form may be inferred from premises of a certain form. It is not the sort of rule which tells you you *must* do something, such as limiting your speed when driving to 55 mph; you will never be *required* to use any given rule. Rather, it is a rule which tells you that you *may* make a certain inference, that, given premises of a certain form, you are permitted to draw a conclusion of a certain form. The rule of Disjunctive Syllogism, for instance, will tell you that, given a disjunction and the negation of one of the disjuncts, you are permitted to infer the other disjunct. In this unit you will be given eight very basic inference rules, all in the following form: there will be one or more statement forms given as premises, and a single statement form given as the conclusion, which can be derived from those premises. A summary of these eight rules will be provided at the end of the unit for easy reference. Remember that *you will be required to memorize the symbolic forms of the inference rules*; otherwise you will simply not be able to do the work in these units. You should also become familiar with, and be able to paraphrase, the informal explanation of the rule, such as the one given above for Disjunctive Syllogism.

It will be easier to learn these rules if you think of them in terms of the operators they contain. Our first three rules, for example, are *horseshoe* rules, because they tell us how we may make inferences with formulas which have horseshoes as their major operators. The next two are *con-*

junction rules, and the two after that are *disjunction* rules. Dilemma, which is presented in subsection *b*, combines the conditional and disjunction operators.

Probably the most fundamental rule in all of logic is the rule of *Modus Ponendo Ponens* (its complete and proper name), more familiarly known as *Modus Ponens*; other rules may come and go, but most systems retain the rule of *M.P.* (as we will abbreviate it). This rule tells you that if you have a conditional, and also the antecedent of the conditional (the "if" part), then you are permitted to infer the consequent (the "then" part). Symbolically, it is stated thus:

MODUS PONENS (M.P.)

$p \supset q$
$\underline{p}$
$/\therefore q$

Given a conditional and the antecedent of that conditional, you are permitted to infer the consequent of the conditional.

It is essential to be aware at the outset that the rules are given in terms of *variables*; that is, they are rule *forms*. You will be applying these to particular instances, and *an instance of a rule will be correct if its premises are a substitution instance of the premises of the rule form and the conclusion is the corresponding instance of the conclusion of the rule form*. The following, for instance, would be a correct application of the rule of M.P.: $(A \lor B) \supset \sim C, (A \lor B) /\therefore \sim C$. The following would *not* be a correct application, because the second premise is not the antecedent of the conditional: $(A \cdot B) \supset \sim C, A /\therefore \sim C$. The following is also not a correct instance, because neither of the premises is in the form of a conditional: $(A \supset B) \lor (C \supset B), A /\therefore B$.[1] The rule of Modus Ponens can *only* be applied in cases where one of the formulas has, as its *major* operator, a conditional. (The other premise, of course, must be the antecedent of the conditional.) A correct application of M.P. occurs in the problem in the preceding section, where step 9, $(P \lor M)$, was inferred from steps 2 and 8, $((W \cdot S) \cdot (R \lor F)) \supset (P \lor M)$ and $((W \cdot S) \cdot (R \lor F))$. As this example illustrates, the applications of the rules—the instances—may be considerably more complex than the rules themselves, which use variables, since a variable may take as a substitution instance *any* formula, no matter how complex. You will need to have a sharp eye for the relationship between a form and its instances, and the second set of exercises at the end of the unit is designed to give you practice in recognizing instances and noninstances of the rules.

---

[1] Not only are these incorrect applications of the rules, they are also truth-functionally invalid. You should use the short truth table method on these and the other instances of incorrect applications given in this unit, constructing counterexamples to convince yourself that they are wrong.

Another very basic rule, which also contains a conditional as a premise, is the rule of *Modus Tollendo Tollens*, more commonly known as *Modus Tollens*. This rule tells us that, given a conditional as one premise, and the *negation of the consequent* of that conditional as the second premise, we may infer the *negation of the antecedent* of the conditional. In symbolic form it looks like this:

MODUS TOLLENS (M.T.)

$p \supset q$
$\sim q$
$/\therefore \sim p$

Given a conditional, and the negation of the consequent of the conditional, you are permitted to infer the negation of the antecedent of that conditional.

An instance of the rule of *M.T.* (as we will abbreviate it) would be the following: $(A \lor B) \supset C$, $\sim C$ $/\therefore \sim (A \lor B)$. This is a very common inference pattern, and now that you are aware of it, you will probably notice instances of it being used continually. An example in English would be "If John broke up with his girlfriend, he is depressed. He isn't depressed. Therefore, he didn't break up with his girlfriend." It is important to remember with this rule that *the second premise must always be the negation of the consequent*. Thus, the following would *not* be a correct application of the rule of M.T.: $\sim (A \cdot B) \supset \sim (C \lor D)$, $\sim (C \lor D)$ $/\therefore \sim (A \cdot B)$. The second premise is a negation, but it is *not* the negation of the consequent, and this is what is required. Another error here is that the conclusion, although it is also a negation, is not the negation of the antecedent. In order to use this conditional in the rule of M.T. we would have to have, as the second premise, the *double negation* $\sim \sim (C \lor D)$, and the correct conclusion would be the double negation $\sim \sim (A \cdot B)$. Notice that the rule of M.T. was used to infer step 11 in the example in the preceding section.

A third rule which uses the conditional is the rule of *Hypothetical Syllogism*, which we will abbreviate as *H.S.* In this rule there are two overlapping conditionals as premises, and the conclusion is the conditional with the middle term eliminated. Its symbolic form is below (notice how much easier it is to say it in symbols than to try to describe it in English):

HYPOTHETICAL SYLLOGISM (H.S.)

$p \supset q$
$q \supset r$
$/\therefore p \supset r$

Given two conditionals, in which the consequent of the first is identical to the antecedent of the second, you may infer the conditional whose antecedent is the antecedent of the first, and whose consequent is the consequent of the second.

What this rule tells us is that the conditional is *transitive*: if the first statement implies the second and the second implies the third, then the first implies the third. An instance of this would be:
$(A \lor B) \supset (C \cdot D), (C \cdot D) \supset (\sim E \cdot F) /\therefore (A \lor B) \supset (\sim E \cdot F)$. It is crucial, in the application of this rule, that the overlapping terms be *exactly identical*. The following would *not* be an instance of Hypothetical Syllogism, because it violates this condition: $(A \lor B) \supset (C \supset D), (C \cdot D) \supset (F \cdot G)$ $/\therefore (A \lor B) \supset (F \cdot G)$. Obviously, $(C \supset D)$ is not the same as $(C \cdot D)$, so the inference is not correct. Not any two conditionals, in other words, can serve as the premises of Hypothetical Syllogism; they must have the very special relationship of sharing a common term. The antecedent of one must be exactly the same as the consequent of the other. An example in English of this rule would be the following: "If either inflation remains high or unemployment increases, Jones will be defeated. If Jones is defeated, either Smith or Andrews will be governor. Therefore, if either inflation remains high or unemployment increases, then either Smith or Andrews will be governor." ("Jones" has been eliminated.)

As noted, it is useful to learn the rules in groups, according to the major operators involved. What you have just learned are three rules which tell us what we can infer from conditionals. The next two rules, which are extremely simple, are rules which tell us how to operate with conjunctions. The first tells us how to *use* a conjunction as a *premise*, that is, what we can derive *from* it, and the second tells us how to *derive* it as a *conclusion*. The rule of *Simplification*, abbreviated as *Simp.*, tells us that, if we have a conjunction as a premise, we may infer either one of its conjuncts as a conclusion. (Thus the rule has two forms.) This is exactly what you would expect from a rule for conjunction; since the conjunction asserts that *both* of the conjuncts are true, we should be able to infer either one separately.

SIMPLIFICATION (SIMP.)

$$\frac{p \cdot q}{/\therefore \quad p} \qquad \frac{p \cdot q}{/\therefore \quad q}$$
From a conjunction as premise, we may infer either of the conjuncts separately as a conclusion.

Again, it will be very important to remember that the rule of Simplification applies *only* when the major operator of the premise is a conjunction. The following would *not* be a correct application of the rule, and is, in fact, invalid: $(A \cdot B) \supset C /\therefore (A \supset C)$. (Again, use the short truth table method to construct a counterexample to verify that this instance is invalid.) The following, however, *is* a correct application of the rule: $((A \cdot B) \supset \sim (C \lor D)) \cdot (F \lor \sim H) /\therefore (F \lor \sim H)$. Notice that the rule of Simp. was used in inferring steps 6 and 10 in the example given in Section 2.

The rule of *Conjunction*, which will be abbreviated as *Conj.*, is also a very easy rule. This rule tells us, as we would expect, that if we have two separate statements, then we may infer their conjunction. The rule is stated as follows:

CONJUNCTION (CONJ.)

p

q

$/\therefore p \cdot q$

Given any two statements, we may infer their conjunction.

This rule should give you little difficulty, except that you must remember that in order to infer the conjunction, you must have *both* components separately first. An example of this rule would be the following: $(A \vee B), \sim (C \supset D) /\therefore (A \vee B) \cdot \sim (C \supset D)$. The rule of Conj. was also used to infer step 8 in the example in the preceding section.

The rules for disjunction are slightly more complex than those for conjunction, but are still quite simple. Again, one rule, Disjunctive Syllogism, tells us how to *use* a disjunction in a premise, while the other, Addition (a confusing name, but standard for this rule), tells us how to *derive* a disjunction as a conclusion. The rule of *Disjunctive Syllogism*, abbreviated as *D.S.*, says that if you are given a disjunction as a premise, and the *negation* of one of the disjuncts as a second premise, you may infer the other disjunct. In this rule, *either* disjunct may be negated as the second premise, so, like the rule of Simplification, it will have two forms. They are given below:

DISJUNCTIVE SYLLOGISM (D.S.)

$p \vee q$ $p \vee q$

$\sim p$ $\sim q$

$/\therefore \quad q$ $/\therefore \quad p$

Given a disjunction, and the negation of one of the disjuncts, you may infer the other disjunct.

This rule is very common in ordinary usage; an example employing the second form of the rule might be: "This tree is either a walnut and will bear nuts in the fall, or it is a cherry true. It is not a cherry tree. Therefore, it is a walnut and will bear nuts in the fall." D.S. was also used to infer the conclusion, that John has mono, in our example in Section 2 of this unit.

In using this rule, you must be sure that the second premise is the *negation of one of the disjuncts*, and not just any old negation. The following, for example, would *not* be an instance of Disjunctive Syllogism, because the second premise is not the negation of the second disjunct (or, or course, of the first). $(A \vee \sim (B \cdot C)), \sim (B \cdot C) /\therefore A$. In order to infer A from the first premise, we would need the *double* negation $\sim \sim (B \cdot C)$. The following, however, would be a correct application of the rule:

$$(\sim A \vee (B \supset C)) \vee \sim D, \sim (\sim A \vee (B \supset C)) /\therefore \sim D.$$

The rule of *Addition*, abbreviated as *Add.*, aside from its rather inapt name, is sometimes confusing to students for several reasons: in the first place, it is not of much use in ordinary discourse, since in using the rule we are going from something more definite to something less definite, which we rarely need to do. In the second place, it looks as if we are just pulling a formula out of thin air, and students often wonder how it is possible to do this. To clarify these comments, we need to look at the rule. The rule will tell us that, *given any statement whatever*, we may infer *any* disjunction with that statement as one of its disjuncts. Since the statement which is used as a premise may appear as either disjunct in the conclusion, the rule again has two forms. The rule is stated symbolically below:

ADDITION (ADD.)

$$\frac{p}{/\therefore \ \ p \vee q} \Big/ \frac{q}{/\therefore \ \ p \vee q}$$

Given any statement, you may infer any disjunction which includes that statement as one of the disjuncts.

What is odd about this rule is that we are going from something stronger to something weaker. From "Tod stole John's wallet," for instance, we can infer that "Either Bill or Tod stole John's wallet," which seems grossly unfair to Bill since we already know that Tod is the culprit! It is true, however, that if Tod stole it, then either Bill or Tod stole it, unfortunate as this may be for Bill. Although we have little use for this rule in ordinary language, it is extremely helpful in formal inferences, in actually constructing proofs. We often need to derive C, for instance, from previous steps of the form $(A \vee B) \supset C$ and A. But we cannot *directly* infer C from these two steps by M.P., since A is not the antecedent of the conditional. From A, however, we can infer $(A \vee B)$, and since this *is* the antecedent, we are then entitled to infer C. We used the rule of Addition in this way, to infer part of a more complex antecedent, in step 7 of the proof in Section 2.

Another rather disconcerting thing about this rule is that we may add on as a disjunct *any* formula, whether it has previously appeared in our proof or not, and no matter how complex it is. From $\sim A$, we may infer by Add. $\sim A \vee [(Z \vee (X \supset Y)) \cdot (\sim B \supset (C \cdot \sim (D \vee F)))]$! This rather odd-looking rule is justified, however, simply by noting that if a given formula is true, then (because of the truth table for disjunction) any formula containing it as a disjunct must be true.

The rule of Addition may be used on any formula; in other words, the premise may be of any form. The *conclusion* of the rule, however, must *always* be in the form of a disjunction; that is, it must always have a disjunction as its major operator, with the premise as one disjunct. It would thus not be correct to infer $((A \lor B) \supset C)$ from $(A \supset C)$ because the major operator of the conclusion is *not* a disjunction, although the formula contains a disjunction as a subformula.

One final reminder: do not confuse the *Addition* rule, which has a *disjunction* as a conclusion, with the *Conjunction* rule, which has a *conjunction* as a conclusion. The disjunctive conclusion, by Addition, requires only a single formula as a premise, while the conjunctive conclusion of Conjunction requires two premises.

We will not, in this unit, introduce a rule for the biconditional. In the next unit we will have a rule which will allow us to replace a biconditional $(p \equiv q)$ with the conjunction of two conditionals, $(p \supset q) \cdot (q \supset p)$, and this is the only biconditional rule we will need.

b. The rule of Dilemma.   Our last rule in this unit is one which combines several operators, and so falls into no particular category; this is the *Dilemma* rule, abbreviated *Dil*. The Dilemma rule tells us that, given two conditionals and the disjunction of their antecedents, we may infer the disjunction of their consequents. This is the only rule which requires more than two premises. The symbolic form is stated below:

DILEMMA (DIL.)

$p \supset q$	Given two conditionals, and the disjunction of
$r \supset s$	the antecedents of those conditionals, we may
$\underline{p \lor r}$	infer the disjunction of the consequences of the
$\therefore \;\; q \lor s$	conditionals.

Notice that this rule is very much like Modus Ponens; it is just a kind of "double-barreled" or "souped up" version of M.P. Instead of one conditional and its antecedent as premises, we have two conditionals and the disjunction of the antecedents as premises, and instead of inferring the single consequent, we infer the disjunction of the consequents. It is essential to remember that, in this rule, the premises must consist of *two conditionals* and one disjunction, and that the conclusion will be a disjunction.

The name of this rule was, no doubt, inspired by instances like the following: "If the United States relies primarily on nuclear power, then waste disposal will be a serious problem and there will someday be a major accident. If the United States relies primarily on coal, then many miners will be killed and air pollution will increase significantly. The United States will rely primarily either on nuclear power or on coal. Therefore, either waste disposal will be a serious problem and there will someday be a major accident, or many miners will be killed and air pollution will increase significantly." Of course, we might look on the bright side of things with the following instance of Dilemma: "If the United States relies primarily on nuclear power, then there will not be an enormous increase in air pollution. If it relies primarily on coal, then there will not be a severe nuclear waste disposal problem. The United States will rely primarily either on nuclear power or on coal. Therefore, either there will not be an enormous increase in air pollution or there will not be a severe nuclear waste disposal problem!"

Since this is our most complicated rule so far, it will be harder to keep straight the premises and conclusion, and easier to fall into error. The following, for instance, would *not* be a correct application of the rule: $((A \vee B) \supset C), ((D \cdot E) \supset F), (A \vee D) /\therefore C \vee F$. Here the problem is that the disjunction, the third premise, is *not* the disjunction of the antecedents. The following would also be incorrect: $A \vee C, B \vee D, A \supset B /\therefore C \supset D$. Here the major operators are reversed. These examples are not only *not* instances of the rule, they are again invalid; you should construct counterexamples to demonstrate this fact. Correct uses of the rule may be rather complex, but just keep in mind that as long as there are two conditionals and the disjunction of their antecedents in the premises, and the disjunction of their consequents in the conclusion, the instance is correct. The following would be a correct application of the rule: $((A \vee B) \supset C) \supset (D \vee F), (F \equiv G) \supset (A \supset F), ((A \vee B) \supset C) \vee (F \equiv G)$ $/\therefore (D \vee F) \vee (A \supset F)$.

c. About the rules.   You now have all the rules for this unit; if you have not memorized them as you went along, you should sit down and do so now. Learn the rules and their names well enough so that you can take out a blank sheet of paper and write them down with little or no hesitation. Learn the verbal descriptions of the rules as well, so that you have a very clear understanding of what each rule can do for you. This will be important when it comes to actually constructing proofs, planning strategies, and thinking up plans of attack. Keep in mind also that the rules you have been given are *forms*. They contain *variables* which can take as substitution instances any statement, simple or complex. You have already seen some examples of fairly complex substitution instances of the rules. Exercises 2, 3, 4, and 5, at the end of the unit, will give you further practice in recognizing complex instances of the rules. *It is very important that you complete all of them.* Don't take shortcuts. It will help to think of the rules in terms of their major operators: M.P., M.T. and H.S. are horseshoe rules; Simp. and Conj. are dot rules; Add. and D.S. are wedge rules. Dilemma combines the horseshoe and the wedge.

You may wonder how we can be so certain that these results are correct and that we will not get into trouble in applying them. The reason, as we have indicated earlier, is that each of these rules can easily be shown to be a truth-functionally valid argument form. This means that there can be no instance of any of these rules that has true premises and a false conclusion. Our rules are all *truth-preserving*, and thus, if our premises are all true, even if we have a very long chain of intermediate steps, we can be sure that in the end our conclusion will be true as well. If you are in doubt about any of these rules, just construct a little truth table to test them out.

You may well wonder why you are allowed just these rules and no others. You will not have a rule, for instance, which will permit you to infer $\sim q$ directly from $(p \equiv q)$ and $\sim p$, though it seems intuitively to be (and is) a perfectly valid inference. Isn't this an arbitrary restriction on your creative powers? Well, yes and no. It is a restriction, certainly, and in some ways it is arbitrary, since we could just as well have included this rule in our system. But on the other hand, there are an infinite number of possible rules we could have included, which would have made the system a bit difficult to learn. (You will gripe enough initially about the few rules you do have to learn.) We need to include *enough* rules to do the job, but not so many that the system becomes unwieldy. What we have here is a reasonable compromise, and these particular rules are included because they are quite powerful and because they are rules which are quite commonly used.

But then, you may ask, why do we need any fixed *system* of rules at all? Why can't we just make those inferences which seem reasonable, like the one mentioned above, and not worry about memorizing a particular list? Well, to take nothing away from your undoubtedly fine reasoning powers, it is quite possible that you might, at some point, be mistaken in what follows from what. You might be tempted, for instance, to infer $\sim A$ from $\sim (A \cdot B)$, which would be incorrect, or to infer $\sim B$ from $(A \lor B)$ and A, which would be just as bad. There is an interesting history of logical mistakes made by some of the best logicians of the twentieth century, and this is, in fact, why there has come to be such an emphasis on *formal* logic, on using only those rules which are tried and true. You can be certain that if you use only the rules given you here (and use them correctly), you will never arrive at an erroneous conclusion. If you try to make up your own, chances are that sooner or later (and most probably sooner) your intuitions will lead you astray.

Finally, it must be emphasized that *you must do the exercises* as well as learn the rules. In learning proofs, as in many other endeavors, practice, although it may not make perfect, is absolutely essential; and lack of practice will virtually guarantee failure. At this point, you should complete Exercises 2 and 3 at the end of the unit; do not even attempt to construct proofs until you can easily do the preliminary exercises. You can't run before you walk, and you can't construct proofs before you can recognize correct and incorrect applications of the rules.

4. Derivations and Proofs

The proof method for determining validity consists, as we have seen, in starting with the premises and deriving a series of intermediate steps until we reach the conclusion. What we do is to construct a *chain* of reasoning, each step a consequence of previous steps, which shows that the conclusion does follow from the premises. But, as noted, not any old step will do; we are only permitted to make an inference if it follows from previous steps according to one of the given rules. We want to be sure that there are no "logical gaps" in our proof construction, no weak links in our chain of reasoning; our sequence of steps must be "logically tight." If an attempted proof contains even one incorrect step, then the chain of reasoning has been broken, so the proof is a failure; we have not shown that the conclusion follows from the premises. To guarantee that the proof sequence is logically tight, each step in the proof which is not a premise must be *justified* by noting that it follows from previous steps (premises or other, intermediate steps) according to one of the given rules of inference. Premises will be justified simply by noting that they are premises, steps given initially from which the others are to follow.

We will require in a proof, then, that every single step be *justified*, either as being a premise or as following from previous steps according to one of the given rules of inference. We will justify a step by writing to the right of it whether it is a premise, or, if it is not, what rule it was derived from, and what steps served as premises for the application of that rule. Fully justified, the proof from Section 2 would look like the following (we will abbreviate "Premise" as "Pr."):

1.	$W \cdot S$	Pr.	7.	$R \lor F$	Add. 6
2.	$((W \cdot S) \cdot (R \lor F))$ $\supset (P \lor M)$	Pr.	8.	$(W \cdot S) \cdot (R \lor F)$	Conj. 1,7
3.	$P \supset B$	Pr.	9.	$P \lor M$	M.P. 2,8
4.	$\sim B \cdot L$	Pr.	10.	$\sim B$	Simp. 4
5.	$\sim R \cdot F$	Pr.	11.	$\sim P$	M.T. 3,10
6.	F	Simp. 5	12.	M	D.S. 9,11

Supplying the justification for each step will be considered essential to the proof, and if the justification is missing, the proof will be considered defective. You should never write down a step in the first place unless you know exactly where it comes from, and by what rule; if you do know this it is a simple matter to jot down to the right of the step the information that justifies it.

We need now to be a little more precise in our definitions. In particular, we have been rather taking for granted what it means for a step to *follow from* previous steps. We can now give the following definition: *A step S follows from previous steps $P_1 \ldots P_n$ according to a given rule of inference R just in case S is a substitution instance of the conclusion of the rule R and $P_1 \ldots P_n$ are the corresponding substitution instances of the premises of the rule R.* Given the definition of "following from" we can now say precisely what it means for a step to be justified: *A justified step is either a premise or a step which follows from previous steps according to one of the given rules of inference.* This definition will be expanded slightly in Unit 9, but for this unit and the next it is adequate.

So far we have been talking about arguments in which the desired conclusion is given, where the problem is to arrive at the stated conclusion by constructing a series of intermediate justified steps. In some cases, however, we do not know what the conclusion is supposed to be. Often we just want to derive the consequences of certain hypotheses, see what follows from them, without any preconceived notions of what the result will be. Such a deduction of consequences, which does not necessarily lead to a particular conclusion, will be called a *derivation. Any* sequence of steps will be a derivation *so long as every step is justified.* This gives us the following short and simple definition of a derivation: *a derivation is a sequence of justified steps.* All sorts of funny things will count as derivations, including just a set of premises (since each step will be justified as being a premise). The following sequence, which doesn't seem to be going anywhere useful, is also a derivation:

1. A Pr.
2. $A \lor B$ Add. 1
3. $(A \lor B) \lor C$ Add. 2
4. $((A \lor B) \lor C) \lor D$ Add. 3
5. $(((A \lor B) \lor C) \lor D) \lor E$ Add. 4
6. $((((A \lor B) \lor C) \lor D) \lor E) \lor F$ Add. 5
7. $(((((A \lor B) \lor C) \lor D) \lor E) \lor F) \lor G$ Add. 6

In contrast to a derivation, a *proof* is a sequence of steps which is going somewhere, in which we arrive at a definite conclusion which we were aiming at ahead of time. The *only* difference between a derivation and a proof is that in a proof we know where we are going (and get there), while in a derivation we simply deduce whatever we can, whatever "falls out" of the premises. It should be noted that the concepts of "derivation" and "proof" are not mutually exclusive. In fact, *every* proof is also a derivation, since each step must be justified. A proof just has something extra—a conclusion towards which the derivation is headed, and with which it must conclude. We can thus define "proof" very simply: *a proof is a derivation in which the last step is the desired conclusion.*

In the next section we will begin to construct proofs, leading to a definite conclusion; but before you go on, you should *do Exercises 4 and 5*, at the end of the unit. In Exercise 4 you will be given derivations, sequences of justified steps, and it will be up to you to supply the justifications. You must find out what steps were used in deriving each line and what rule was applied. In Exercise 5 you will first be given premises and a rule, and will be asked to draw the appropriate conclusion; you will also be asked to supply the missing premise used in drawing a particular conclusion from a particular rule; and in some cases you will be asked simply to draw a conclusion, and you will have to figure out yourself what rule is applicable. These exercises will prepare you for the task of constructing proofs yourself.

5. Constructing Simple Proofs

It is at this point in the course, where you begin to construct your own proofs, that you may begin to feel some frustration. Learning the rules may be tedious, but at least it requires nothing more than perseverance, and supplying the justification for derivations which are already worked out is a purely mechanical procedure, requiring nothing more than a thorough knowledge of the rules and an understanding of the concept of a substitution instance. Constructing proofs from scratch, however, as noted in the introduction to this unit, is not a mechanical procedure; there is no precise method, no set of instructions, for coming up with a series of steps which leads to the desired conclusion. Constructing proofs often requires insight and ingenuity, and even if you know the rules well you may often find that you simply don't know how to get started, or once started, you may not be able to see how to continue. This is a very common experience with beginning logic students, and should be no cause for alarm, at least initially. What we will do in this section is to give you some idea of how to attack proof problems and some practical hints for learning the process of constructing proofs.

What is absolutely essential, of course, is to know the rules thoroughly; you should have them all clearly in your head, so that you can look at a pair of formulas such as $(A \supset (B \cdot C))$, $\sim (B \cdot C)$, and see immediately that it is an instance in which you can apply the rule of Modus Tollens. You should also be sure you have done all the preliminary exercises, numbers 1 through 5, at the end of the unit; these will help make the rules second nature to you, and will help you get started in applying them. Finally, you should make it a habit to *start with the easiest problems first*, and gradually work your way to up to the more difficult ones. And do persevere; even though it may seem that you are making no progress, if you follow the instructions, you *will* eventually come to the point where what once seemed impossible now looks rather trivial. It often takes a couple of weeks to catch on to the method, but sooner or later the light will dawn. Once it does, you will have a real sense of accomplishment, and may very well find that this is the most enjoyable part of the course. Constructing proofs is like doing puzzles, a challenging intellectual exercise, and if you take it in this spirit you should find it much easier and more enjoyable. Don't expect miracles the first time you sit down to work; it sometimes takes time for everything to sink in, but your patience will eventually be rewarded.

Having given the sermon, let us look at a very simple example: suppose we have as premises $(A \cdot B) \cdot \sim C$, and $D \supset C$, and we want to prove $B \cdot \sim D$. Whether we list the premises vertically or horizontally, we will generally write the conclusion, for reference, to the right of the last premise, preceded by our "therefore" sign /∴.. We might, then, set up our problem as follows: (1) $(A \cdot B) \cdot \sim C$, (2) $D \supset C$, /∴ $B \cdot \sim D$. We will also "number" our problems in this unit with small letters (this one is a). Try to do the proof yourself before you go on. One method you can use, which will not always work but does work here, is to treat the problem like a derivation, and simply try to deduce whatever can be derived from the premises. We will call this the method of working from the "top down" (as opposed to the "bottom up" method which will be applied in more complicated problems). Let us see, then, what can be deduced from these premises. When you have a conjunction in the premises, of course, you can always apply the rule of Simplification, and we can here apply it to premise 1. This would give us two intermediate steps, from the two forms of the rule:

3. $A \cdot B$ Simp. 1
4. $\sim C$ Simp. 1

Now, given the conjunction in step 3, we could apply the Simp. rule once again to get two more intermediate steps:

5. A Simp. 3
6. B Simp. 3

At this point we can check out our other premise and see what can be deduced from it. This premise is a conditional, $D \supset C$, and we note that, in step 4, we have the negation of its consequent. This allows us to derive the negation of the antecedent, according to the rule of Modus Tollens. Thus our next step in the deduction would be:

7. $\sim D$ M.T. 2,4

This seems to exhaust the resources of our premises, but it is all we need for the conclusion. The conclusion, if you remember, was simply the conjunction $B \cdot \sim D$. We now have both of the conjuncts separately, so we are entitled to join them together with the conjunction rule, and the last step in our derivation will be the desired conclusion.

    8. $B \cdot \sim D$ Conj. 6,7

You may have noticed here that we could just as well have applied the rule of M.T. to get $\sim D$ at step 5; there was no need to do the other Simplification step first. In general, *there will be many different ways in which a proof can be constructed*; the order makes no difference except that, of course, every step must be validly derived from previous steps, so that you must first have the steps you need to apply the rules. We could not have derived $\sim D$ immediately after step 3, for instance, because we did not yet have $\sim C$. Another thing which needs to be mentioned, although it does not come up in this problem, is that *the order of the premises in the application of a rule is irrelevant*. If we first had $\sim C$ and only then derived $D \supset C$, we could still have applied the M.T. rule.

Let us take a slightly more complicated example, but one in which we can still work from the "top down." Our premises will be the following, numbered 1 through 4, and the conclusion will be listed following the "therefore" sign to the right of the last premise:

  b.   1. $(A \lor B) \supset (C \lor D)$ Pr.
       2. $(C \lor F) \supset H$ Pr.
       3. $E \cdot \sim D$ Pr.
       4. $E \supset A$ Pr. /∴ $H \lor I$

(*Hint*: When you have a letter in the conclusion which appears nowhere in the premises, such as the I in this proof, it will have to be "imported" by the use of Addition. Here we will need to get H first, and then $H \lor I$ by Add.) Again we have a conjunction in the premises, at step 3, so we may as well begin there by applying the Simplification rule. This gives us:

    5. E Simp. 3
    6. $\sim D$ Simp. 3

Now, if you know your rules well the next step should be obvious: the rule of Modus Ponens will give us:

    7. A M.P. 4,5

Now what? Here you should notice that we could apply the rule of Modus Ponens to step 1 if we had $A \lor B$ (this is, again, a matter of knowing your rules well). We do have A, and if you think for a minute you will realize that you can derive $A \lor B$ from A by Addition. This gives:

    8. $A \lor B$ Add. 7

At this point we can apply M.P., as we saw earlier.

    9. $C \lor D$ M.P. 1,8

We now have quite a few steps, and you may need to pause to look them over to see what else you can do. As we scan the list, we see that we have a disjunction and the negation of one of the disjuncts in steps 9 and 6; thus, we can apply the rule of Disjunctive Syllogism. (Notice here that the order of the premises is reversed.) This gives us:

    10. C D.S. 9,6

We can now apply our Addition trick again to get:

    11. $C \lor F$ Add. 10

This we need in order to apply M.P. to step 2, which will yield:

    12. H M.P. 2,11

Now what? We seem to have used up all our ammunition. Fortunately, the conclusion is only a short hop away: from H we can derive $H \lor I$ by Addition, and this is the end of the proof.

    13. $H \lor I$ Add. 12

At this point it would be a good idea to take a fresh sheet of paper, write down the last two problems, close your book, and try to reconstruct the proofs. If you get stuck, go back and reread the explanation and then try to do the proofs again. Keep trying and rereading until you can do them.

Although there are many things you *cannot* do in a proof, such as make up your own rules, there are a number of things that are permitted, that make no difference to the proof. You may use premises more than once, or you may not use some of them at all. Nothing in the definition of "proof" says that each premise must be used once and only once. It has already been noted that the premises may be in any order. Perhaps surprisingly, it also does not matter if you have extra, unneeded steps. If in doing a proof you go down the wrong road and reach a dead end, it is not necessary to erase all the useless steps and start over. Once you discover the correct route, you can simply write down those useful steps immediately following the useless steps. There is nothing in the definition of a proof that says it must be optimally short, so it is not necessary to "purge" a proof of unnecessary steps. A proof may contain all sorts of useless garbage as long as the two essential conditions are met: (1) the conclusion is eventually reached, and (2) every step in the proof sequence, whether useful or not, is *justified*. You should keep these things in mind as you attempt to do the following proof:

c. 1. ~ *A*   Pr.
   2. *D* ⊃ *A*   Pr.
   3. *B* ⊃ *A*   Pr.
   4. *A* ⊃ *C*   Pr.
   5. *C* ⊃ (*D* ∨ *E*)   Pr.
   6. *B* ∨ *F*   Pr.
   7. *D* ∨ *G*   Pr. /∴ *F* · *G*

Suppose you were given the following problem; would you consider it a misprint?

d. 1. *A* ∨ *B*   Pr.
   2. *B* ⊃ *C*   Pr.
   3. ~ *A* · ~ *C*   Pr. /∴ *F* ⊃ *G*

It certainly looks as if there is no possible way to derive the conclusion; *F* and *G* don't appear anywhere in the premises. But being a conscientious student, you would no doubt give it a try anyway, perhaps in the following manner:

  4. ~ *A*   Simp. 3
  5. ~ *C*   Simp. 3
  6. *B*   D.S. 1,4
  7. ~ *B*   M.T. 2,5

At this point you need to know that given a contradiction, you may derive *anything*, including the desired conclusion (or its negation!). Simply use addition on *B*, to get *B* ∨ (*F* ⊃ *G*), and then D.S., with ~ *B*, to get the conclusion.

  8. *B* ∨ (*F* ⊃ *G*)   Add. 6
  9. *F* ⊃ *G*   D.S. 7,8

In general, if you have a contradiction (*p* and also ~ *p*), you can derive *any* formula *q* at all, by getting *p* ∨ *q* by Addition from *p*, and then

q by the application of Disjunctive Syllogism to $p \lor q$ and $\sim p$. Some of the proof problems at the end of the unit may require this little trick.

Having seen several examples, and assuming you know your rules, you should now be in a position to try the exercises in 6 at the end of the unit. These are graduated, starting with some fairly trivial problems and then moving on to more complex ones. They should all be solvable by the application of the "top down" method, where you simply try to deduce whatever you can until you reach the conclusion. More complex problems, which will require a little more planning and strategy, will be given in Exercise 7. For these you will sometimes need to use the method of "working backwards," or "from the bottom up." This method, and other strategy hints for more complex problems, will be developed in the next section.

6. Constructing More Complex Proofs

In some proofs, even relatively easy ones on occasion, you may simply not see how to get started; you may not be able to use the "top down" method because you don't see what things can be deduced from the premises to begin with. In such cases it may be helpful to use the "bottom up" method, or the method of *working backwards*. What this entails is looking first at the conclusion, then checking the premises to see how the conclusion could "come out" of them, and then working out step by step, *from the bottom up*, what will be needed to obtain the conclusion. In order to use this method, you have to not only know your rules by heart but also know what they can do for you. If your conclusion, for instance, is B, and you have a formula $(C \supset (A \lor B))$, you must know your rules well enough to reason in the following way: "If I had $A \lor B$ and $\sim A$, I could derive B. I could get $A \lor B$ by the rule of Modus Ponens from $(C \supset (A \lor B))$, provided I had the antecedent C." This bit of reasoning would have told you that you need C and $\sim A$ in order to derive B from $(C \supset (A \lor B))$. This kind of reasoning will be used constantly in the "bottom up" method, because you will be asking "what do I need to get the conclusion?" and, having discovered that you need a certain intermediate formula to get it, you will then have to ask "how can I get that intermediate formula?" and so on until you reach back into the premises. We will take a relatively simple problem (which actually could be solved by the direct "top down" method) to illustrate this procedure.

e. 1. $(A \lor B) \supset (C \lor (F \supset H))$ Pr.
    2. $A \cdot \sim C$ Pr.
    3. $H \supset C$ Pr. $/\therefore \ \sim F$

Since our conclusion is $\sim F$, we look into the premises to see where it might come from. F occurs in the rather complex first premise, and what we must do is to figure out how to pull $\sim F$ out of that formula. For this you need to know, again, what each rule can do for you. F appears as a subformula of the consequent of a conditional; thus you must first get that consequent, $C \lor (F \supset H)$. You should know that *you can derive the consequent of a conditional by M.P., provided you can get its antecedent.* Thus, you will need $(A \lor B)$. Once you get $C \lor (F \supset H)$, you should note that you need $(F \supset H)$, and that this could be derived by D.S. if you could get $\sim C$. $\sim C$, then, is another formula you will need. Once you have $F \supset H$ by this process, clearly you will need $\sim H$ to get $\sim F$ by M.T. Thus you will need the three formulas $(A \lor B)$, $\sim C$, and $\sim H$ to get $\sim F$ from the first premise. Having figured this out, the obvious next move is to try to figure out how to get these three formulas. You look back into the premises and note that one is immediate: $\sim C$ occurs as a part of a conjunction in step 2, and so can be derived just by Simp. We might as well do A at the same time, for the following two steps:

4. $\sim C$ Simp. 2
5. A Simp. 2

At this point it's no mystery how to get $A \lor B$, since we already have A.

6. $A \lor B$ Add. 5

The only thing left is to get $\sim H$. By this time that is no trick either, since we have $H \supset C$ as premise 3, and we already have $\sim C$ at step 4. Thus we get:

7. $\sim H$ M.T. 3,4

We now have the steps we said we needed in order to pull $\sim F$ out of the first premise, so let us apply them as needed:

8. $C \lor (F \supset H)$ M.P. 1,6
9. $F \supset H$ D.S. 8,4
10. $\sim F$ M.T. 7,9

In the "bottom up," or "working backwards" method, as the names imply, you start out by considering the *conclusion* in your reasoning process, rather than the premises. Instead of asking "what can be derived from these premises?" as in the last section, you ask "given that this is the conclusion I want, how can I get it; that is, what formulas do I need to derive it, and what rules will be applied?" If your conclusion is a conjunction, for instance, in general the thing to do will be to get both conjuncts separately, and if the conclusion is a disjunction, it can be derived either by Addition or Dilemma.

Let us take a slightly more complex problem which requires more steps and a more extended use of the working-backwards method:

f.   1. $A \supset (B \lor C)$ Pr.
    2. $B \supset F$ Pr.
    3. $A \cdot \sim F$ Pr.
    4. $(A \lor H) \supset D$ Pr. $/\therefore$ $C \cdot D$

Here our conclusion is a conjunction, so the thing to do is work on each conjunct separately, and then put them together with the rule of Conj. D should not be hard to derive, since it occurs as the consequent of a conditional in Premise 4 and thus can be derived by M.P., provided we have the antecedent. The antecedent is $A \lor H$, and to get this, we will probably use either Add. or Dil. Checking out the premises, we see that we have A in Premise 3, which occurs as a part of a conjunction. Thus we can get A by Simp., $A \lor H$ by Add., and D by M.P. This is one-half of our conclusion.

    5. A Simp. 3
    6. $A \lor H$ Add. 5
    7. D M.P. 4,6

The other half is C, and this will be a little more involved. C occurs in Premise 1 in the consequent of a conditional, as one disjunct. Thus we could derive C from Premise 1 if we could get A, the antecedent of the conditional, and $\sim B$, the negation of the other disjunct. Of course, we already have A, so we can use M.P. to get:

    8. $B \lor C$ M.P. 1,5

Now we need $\sim B$, and at this point it is easy to get. Since B appears in Premise 2 as the antecedent of a conditional, we could drive $\sim B$ by M.T. provided we could get the negation of the consequent, $\sim F$. But we have this ready and waiting in Premise 3, so we are almost there.

    9. $\sim F$ Simp. 3
  10. $\sim B$ M.T. 2,9

We can now apply D.S. to steps 8 and 10 to get our other conjunct, and we can then conjoin the two conjuncts to get our conclusion.

11. C D.S. 8,10
12. $C \cdot D$ Conj. 7,11

You should read through these problems until you fully understand every step; keep in mind also, however, that there may be other ways of doing the proofs. Especially with complex problems, there are usually a number of different and equally correct ways to derive the conclusions. As long as every step is justified and you eventually get to the conclusion, it doesn't really matter how you got there, even if your route is rather circuitous. In doing your problems, start with the easiest ones first and gradually work your way up to the more difficult ones. Take your time and don't get discouraged if you don't immediately succeed. If you know your rules well, apply them correctly, study the previous examples, and persevere, you should eventually be able to complete most of the proofs. And again, think of them as puzzles, as intellectual challenges; proofs are rather like mazes, where you have to reach the conclusion by working through to the right path. If you take it in this spirit, the process of learning to construct proofs should prove to be intellectually rewarding.

SUMMARY OF RULES OF INFERENCE

MODUS PONENS (M.P.)   **MODUS TOLLENS (M.T.)**   **HYPOTHETICAL SYLLOGISM (H.S.)**

$$p \supset q$$
$$\underline{p}$$
$$/\therefore q$$

$$p \supset q$$
$$\underline{\sim q}$$
$$/\therefore \sim p$$

$$p \supset q$$
$$\underline{q \supset r}$$
$$/\therefore p \supset r$$

SIMPLIFICATION (SIMP.)   **CONJUNCTION (CONJ.)**   **DILEMMA (DIL.)**

$$\underline{p \cdot q}$$
$$/\therefore p$$

$$\underline{p \cdot q}$$
$$/\therefore q$$

$$p$$
$$\underline{q}$$
$$/\therefore p \cdot q$$

$$p \supset q$$
$$r \supset s$$
$$\underline{p \vee r}$$
$$/\therefore q \vee s$$

DISJUNCTIVE SYLLOGISM (D.S.)   **ADDITION (ADD.)**

$$p \vee q$$
$$\underline{\sim p}$$
$$/\therefore q$$

$$p \vee q$$
$$\underline{\sim q}$$
$$/\therefore p$$

$$\underline{p}$$
$$/\therefore p \vee q$$

$$\underline{q}$$
$$/\therefore p \vee q$$

DEFINITIONS

1. A **constant** is a term that has a definite, particular value.
2. A **variable** is a term that may represent any value.
3. A **statement variable** is a letter that can take as substitution instances

any particular statement, simple or complex. (We will use lowercase letters from the middle of the alphabet, $p, q, r, \ldots$ as our sentential variables.)

4. A **statement constant** is a capital letter that is used as an abbreviation for a particular truth-functionally simple English sentence.

5. A **statement** is a formula (simple or complex) that has statement constants as its smallest components.

6. A **statement form** is a formula (simple or complex) that has statement variables as its smallest components.

7. A **substitution instance** (s.i.) of a statement form is a statement obtained by substituting (uniformly) some statement for each variable in the statement form. (We *must* substitute the same statement for repeated occurrences of the same variable, and we *may* substitute the same statement for different variables. Thus, both $A \lor B$ and $A \lor A$ *are s.i.'s of $p \lor q$, but $A \lor B$ is *not* an s.i. of $p \lor p$.)

8. A **justified step** is either a premise or a step which follows from previous steps according to one of the given rules of inference.

9. A **derivation** is a sequence of justified steps.

10. A **proof** is a derivation (sequence of justified steps) in which the last step is the desired conclusion.

EXERCISES

*1.   Using the definition of substitution instance in the glossary, decide whether the statements under each form are substitution instances of that form.

(1) $p \supset \sim q$

a. $A \supset \sim A$
b. $\sim A \supset \sim B$
c. $\sim A \supset B$
d. $A \supset \sim (B \cdot C)$
e. $A \supset (\sim B \lor \sim C)$
f. $\sim (A \supset B) \supset \sim (A \supset B)$

(2) $(p \lor q) \supset r$

a. $(\sim A \lor \sim B) \supset C$
b. $(A \lor A) \supset B$
c. $A \lor (B \supset C)$
d. $(A \lor B) \supset \sim (C \lor D)$
e. $\sim (A \lor B) \supset \sim C$
f. $(\sim (A \lor B) \lor \sim (A \lor B)) \supset (A \lor B)$

(3) $p \supset \sim (q \cdot r)$

a. $A \supset (\sim B \cdot \sim C)$
b. $A \supset \sim (B \cdot \sim C)$
c. $A \supset \sim \sim (B \cdot C)$
d. $\sim (A \cdot B) \supset \sim (A \cdot B)$
e. $(A \lor B) \supset \sim ((A \lor B) \cdot C)$
f. $(A \lor B) \supset \sim (A \lor (B \cdot C))$

(4) $p \supset (q \supset r)$

a. $A \supset (\sim A \supset A)$
b. $(A \supset B) \supset C$
c. $C \supset (\sim B \supset \sim B)$
d. $(A \supset B) \supset (A \supset B)$
e. $(A \supset (B \supset C)) \supset ((A \supset B) \supset (A \supset C))$
f. $(A \supset B) \supset (C \supset ((A \supset B) \supset (A \supset C)))$

(5) $\sim p \supset \sim q$ | (6) $(p \supset q) \supset r$

(5) $\sim p \supset \sim q$

a. $\sim (A \supset \sim B)$
b. $\sim A \supset \sim \sim B$
c. $\sim (A \supset B) \supset \sim (B \supset A)$
d. $\sim A \supset (\sim B \supset \sim C)$
e. $\sim (B \supset \sim C) \supset \sim A$
f. $\sim (\sim (B \supset \sim C) \supset \sim (\sim C \supset B))$

(6) $(p \supset q) \supset r$

a. $(A \supset B) \supset (A \supset B)$
b. $(A \supset (B \supset C)) \supset ((A \supset B) \supset (A \supset C))$
c. $A \supset ((B \supset C) \supset ((A \supset B) \supset (A \supset C)))$
d. $\sim (A \supset B) \supset (C \supset C)$
e. $(\sim (A \supset B) \supset C) \supset C$
f. $((A \supset B) \supset C) \supset C$

*2. Which of the following are correct applications of the rule cited? For those which are not, say why not, and what would be needed in order for them to be correct applications. The order of the premises does not matter.

a. $(A \lor \sim B) \supset \sim C$, $(A \lor \sim B)$ /∴ $\sim C$ — M.P. — yes.

b. $(B \supset \sim C) \lor (A \cdot D)$, B /∴ $\sim C \lor (A \cdot D)$ — M.P. — no

c. $\sim (A \lor B) \supset C$, $\sim C$ /∴ $\sim (A \lor B)$ — M.T. — no

d. $A \supset (B \supset C)$, $B \supset (C \supset D)$ /∴ $A \supset (C \supset D)$ — H.S. — no.

e. $\sim A \supset \sim (B \lor C)$, $(D \cdot E) \supset \sim A$ /∴ $(D \cdot E) \supset \sim (B \lor C)$ — H.S.

f. $((A \lor B) \supset \sim C) \cdot \sim D$ /∴ $\sim D$ — Simp.

g. $\sim B$ /∴ $(\sim A \cdot (D \supset (\sim C \lor E))) \lor \sim B$ — Add.

h. $(A \cdot \sim B) \supset (C \cdot \sim D)$ /∴ $A \supset (C \cdot \sim D)$ — Simp.

i. $\sim A \lor \sim B$, $\sim A$ /∴ $\sim B$ — D.S.

j. $(\sim C \lor \sim D) \supset (A \supset B)$, $(A \supset B)$, /∴ $\sim C \lor \sim D$ — M.P.

k. $\sim (A \lor B) \lor (C \supset \sim D)$, $\sim (C \supset \sim D)$ /∴ $\sim (A \lor B)$ — D.S.

l. $(A \cdot B) \supset (C \cdot D)$, $\sim (A \cdot B)$ /∴ $\sim (C \cdot D)$ — M.P.

m. $\sim (A \lor B) \supset (C \lor D)$, $\sim (C \lor D)$ /∴ $\sim (A \lor B)$ — M.T.

n. $A \lor B$ /∴ $(A \lor B) \cdot C$ — Conj.

o. $(\sim A \lor B) \supset (\sim A \lor B)$, $\sim A \lor B$ /∴ $\sim A \lor B$ — M.P.

p. $\sim \sim (A \lor \sim B) \supset \sim (\sim C \lor \sim D)$, $\sim \sim (\sim C \lor \sim D)$
/∴ $\sim \sim \sim (A \lor \sim B)$ — M.T.

q. $(A \cdot (B \lor C)) \lor (B \lor C)$ /∴ $(B \lor C) \lor (B \lor C)$ — Simp.

r. $\sim A \lor (B \supset C)$, $(B \supset C) \supset (C \supset D)$, $\sim A \supset (C \supset D)$
/∴ $(C \supset D) \lor (C \supset D)$ — Dil.

s. $\sim A \lor A$, $\sim A \lor \sim A$ /∴ $(\sim A \lor A) \cdot (\sim A \lor \sim A)$ — Conj.

t. $A \supset (\sim B \cdot C)$ /∴ $(A \lor D) \supset (\sim B \cdot C)$ — Add.

u. $(\sim B \lor \sim C) \supset \sim E$, $(\sim C \lor \sim D) \supset \sim F$, $\sim B \lor \sim D$
/∴ $\sim E \lor \sim F$ — Dil.

v. $\sim (A \lor \sim (B \lor C)) \lor (D \lor E)$, $\sim \sim (A \lor \sim (B \lor C))$
/∴ $\sim (D \lor E)$ — D.S.

w. $\sim (A \equiv (E \cdot \sim F)) \supset \sim (F \equiv (E \cdot A))$, $\sim (F \equiv (E \cdot A))$
/∴ $\sim (A \equiv (E \cdot \sim F))$ — M.T.

x. $((A \supset \sim B) \supset \sim C) \supset (D \lor E)$, $D \lor (E \supset (F \lor G))$
/∴ $((A \supset \sim B) \supset \sim C) \supset (F \lor G)$ — H.S.

y. $\sim A \supset (\sim A \cdot (\sim B \supset \sim C))$, $(\sim A \vee ((\sim B \cdot \sim C) \supset \sim B)) \supset \sim B$,
$\sim A \vee (\sim A \vee ((\sim B \cdot \sim C) \supset \sim B))$ /∴$(\sim A \cdot (\sim B \supset \sim C)) \vee \sim B$ Dil.

z. $(A \vee \sim B) \supset (C \vee \sim D)$, $\sim (A \vee \sim B)$ /∴ $\sim (C \vee \sim D)$ M.T.

*3. Cite the rule which was used in the following valid inferences. DS

a. $(\sim A \vee \sim B) \vee C$, $\sim C$ /∴ $\sim A \vee \sim B$

b. $(A \vee \sim B) \supset \sim (C \vee D)$, $A \vee \sim B$ /∴ $\sim (C \vee D)$

c. $\sim (A \vee B) \supset (\sim A \vee \sim B)$, $\sim (\sim A \vee \sim B)$ /∴ $\sim \sim (A \vee B)$

d. $(\sim (A \vee B) \supset \sim C) \cdot \sim C$ /∴ $\sim C$

e. $(\sim B \vee \sim C) \supset \sim D$, $\sim A \supset (\sim B \vee \sim C)$ /∴ $\sim A \supset \sim D$

f. $\sim A \supset \sim B$, $\sim A \vee \sim B$, $\sim B \supset \sim A$ /∴ $\sim B \vee \sim A$

g. $\sim (\sim B \supset \sim A)$, $A \supset (\sim B \supset \sim A)$ /∴ $\sim A$

h. $(\sim A \supset \sim B) \supset (\sim C \supset \sim B)$, $\sim A \supset \sim B$ /∴ $\sim C \supset \sim B$

i. $(\sim B \supset \sim C) \supset (\sim A \supset \sim B)$, $\sim A \supset (\sim B \supset \sim C)$ /∴ $\sim A \supset (\sim A \supset \sim B)$

j. $(B \supset \sim C) \supset (A \supset \sim B)$, $\sim (A \supset \sim B)$ /∴ $\sim (B \supset \sim C)$

k. $(A \supset \sim B) \vee \sim C$, $\sim (A \supset \sim B)$ /∴ $\sim C$

l. $A \vee B$, $\sim A \vee \sim B$ /∴ $(\sim A \vee \sim B) \cdot (A \vee B)$

m. $\sim B \supset (\sim C \equiv \sim D)$, $\sim B$ /∴ $\sim C \equiv \sim D$

n. $(\sim A \vee \sim B) \vee (\sim A \cdot \sim B)$ /∴ $((\sim A \vee \sim B) \vee (\sim A \cdot \sim B)) \vee \sim A$

o. $\sim (A \equiv B)$, $(A \equiv (\sim B \equiv \sim C)) \supset (A \equiv B)$ /∴ $\sim (A \equiv (\sim B \equiv \sim C))$

*4. Supply the justifications for each step which is not a premise in the following derivations, in other words, indicate the rule used to derive the step, and the previous steps which were used to get it.

a.
1. $(A \vee \sim B) \supset \sim (C \vee \sim D)$ — Pr.
2. $(F \cdot \sim H) \vee A$ — Pr.
3. $\sim (F \cdot \sim H) \cdot (\sim O \cdot \sim P)$ — Pr.
4. $(\sim Z \cdot X) \supset O$ — Pr.
5. $\sim (F \cdot \sim H)$ — (3, Simp)
6. A — (2,5 DS)
7. $\sim O \cdot \sim P$ — (3, Simp)
8. $\sim O$ — (7, Simp)
9. $\sim (\sim Z \cdot X)$ — (4,8, MT)
10. $A \vee \sim B$ — (6, Add)
11. $\sim P$ — (7, Simp)
12. $\sim (C \vee \sim D)$ — (1,10, MP)
13. $\sim P \vee (\sim F \cdot \sim H)$ — (11, Add)

b.
1. $(\sim A \vee (A \vee B)) \supset (A \vee B)$ — Pr.
2. $\sim A \supset \sim (B \vee C)$ — Pr.

3. $\sim \sim (B \lor C) \cdot (\sim B \lor \sim A)$ Pr.

4. $(\sim \sim A \lor B) \supset \sim (A \lor B)$ Pr.

5. $\sim \sim (B \lor C)$ (3, Simp)

6. $\sim \sim A$ (2, 5, MT)

7. $\sim B \lor \sim A$ (3, Simp)

8. $\sim \sim A \lor B$ (6, Add)

9. $\sim B$ (7, DS)

10. $\sim (A \lor B)$ (4, 8, MT)

11. $\sim A \lor \sim (A \lor B)$ (

12. $A \lor B$

13. $\sim B \cdot (A \lor B)$

c.   1. $((C \lor D) \cdot (\sim C \lor \sim D)) \cdot ((\sim C \lor \sim D) \supset \sim (D \lor E))$ Pr.

2. $((C \lor D) \lor \sim (C \lor E)) \supset ((\sim C \lor \sim D) \supset \sim C)$ Pr.

3. $((D \lor E) \supset C) \cdot (\sim D \lor (D \lor E))$ Pr.

4. $(\sim D \cdot (\sim C \lor \sim D)) \supset (\sim (D \lor E) \supset (\sim D \supset (\sim C \supset \sim E)))$ Pr.

5. $(C \lor D) \cdot (\sim C \lor \sim D)$

6. $\sim C \lor \sim D$

7. $C \lor D$

8. $(\sim C \lor \sim D) \supset \sim (D \lor E)$

9. $(C \lor D) \lor \sim (C \lor E)$

10. $\sim (D \lor E)$

11. $(\sim C \lor \sim D) \supset \sim C$

12. $\sim C$

13. $(D \lor E) \supset C$

14. $\sim D \lor (D \lor E)$

15. $\sim D$

16. $\sim D \cdot (\sim C \lor \sim D)$

17. $\sim (D \lor E) \supset (\sim D \supset (\sim C \supset \sim E))$

18. $\sim D \supset (\sim C \supset \sim E)$

19. $\sim C \supset \sim E$

20. $\sim E$

21. $((\sim C \supset \sim D) \supset \sim E) \lor \sim E$

d.   1. $(\sim ((C \lor D) \supset E) \cdot (\sim (A \lor B) \equiv F)) \supset (\sim (A \equiv E) \lor \sim (C \equiv F))$ Pr.

2. $\sim ((C \lor D) \supset E) \supset (((A \equiv F) \equiv (C \equiv E)) \lor \sim (C \lor D))$ Pr.

3. $(\sim (A \equiv E) \cdot (C \lor D)) \cdot (((A \equiv F) \equiv (C \equiv E)) \supset (A \equiv E))$ Pr.

4. $(C \lor D) \supset ((C \equiv F) \supset (A \equiv E))$ Pr.

5. $((C \equiv F) \supset (A \equiv E)) \supset \sim ((C \lor D) \supset E)$ Pr.

6. $\sim (A \equiv E) \cdot (C \vee D)$

7. $\sim (A \equiv E)$

8. $((A \equiv F) \equiv (C \equiv E)) \supset (A \equiv E)$

9. $C \vee D$

10. $\sim (A \equiv E) \vee \sim (C \equiv F)$

11. $(C \equiv F) \supset (A \equiv E)$

12. $(C \vee D) \supset \sim ((C \vee D) \supset E)$

13. $(C \vee D) \vee ((C \vee D) \supset E)$

14. $\sim ((C \vee D) \supset E)$

15. $((A \equiv F) \equiv (C \equiv E)) \vee \sim (C \vee D)$

16. $\sim ((A \equiv F) \equiv (C \equiv E))$

17. $\sim (C \vee D)$

18. $(C \vee D) \vee (\sim (A \vee B) \equiv F)$

19. $\sim (A \vee B) \equiv F$

20. $\sim ((C \vee D) \supset E) \cdot (\sim (A \vee B) \equiv F)$

21. $\sim (A \equiv E) \vee \sim (C \equiv F)$

*5. Answer the following:

a. Indicate what conclusion would follow from the application of the rule cited to the given premises.

1. $(\sim A \vee \sim B) \vee (C \cdot D)$, $\sim (C \cdot D)$ /∴		D.S.
2. $(A \vee (B \supset C)) \supset D$, $(E \supset A) \supset (A \vee (B \supset C))$ /∴		H.S.
3. $\sim (A \vee B) \supset (C \supset (D \vee F))$, $\sim (C \supset (D \vee F))$ /∴		M.T.
4. $(P \supset G) \supset (G \supset R)$, $P \supset G$ /∴		M.P.
5. $\sim (\sim B \equiv \sim C)$, $(B \equiv C) \supset (\sim B \equiv \sim C)$ /∴		M.T.
6. $\sim A \vee (B \supset C)$, $\sim A \supset (A \supset D)$, $(B \supset C) \supset C$ /∴		Dil.
7. $\sim (A \vee \sim B) \vee (B \vee \sim A)$, $\sim (B \vee \sim A)$ /∴		D.S.
8. $\sim (D \supset \sim F)$, $\sim (A \vee (B \supset \sim C)) \supset (D \supset \sim F)$ /∴		M.T.
9. $(A \supset B) \supset (D \supset F)$, $(D \supset (D \supset F)) \supset (A \supset B)$ /∴		H.S.
10. $(A \supset B) \supset (\sim D \vee F)$, $(B \vee A) \supset (D \vee \sim F)$, $(B \vee A) \vee (A \supset B)$ /∴		Dil.

b. State what additional premises would be needed in order to derive the indicated conclusion according to the rule cited.

1. $\sim C \vee (\sim D \cdot F)$ /∴ $\sim C$		D.S.
2. $(A \supset (A \supset B)) \supset C$ /∴ $(A \supset B) \supset C$		H.S.
3. $\sim (\sim F \supset E)$ /∴ $\sim (E \supset \sim F)$		M.T.
4. $\sim F$ /∴ $A \equiv \sim F$		M.P.

5. $(A \cdot B) \supset (B \vee C)$ $/\therefore A \supset (B \vee C)$ M.P.

6. $(A \vee C) \supset \sim (B \vee C), (\sim B \vee \sim C) \vee (A \vee C)$
   $/\therefore \sim A \vee \sim (B \vee C)$ Dil.

7. $\sim (F \supset (G \vee H))$ $/\therefore \sim (A \vee B)$ D.S.

8. $\sim (\sim B \vee (A \supset C))$ $/\therefore \sim \sim \sim (\sim A \supset B)$ M.T.

9. $(B \supset (A \supset B)) \supset (\sim A \supset B)$ $/\therefore (B \supset \sim A) \supset (\sim A \supset B)$ H.S.

10. $(\sim B \supset \sim A) \supset \sim A, (\sim A \supset B) \supset (B \supset \sim A)$
    $/\therefore (B \supset \sim A) \vee \sim A$ Dil.

c. What conclusion could be drawn from the following premises, and by what rule (*excluding* Add., Simp., and Conj.)?[2]

   1. $(\sim A \vee \sim B) \supset \sim (C \vee \sim D), \sim A \vee \sim B$ $/\therefore$ $\sim\sim (C \vee D)$ MP

   2. $(A \vee \sim B) \supset (\sim B \vee C), \sim (\sim B \vee C) \supset (A \vee \sim B)$ $/\therefore$ $(\sim B \vee C) \vee \sim (B \vee C)$ HS

   3. $\sim (A \vee \sim B), (\sim A \supset \sim B) \vee (A \vee \sim B)$ $/\therefore$ $(\sim A \supset \sim B)$ DS

   4. $\sim (C \vee \sim D), (\sim (C \vee \sim D) \vee \sim C) \supset (C \vee \sim D)$ $/\therefore$ $\sim \sim (C \vee \sim D) \vee C)$ MT

   5. $(E \equiv \sim F) \vee (F \equiv (\sim E \equiv \sim F)), \sim (E \equiv \sim F)$ $/\therefore$ $(F \equiv (\sim E \equiv \sim F))$ DS

   6. $\sim A \supset (B \vee \sim C), \sim A \vee (\sim B \vee \sim C),$
   $\quad (\sim B \vee \sim C) \supset (\sim A \vee (C \vee \sim B))$ $/\therefore$

   7. $(\sim A \supset (\sim B \vee \sim A)) \supset (A \supset \sim B), \sim (A \supset \sim B)$ $/\therefore$

   8. $(A \supset B) \supset (A \supset (C \supset A)), (C \supset (A \supset C)) \supset (A \supset B)$ $/\therefore$

   9. $\sim (C \vee (A \vee C)) \supset \sim (A \supset (C \supset A)), \sim (C \vee (A \vee C))$ $/\therefore$

  10. $\sim (C \vee (A \vee C)) \vee (\sim A \vee \sim C), \sim (\sim A \vee \sim C)$ $/\therefore$

6. For each of the following arguments, construct a proof of the conclusion from the given premises, and justify every step which is not a premise.

  a. $D \supset (A \vee C), D \cdot \sim A$ $/\therefore C$

  b. $(A \vee B) \supset C, (C \vee D) \supset (E \vee F), A \cdot \sim E$ $/\therefore F$

  c. $(F \vee G) \supset \sim A, A \vee W, F \cdot T$ $/\therefore W$

  d. $(A \vee B) \supset T, Z \supset (A \vee B), T \supset W, \sim W$ $/\therefore \sim Z$

*e. $\sim A \supset \sim B, A \supset C, Z \supset W, \sim C \cdot \sim W$ $/\therefore \sim B \vee W$

  f. $(A \vee B) \supset (C \vee D), C \supset E, A \cdot \sim E$ $/\therefore D \vee W$

*g. $(A \cdot B) \supset \sim C, C \vee \sim D, A \supset B, E \cdot A$ $/\therefore \sim D$

  h. $(\sim A \vee \sim B) \supset \sim G, \sim A \supset (F \supset G), (A \supset D) \cdot \sim D$ $/\therefore \sim F$

*i. $F \supset (G \supset \sim H), (F \cdot \sim W) \supset (G \vee T), F \cdot \sim T, W \supset T$ $/\therefore \sim H$

*j. $P \supset (Q \supset (R \vee S)), P \cdot Q, S \supset T, \sim T \vee \sim W, \sim \sim W$ $/\therefore R$

7. Construct proofs for the following more challenging problems, justifying each step which is not a premise.

  a. $(F \vee G) \vee H, (F \supset H) \cdot (H \supset T), \sim T \vee W, \sim W \cdot S$ $/\therefore G \cdot S$

[2] There are, of course, an infinite number of conclusions that would follow from various applications of the rule of Addition.

b. $(B \lor C) \supset A, A \supset (S \lor T), B \cdot \sim S, (T \cdot A) \supset (W \supset S)$ $/\therefore \sim W$

c. $(X \lor Y) \supset (Y \lor Z), X \cdot (Y \supset \sim Y), Z \supset \sim Z$ $/\therefore \sim Y \lor \sim Z$

d. $(A \supset B) \supset (C \supset D), (F \supset A) \supset (A \supset B), A \supset (F \supset A), A \cdot C$ $/\therefore D \lor F$

*e. $A \supset ((C \lor D) \supset B), (\sim W \lor \sim T) \supset (A \cdot C), W \supset (S \lor P),$
   $\sim H \lor \sim (S \lor P), \sim H \supset Z, \sim Z \cdot \sim Y$ $/\therefore B \cdot \sim Y$

*f. $(\sim A \cdot \sim B) \supset (\sim C \lor \sim D), (E \lor \sim F) \supset \sim A, \sim H \supset (B \supset J),$
   $(\sim F \cdot \sim H) \supset (\sim \sim C \cdot \sim J), \sim H \cdot (F \supset H)$ $/\therefore \sim D$

g. $(A \lor B) \supset (C \lor D), (C \supset W) \cdot (D \supset \sim A), A \cdot \sim W$ $/\therefore \sim (C \lor D)$

*8. Symbolize and construct proofs for the following valid arguments.

a. Either Plato or Democritus believed in the theory of forms. Plato believed in the theory of forms only if he was not an atomist, and Democritus was an atomist only if he did not believe in the theory of forms. Democritus was an atomist. Therefore, Plato was not an atomist.

b. If I smoke or drink too much then I don't sleep well, and if I don't sleep well or don't eat well then I feel rotten. If I feel rotten I don't exercise and don't study enough. I do smoke too much. Therefore, I don't study enough.

c. If the Bible is literally true then the Earth was created in six days. If the earth was created in six days, then carbon dating techniques are useless and scientists are frauds. Scientists are not frauds. The Bible is literally true. Therefore, God does not exist.

d. If nuclear power becomes our chief source of energy, then either there will be a terrible accident or severe waste disposal problems. If there are severe waste disposal problems and an increase in uranium costs, then Americans will cut their energy consumption. There will be a terrible accident only if safeguards are inadequate. Nuclear power will become our chief source of power, and uranium costs will increase, but safeguards are not inadequate. Therefore, Americans will cut their energy consumption.

e. Either scientists don't know what they are talking about, or the sun will eventually burn out and the Earth will become dark and cold. If scientists don't know what they are talking about then Mars is teeming with life. If the Earth becomes dark and cold, then either the human race will migrate to other planets or will die out. Mars is not teeming with life, but the human race will not die out. Therefore, the human race will migrate to other planets.

UNIT 8
Replacement Rules

A. INTRODUCTION

In Unit 7 you learned some very basic inference rules in which, given certain premises, you could derive a conclusion. In this unit you will learn some rules of a slightly different form, which we will call "replacement rules." These rules will be written as two statement forms separated by four dots $::$, which we will call the "replacement sign." The rule of Contraposition, for example, will be stated as $(p \supset q) :: (\sim q \supset \sim p)$. In each of the replacement rules the formula to the right of the four dots is logically equivalent to the formula to the left, and that is why we are permitted to replace the one with the other.

In using these rules, if you have an instance of the statement form on one side of the four dots, you may *replace* it with the corresponding instance of the statement form on the other side. A correct application of the rule of Contraposition, cited above, would be the following: $((A \cdot B) \supset C) /\therefore (\sim C \supset \sim (A \cdot B))$. Two very important ways in which these rules differ from the ones you learned in the last unit are that the replacement rules may be used from right to left as well as from left to right, and they may be used on subformulas. Thus, the following would also be a correct application of the rule of Contraposition:

$$(A \equiv B) \supset (\sim C \supset \sim D) /\therefore (A \equiv B) \supset (D \supset C)$$

You will be expected to learn, that is, *memorize* these rules, just as in the last unit, and for the same reason: it is absolutely impossible to work out the proofs without a thorough knowledge of the rules, and this can come only by memorization. You will also be expected, as in the last unit, to learn to do proofs with these rules, starting with rather simple proofs and working up to the more difficult ones, and you will also learn more about plotting proof strategies. On the whole, this unit is just a continuation of the last, with a new batch of rules for your edification and enjoyment.

B. UNIT 8 OBJECTIVES

1. Learn, that is, memorize, the symbolic forms of the rules.
2. Be able to explain the difference in structure and usage between the replacement rules and the basic rules which you learned in Unit 7.
3. Be able to identify correct applications (substitution instances) of the rules, and be able to spot incorrect applications.
4. Given a derivation which uses both the basic rules and the replacement rules, be able to supply the justification for each step.
5. Given a set of premises and the conclusion, be able to construct a short proof (up to 5 intermediate steps), using both the basic rules and the replacement rules.
6. Learn the "mini-proofs."
7. Learn more about plotting proof strategies.
8. Given a set of premises and a conclusion, be able to construct longer proofs (up to 25 or 30 intermediate steps).

C. UNIT 8 TOPICS

1. The Structure of Replacement Rules

The rules which you will be learning in this unit have a fundamentally different structure from the basic inference rules you learned in Unit 7. The basic rules all consist of one or more premises plus a conclusion. The *replacement rules*, by contrast, will be given as two statement forms separated by four dots $::$, the replacement sign. The rule of Exportation, for example, will be stated as $((p \cdot q) \supset r) :: (p \supset (q \supset r))$. In using these rules, if you have a statement which is an instance of one side of the rule, you may replace it with the corresponding instance of the other side of the rule.

There are two important ways in which the application of the replacement rules differs from the application of the basic rules you learned in Unit 7. First, the replacement rules are *symmetric*, that is, *they go both ways.* Given an instance of the left side, you may replace it with the corresponding instance of the right side and *also*, given an instance of the right side, you may replace it with the corresponding instance of the left side. Our basic rules, by contrast, were a one-way street; we could infer the conclusion, given the premises, but *not* the other way around. We may infer B from $(A \cdot B)$ by Simp., for instance, but we most certainly may *not* infer $(A \cdot B)$ from B.

The second difference is that the *replacement rules may be applied to any part of a formula*, as well as to the formula as a whole; that is, *they may be used on subformulas.* This is certainly not true for the basic inference rules; as we have seen, Simplification may only be used on a formula whose *major* operator is a conjunction, such as $(A \vee B) \cdot (C \vee D)$. It may *not* be used on a subformula of a larger formula. We are not permitted, for instance, to infer $(B \supset C)$ from $((A \cdot B) \supset C)$. With the replacement rules, however, if we have an instance of one side of the rule appearing as a *part* of another formula, we may replace *that part* with the instance of the other side. In using Contraposition, for instance, if we had the formula $(B \cdot D) \vee (\sim A \supset \sim C)$, we would be able to apply the rule to the second disjunct only, and derive $(B \cdot D) \vee (C \supset A)$.

The reason for these differences in applications is that *with the replacement rules the statement forms on either side of the replacement sign are always logically equivalent*, that is, they always have exactly the same truth value for every possible instance. This means that the conclusion which we derive by using one of the replacement rules will always have the same truth value as the premise, and thus it will never be possible, in using these rules, to go from a true premise to a false conclusion. It doesn't matter if we go from right to left or from left to right, since the two sides are the same in truth value; and if we use the rule on only part of a formula, the part which is replacing the original will have the same truth value as the original. Since the truth values of the parts are the same, and nothing else is being changed, the truth values of the wholes will be the same. Thus, again, it will never be possible in using these rules to go from true premises to a false conclusion, since the truth value of the conclusion will always be the same as the truth value of the premise.

In the next section, we will explain each of the ten replacement rules and will illustrate how they are used in proofs. These rules are listed at the end of the unit for easy reference, and are also given on the inside of the front cover.

2. The Ten Replacement Rules

a. Four simple rules. Again you will have to memorize these rules, but it may be easier in this unit because many of the rules are rather simpleminded, and many others you have already encountered in the unit on symbolization. In any case, it will be easier to learn them by starting with the more elementary ones first, and this is the order in which we will introduce them.

By far the simplest of the ten rules is *Double Negation*, abbreviated as *D.N.*, which tells you that given a formula preceded by two contiguous negation signs, you may replace it with the same formula without the negation signs, and vice versa. In other words, *double negations may be dropped from a formula or added on whenever needed*. In symbols:

$p :: \sim \sim p$ Double Negation (D.N.)

Instances of this would be $\sim \sim A /\therefore A$ and $(A \cdot B) /\therefore \sim \sim (A \cdot B)$. Of course, since replacement rules may be used on subformulas, the following would also be an instance: $A \supset B /\therefore A \supset \sim \sim B$. It is essential to this rule, however, that the two negation signs occur together, with *nothing in between*; the outer negation must be negating a *negation*, and not, for instance, a conjunction. The following would *not* be an instance of D.N.: $\sim (\sim A \cdot B) /\therefore (A \cdot B)$. We can see that this is not correct because there is a parenthesis intervening between the first and second negations. It is absolutely essential to remember that *not any two negations can be dropped* in using D.N., but *only* those which occur immediately next to each other.

The *Commutation* rule (*Com.*) is almost as simple, and tells us that *we are permitted to reverse the order of the components in conjunctions and disjunctions*. From $(A \equiv B) \lor (C \equiv D)$ we may infer $(C \equiv D) \lor (A \equiv B)$, for instance, and from $((A \supset B) \supset (C \cdot \sim D))$ we may infer $((A \supset B) \supset (\sim D \cdot C))$. Notice, however, that this rule holds *only* for these two operators, and in particular, it does *not* hold for conditionals. $(A \supset B)$ means something very different from $(B \supset A)$, as you saw in Unit 4. The Commutation rule will be stated in the following way:

$(p \lor q) :: (q \lor p)$ Commutation (Com.)
$(p \cdot q) :: (q \cdot p)$

The rule, of course, must have two forms, since it holds for both conjunction and disjunction. This is one of the rules (there are several) which have analogues in arithmetic; commutation also holds for addition and multiplication, as you no doubt learned some time ago. We have $(x + y) = (y + x)$ and $(x \cdot y) = (y \cdot x)$.

Another rule which has an analogue in arithmetic is the rule of *Association* (*Assoc.*), which tells us that *if we have a string of conjuncts or a string of disjuncts, it does not matter how they are grouped*. More precisely, the rule is stated:

$$((p \vee q) \vee r) :: (p \vee (q \vee r))$$
$$((p \cdot q) \cdot r) :: (p \cdot (q \cdot r))$$ Association (Assoc.)

The analogues in arithmetic, of course, are $((x + y) + z) = (x + (y + z))$ and $((x \cdot y) \cdot z) = (x \cdot (y \cdot z))$. Notice again that the rule holds only for the two operators, conjunction and disjunction, and *not* for the conditional. $((p \supset q) \supset r)$ means something very different from $(p \supset (q \supset r))$. Notice also that it holds only where *all* the operators in a string are disjunctions, or *all* are conjunctions. Association is *incorrect* if there is a mixture of operators; one of your truth table problems, in fact, showed you that, in general, $((p \cdot q) \vee r)$ is *not* logically equivalent to $(p \cdot (q \vee r))$. Some applications of this rule would be the following:

$\sim B \supset (A \vee (\sim C \vee D)) /\therefore \sim B \supset ((A \vee \sim C) \vee D)$ and
$((\sim A \cdot \sim B) \cdot \sim C) \supset \sim D /\therefore (\sim A \cdot (\sim B \cdot \sim C)) \supset \sim D.$

Another very simple rule, and again one that holds only for conjunction and disjunction, is the *Duplication* rule (*Dup.*), which tells us that if, in a disjunction or a conjunction, both the disjuncts or the conjuncts are the same, then the disjunction or conjunction is equivalent simply to the one disjunct or conjunct. We could also say that *p is replaceable by its own disjunction and its own conjunction*. It is much easier to say this in symbols:

$$p :: (p \vee p)$$
$$p :: (p \cdot p)$$ Duplication (Dup.)

Instances of this rule would be the following:

$((A \cdot B) \vee (A \cdot B)) \supset (B \vee C) /\therefore (A \cdot B) \supset (B \vee C)$ and
$(A \cdot B) \supset (C \equiv (\sim D \cdot \sim D)) /\therefore (A \cdot B) \supset (C \equiv \sim D).$

Again notice that this rule does *not* hold for the conditional. $p \supset p$ will be a tautology, while p will have one T and one F in its truth table. The two are *not* logically equivalent, and thus are not mutually replaceable.

Given these four rules, many inferences are possible which could not be carried out with just the first eight basic inference rules. Given just those eight rules, for instance, we may *not* infer $\sim A$ from $(A \supset \sim B)$ and B, even though this is a truth-functionally valid argument. We could not use M.T. here because our second premise B is *not* the negation of the consequent. This may seem picky, because it is certainly *equivalent* to the negation of the consequent, but as it stands the inference is *not* an instance of M.T. With the Double Negation rule at hand, however, we can infer $\sim \sim B$ from B, and then use $\sim \sim B$ as our second premise, from which we can legitimately infer $\sim A$. You will often use D.N. in this way, to infer a statement which is the negation of another negation, for use in rules such as M.T. and D.S., where you need the negation of the consequent and the negation of one disjunct as premises. A rather simple proof which uses all these new rules is the following:

a. 1. $((A \vee \sim B) \vee \sim B)$ Pr.   6. D M.P. 2,5
            $\vee \sim D$ 7. $\sim \sim D$ D.N. 6

2. $(E \cdot F) \supset D$ Pr.   8. $(A \vee \sim B) \vee \sim B$ D.S. 1,7

3. $\sim A$ Pr.   9. $A \vee (\sim B \vee \sim B)$ Assoc. 8

4. $F \cdot E$ Pr. /∴ $\sim B$ 10. $\sim B \vee \sim B$ D.S. 3,9

5. $E \cdot F$ Com. 4   11. $\sim B$ Dup. 10

At this point, you should commit these first four rules to memory, if you haven't already done so just by reading through the material. Then go on to the more complex rules.

b. Three intermediate rules.   The next three rules ought to be fairly easy to learn, because they should all be familiar to you from the symbolization material. We saw in Unit 4 that there were certain symbolizations that were equivalent to each other and could thus be used interchangeably. Some of these equivalences are so common, and so important, that they have been elevated to the status of rules; we will discuss three of these rules in this subsection.

As we saw in Unit 4, "Neither A nor B" can be properly symbolized as either $\sim (A \vee B)$ or $(\sim A \cdot \sim B)$, and "Not both A and B" can be symbolized as either $\sim (A \cdot B)$ or $(\sim A \vee \sim B)$. These equivalences are especially important in logic, and will be the first set of rules in this subsection. The rules tell us that we may always replace a *negated disjunction* by a corresponding conjunction with both conjuncts negated, and a *negated conjunction* by a disjunction with both disjuncts negated. They are called *DeMorgan's Rules* after Augustus DeMorgan, one of the founders of modern symbolic logic. The symbolic form of the rules is the following:

DeMorgan's (DeM.)
     $\sim (p \vee q) :: (\sim p \cdot \sim q)$ DeMorgan's (DeM.)
     $\sim (p \cdot q) :: (\sim p \vee \sim q)$

Again, our rules are stated in terms of *variables*, which can take any statements as substitution instances. Thus the following would all be instances of DeMorgan's Rules:

$\sim (A \vee \sim B) /\therefore \sim A \cdot \sim \sim B$ (first form)

$\sim (A \supset B) \cdot \sim (B \supset A) /\therefore \sim ((A \supset B) \vee (B \supset A))$ (first form)

$A \equiv \sim (B \cdot \sim C) /\therefore A \equiv (\sim B \vee \sim \sim C)$ (second form)

$\sim \sim A \vee \sim \sim B /\therefore \sim (\sim A \cdot \sim B)$ (second form)

It is very important using DeMorgan's Rules to be sure it is the entire conjunction or disjunction which is being negated. This will be indicated by having the conjunction or disjunction in parentheses, as in the instances we have given. The following would *not* be a correct application of DeM.: $\sim (A \supset B) \vee (B \supset A) /\therefore \sim (A \supset B) \cdot \sim (B \supset A)$, since it is not the entire disjunction which is being negated, but only *the first disjunct of it*, a conditional. This formula *will* be a candidate for one of our other rules, Conditional Exchange, and it will be important not to confuse the two forms. We will say more about this when we discuss Conditional Exchange.

Finally, keep clearly in mind that *the negation of a conjunction will be a disjunction*, not another conjunction, and *the negation of a disjunction will be a conjunction*. We will see another example of this "cross-relation" between conjunction and disjunction in our rule of Distribution, in Subsection c.

Another familiar equivalence is that between the *biconditional* and the *conjunction of two conditionals*. The biconditional $p \equiv q$ means "*p* if and only if *q*," which, as we saw in Unit 4, really means just "if *p* then *q* and if *q* then *p*." This equivalence will be reflected in our rule of *Biconditional Exchange (B.E.)*, which will tell us that the biconditional may be replaced with the conjunction of two conditionals, and vice versa.

$(p \equiv q) :: ((p \supset q) \cdot (q \supset p))$ Biconditional Exchange (B.E.)

Given this rule, we can finally make inferences from formulas containing the triple bar, and the inferences we can make are just what one would expect, given the meaning of the biconditional. If we have a biconditional and the formula on one side of the triple bar, we may derive the formula on the other side of the triple bar by using B.E., Simp., and M.P. An example of this kind of derivation is the following:

b. 1. $A \equiv (B \vee \sim C)$ Pr.

2. $B \vee \sim C$ Pr. $/\therefore A$

3. $(A \supset (B \vee \sim C)) \cdot ((B \vee \sim C) \supset A)$ B.E. 1

4. $(B \vee \sim C) \supset A$ Simp. 3

5. A M.P. 2,4

Also, given a biconditional and the *negation* of one side, we may derive the negation of the other side by using B.E., Simp. and M.T. (instead of M.P.). An example of this sort is the following:

c. 1. $A \equiv (B \vee \sim C)$ Pr.
    2. $\sim A$ Pr. /∴ $\sim (B \vee \sim C)$
    3. $(A \supset (B \vee \sim C)) \cdot ((B \vee \sim C) \supset A)$ B.E. 1
    4. $(B \vee \sim C) \supset A$ Simp. 3
    5. $\sim (B \vee \sim C)$ M.T. 2,4

Our third familiar rule is the rule of *Contraposition* (*Contrap.*), which tells us that given a conditional we can replace it with another conditional in which antecedent and consequent have been reversed, and both have been negated. Again, it is much easier to say it in symbols:

$(p \supset q) :: (\sim q \supset \sim p)$ Contraposition (Contrap.)

As we have seen in Unit 4, the sentence "John will win only if he tries harder" could be symbolized *either* as $(J \supset T)$ *or* as $(\sim T \supset \sim J)$. Again, the reason either of these symbolizations will do equally well is that they are logically equivalent. It is important to notice here that we not only negate antecedent and consequent; we must reverse the order as well. Thus $(\sim A \supset \sim B)$ /∴ $(A \supset B)$ would *not* be an instance of Contrap.; it is not even a valid argument. From $(\sim A \supset \sim B)$ we *could* infer $(B \supset A)$, however. Other slightly more complicated applications would be the following:

$(\sim (A \cdot B) \supset (\sim C \equiv D))$ /∴ $(\sim (\sim C \equiv D) \supset \sim \sim (A \cdot B)$ and

$(\sim (\sim A \supset (B \vee C))) \supset \sim (B \equiv (C \cdot D))$ /∴$(B \equiv (C \cdot D)) \supset (\sim A \supset (B \vee C))$.

Notice that this rule is reminiscent of Modus Tollens; going from left to right, it says that given the conditional $(p \supset q)$, then, if $\sim q$ then $\sim p$. This rule is more inclusive than M.T., however, since it goes from right to left as well as from left to right, and can be used on subformulas.

At this point, before we go on to introduce our last three rules, let us illustrate the ones we have so far in a proof of some length.

d. 1. $\sim (\sim A \cdot B) \cdot B$ Pr.
    2. $\sim (\sim B \equiv C) \supset \sim A$ Pr.
    3. $(A \cdot D) \supset C$ Pr. /∴ $\sim D$

Since we have a conjunction in the first premise, that is the natural place to start, so we have as our first intermediate steps the following:

    4. $\sim (\sim A \cdot B)$ Simp. 1
    5. B Simp. 1

One of the things you may find difficult, now that you have so many rules, is figuring out what to do next. A good bet is to *apply the rules whose applications are obvious*, so here we would look to step 2, where we could apply Contrap., and step 4, which is a candidate for DeM. (Notice again that this means knowing your rules extremely well, so you can spot the obvious applications.)

$$6.\ A \supset (\sim B \equiv C) \qquad \text{Contrap. 2}$$
$$7.\ \sim \sim A \lor \sim B \qquad \text{DeM. 4}$$

At this point you should notice that you almost have an application of D.S., at steps 5 and 7, except that you need the *negation* of one of the disjuncts. We can get this simply by double negating B, however, so let us do this and then apply D.S.

$$8.\ \sim \sim B \qquad \text{D.N. 5}$$
$$9.\ \sim \sim A \qquad \text{D.S. 7,8}$$

Now at step 6 we have a conditional with A as the antecedent, so if we had A we could derive $(\sim B \equiv C)$. But we can get A simply by using D.N. on step 9, so we have:

$$10.\ A \qquad \text{D.N. 9}$$
$$11.\ \sim B \equiv C \qquad \text{M.P, 6,10}$$

At this point we need to pause and take stock of the situation. We want $\sim D$ as our conclusion, and that is apparently going to come from Premise 3, after an application of M.T. (and some other steps). For this we need $\sim C$. We should be able to get this from $\sim B \equiv C$ and $\sim \sim B$, but we cannot do this directly. Remember from our discussion of B.E. that we need to make use of B.E., Simp., and M.T. in order to get the negation of one side of a biconditional. Thus we have the following steps:

$$12.\ (\sim B \supset C) \cdot (C \supset \sim B) \qquad \text{B.E. 11}$$
$$13.\ C \supset \sim B \qquad \text{Simp. 12}$$
$$14.\ \sim C \qquad \text{M.T. 8,13}$$

Now given $\sim C$, we can use M.T. on Premise 3, to get:

$$15.\ \sim (A \cdot D) \qquad \text{M.T. 3,14}$$

A negated conjunction is always a candidate for DeM., and we have nothing to lose, so let us apply this rule and see the results:

$$16.\ \sim A \lor \sim D \qquad \text{DeM. 15}$$

We are now only a step away from our conclusion; all we need is the double negation of A, in combination with step 16. We do have $\sim \sim A$, in step 9, so we are home.

       17. $\sim D$ D.S. 9,16

      c. Three final replacement rules.   The last three rules are the ones you will probably find the hardest to learn, since they are a little less obvious than the others, but if you read the explanations carefully, and try to understand what the rules *mean*, you should have no real difficulty.

      The rule of *Conditional Exchange* (abbreviated *C.E.*) will be used frequently in this unit, and (until you get the rule of Conditional Proof in the next unit) is the rule that will be used to derive conditionals. Stated symbolically, it looks like this:

      $(p \supset q) :: (\sim p \lor q)$ Conditional Exchange (C.E.)

      This rule tells us that, given a *conditional*, we may replace it by a *disjunction* which has as its left disjunct the negation of the antecedent, and as its right disjunct the consequent of the conditional. An instance in English of the *right* side of the rule would be "Either inflation does not remain high or my savings will be wiped out," which could be symbolized as $(\sim I \lor S)$. The corresponding instance of the *left* side of the rule would be "If inflation remains high, then my savings will be wiped out," symbolized as $(I \supset S)$. It should be clear that these two sentences mean just the same.

      As noted, the Conditional Exchange rule is the one you will be using in this unit to derive *conditionals*; you will first get the corresponding *disjunction* and then use C.E. to get the desired conditional. A very simple proof which uses this stragegy is below; note that the conclusion could *not* be derived in just one step by Addition, since Add. may not be used on just part of a formula.

    e.   1. $A \supset B$ Pr. $/\therefore A \supset (B \lor C)$
          2. $\sim A \lor B$ C.E. 1
          3. $(\sim A \lor B) \lor C$ Add. 2
          4. $\sim A \lor (B \lor C)$ Assoc. 3
          5. $A \supset (B \lor C)$ C.E. 4

      Finally, be very careful not to confuse statements of the form $\sim p \lor q$, on which C.E. can be used, with statements of the form $\sim (p \lor q)$, on which you can use DeM. but *not* C.E. The following, for instance, would *not* be a correct application of C.E.:

      $\sim (A \lor (B \supset C)) /\therefore A \supset (B \supset C).$

What we could infer from that premise, but by DeM., would be $\sim A \cdot \sim (B \supset C)$. In C.E., remember, the premise (in going from right to left) will be a *disjunction*, with a negated left disjunct, while in using DeM., the premise will be a *negated disjunction*. This difference is crucial.

Our next rule is the rule of *Exportation*, and is really quite sensible. The left-hand formula says that if two things, both p and q, happen, then r will occur, and the equivalent right-hand side says that *if p* happens, then, *if q* happens (as well), r will occur. Both formulas, in slightly different ways, say that *if* the two things, p and q, both happen, then so will r. In both cases we have two antecedents, p and q. On the left side, the two antecedents are *conjoined* into a single antecedent, while on the right, p is the antecedent of the first conditional and q is the antecedent of the second conditional. Stated symbolically, the rule is:

$$((p \cdot q) \supset r) :: (p \supset (q \supset r)) \qquad \text{Exportation (Exp.)}$$

Notice that in the first formula the parentheses are around the conjunction, to the left, while in the second formula the parentheses are around the second conditional, to the right. It would *not* be correct to go from $((p \cdot q) \supset r)$ to $((p \supset q) \supset r)$. Some applications of the rule would be the following: $((A \supset B) \cdot (C \supset D)) \supset F$ /∴ $(A \supset B) \supset ((C \supset D) \supset F)$ and $\sim A \supset (\sim (B \vee C) \supset \sim (F \vee A))$ /∴ $(\sim A \cdot \sim (B \vee C)) \supset \sim (F \vee A)$.

A very simple proof that illustrates the use of this rule follows:

f. 1. $A \supset (B \supset C)$ Pr.   /∴ $B \supset (A \supset C)$
    2. $(A \cdot B) \supset C$ Exp. 1
    3. $(B \cdot A) \supset C$ Com. 2
    4. $B \supset (A \supset C)$ Exp. 3

Last but not least is the *Distribution* rule, (*Dist.*), which tells us that conjunction "distributes over" disjunction and disjunction distributes over conjunction. In symbols:

$$(p \cdot (q \vee r)) :: ((p \cdot q) \vee (p \cdot r))$$
$$(p \vee (q \cdot r)) :: ((p \vee q) \cdot (p \vee r)) \qquad \text{Distribution (Dist.)}$$

Notice that in the Distribution rule we have the same kind of "cross relationship" between conjunction and disjunction as we had for DeMorgan's Rules: the *conjunction* is equivalent to a *disjunction*, and the *disjunction* is equivalent to a *conjunction*. In the first case, if we *conjoin* p to $(q \vee r)$, we end up with the *disjunction* which has $(p \cdot q)$ on the left and $(p \cdot r)$ on the right.

Just as in DeMorgan's Rules, a conjunction will *not* be equivalent to another conjunction, and a disjunction will not be equivalent to another disjunction. You will never have, for instance, $(A \lor B) \cdot (A \lor C)$ /∴ $A \cdot (B \lor C)$. Nor would you have $A \lor (B \cdot C)$ /∴ $(A \cdot B) \lor (A \cdot C)$. *Conjunctions* will always turn into *disjunctions*, and vice versa. Distribution is your most complex rule, and you will need to work with lots of examples to get the hang of it. A symbolic example would be the following:

$$(\sim A \supset \sim B) \cdot (\sim C \lor \sim D) /∴ ((\sim A \supset \sim B) \cdot \sim C) \lor ((\sim A \supset \sim B) \cdot \sim D).$$

There will be many exercises involving Distribution to give you practice in recognizing its uses, and you should now turn to the exercises at the end of the unit and complete 1, 2, and 3, which will give you practice in recognizing substitution instances of all the rules we have discussed in this section.

3. Constructing Simple Proofs with Replacement Rules

Now that you have so many more rules, your proofs, and proof strategies, will in general be much more complex. You will have so many possible moves to choose from that you may well have trouble getting started. For many students, in fact, this proves to be the hardest unit, but with the hints we will provide in this section, it should be simplified. Again, you will have to know your rules *very* thoroughly, because a major part of proof strategy is knowing what rule can be used where, being able to spot a fruitful application of a rule. We will discuss strategy in general, and in particular the method of "working backwards" again, in the next section; in this section we will concentrate on some very elementary proofs which we will call "miniproofs." You will find, for example, that you can always derive a *negated conjunction*, $\sim (p \cdot q)$, given one negated conjunct, $\sim p$ (or, of course, $\sim q$). These little proofs will often be needed within the context of a more complex proof, and can always be used to derive a conclusion of a certain form from premises of a certain form.[1]

Two very simple examples of mini-proofs are the following, which can be described by saying that (1) *given the consequent of a conditional, you can always prove the conditional*, and (2) *given the negation of the antecedent of a conditional, you can always prove the conditional*. (We will number our mini-proofs as (1), (2), (3), etc., to keep them separate from the other proof examples in this unit.)

(1)	1. q	Pr. /∴ $p \supset q$	(2)	1. $\sim p$	Pr. /∴ $p \supset q$
	2. $\sim p \lor q$	Add. 1		2. $\sim p \lor q$	Add. 1
	3. $p \supset q$	C.E. 2		3. $p \supset q$	C.E. 2

[1] We will construct these proofs with variables, so that it is clear to you that *any* instance of the mini-proof is valid.

It will be important to remember that there is an exact parallelism between the little proofs presented here and the results of the truth table method. Whatever is valid according to the truth tables can be proved by using our rules, and vice versa. In particular, we can always describe the little mini-proofs in truth-table terminology, and this may make it easier to remember them. The parallelism for the examples above should be clear; the truth table tells us that *whenever the consequent is true, the conditional is true (1)* and that *whenever the antecedent is false, the conditional is true (2)*. Further examples of these "mini-arguments" are given below, along with their truth table descriptions.

(3)	$\sim p \;/\therefore \sim (p \cdot q)$	If one of the conjuncts is false, the conjunction
(4)	$\sim q \;/\therefore \sim (p \cdot q)$	as a whole is false. (3) and (4)
(5)	$\sim (p \lor q) \;/\therefore \sim p$	If a disjunction is false, each disjunct must be
(6)	$\sim (p \lor q) \;/\therefore \sim q$	false. (5) and (6)
(7)	$p, q \;/\therefore (p \equiv q)$	If two formulas are both true, or both false, then
(8)	$\sim p, \sim q \;/\therefore (p \equiv q)$	the biconditional is true. (7) and (8)
(9)	$\sim (p \supset q) \;/\therefore p$	If a conditional is false, then the antecedent is
(10)	$\sim (p \supset q) \;/\therefore \sim q$	true and the consequent is false. (9) and (10)
(11)	$p, \sim q \;/\therefore \sim (p \supset q)$	If the antecedent is true and the consequent is false, then the conditional is false. (11)
(12)	$p, \sim q \;/\therefore \sim (p \equiv q)$	If one formula is true and the other false, the
(13)	$\sim p, q \;/\therefore \sim (p \equiv q)$	biconditional is false. (12) and (13)

Proofs of some of these little arguments will be given here; the rest will be included as exercises at the end of the unit, along with a few other simple proofs. You should not only do these exercises, construct the proofs, but also *remember the results*, so that you will be able to apply them fruitfully when needed in longer proofs.

We can prove (3) and (5), and analogously (4) and (6), by a simple application of DeMorgan's Rules. These little arguments will find frequent application, so you should learn them well and always keep them in mind.

(3)	1. $\sim p$	Pr. $/\therefore \sim (p \cdot q)$	(5)	1. $\sim (p \lor q)$	Pr. $/\therefore \sim p$
	2. $\sim p \lor \sim q$	Add. 1		2. $\sim p \cdot \sim q$	DeM. 1
	3. $\sim (p \cdot q)$	DeM. 2		3. $\sim p$	Simp. 2

To prove a biconditional given two formulas p and q, as in problem (7), we simply derive the two conditionals $(p \supset q)$ and $(q \supset p)$, conjoin them, and use BE.

(7)

1. p	Pr.	5. $\sim p \lor q$	Add. 2
2. q	Pr. /∴ $p \equiv q$	6. $p \supset q$	C.E. 5
3. $\sim q \lor p$	Add. 1	7. $(p \supset q) \cdot (q \supset p)$	Conj. 6,4
4. $q \supset p$	C.E. 3	8. $p \equiv q$	B.E. 7

It should be clear from this how we would prove the biconditional given the *negated* formulas $\sim p$ and $\sim q$, and this will be left as an exercise. To prove the facts about conditionals, in problems (9), (10), and (11), we need to make use of both C.E. and DeM. We will do two of these to illustrate:

(9)

			(11)		
1. $\sim (p \supset q)$	Pr. /∴ p			1. p	Pr.
2. $\sim (\sim p \lor q)$	CE 1			2. $\sim q$	Pr. /∴ $\sim (p \supset q)$
3. $\sim \sim p \cdot \sim q$	DeM. 2			3. $p \cdot \sim q$	Conj. 1,2
4. $\sim \sim p$	Simp. 3			4. $\sim \sim p \cdot \sim q$	D.N. 3
5. p	D.N. 4			5. $\sim (\sim p \lor q)$	DeM. 4
				6. $\sim (p \supset q)$	C.E. 5

Aside from the mini-proofs, we need to say something about proving conditionals in general. Until the next unit, when you will learn the rule of Conditional Proof, the best way to derive a conditional will be to use C.E., after deriving the right corresponding disjunction. (Of course, in some cases you may get a conditional by H.S., if you have the right premises.) We did one of these little proofs earlier; now that we have Distribution, we can look at some slightly more complex examples. Although we cannot use Simp. on a part of a formula, the following argument is valid, and can be proved by means of some of our replacement rules.

    g. $A \supset (B \cdot C) /∴ A \supset B$

Here both premise and conclusion are conditionals, so what we can do is to use C.E. on both ends, that is, apply C.E. first to the premise, and then try to derive the disjunction which will be equivalent to the conclusion, so that we can use C.E. at the last step. In other words, we want our proof to look like this at the beginning and the end:

g.			
1. $A \supset (B \cdot C)$	Pr. /∴	$A \supset B$
2. $\sim A \lor (B \cdot C)$	C.E. 1	
	.	
	.	
	.	
n-1. $\sim A \lor B$	?	
n. $A \supset B$	C.E.	

The question is, how do we fill in the intervening steps; that is, how do we derive the next to last step $\sim A \vee B$? If you know your rules well, you should be able to see it; we have an application of Dist. ready and waiting at step 2. All we need to do is use that rule, and we have what we want. Step 3 would thus be: $(\sim A \vee B) \cdot (\sim A \vee C)$. The completed proof would look like this:

g. 1. $A \supset (B \cdot C)$ Pr. /∴ $A \supset B$
    2. $\sim A \vee (B \cdot C)$ C.E. 1
    3. $(\sim A \vee B) \cdot (\sim A \vee C)$ Dist. 2
    4. $\sim A \vee B$ Simp. 3
    5. $A \supset B$ C.E. 4

There will be several problems like this in the exercises, and again, you should not only do the proofs, but also keep in mind that these moves can always be made, and that the arguments are valid. Some other examples of valid conditional arguments are the following:

$$(A \supset B) \cdot (A \supset C) /∴ (A \supset (B \cdot C))$$
$$(A \supset B) \cdot (C \supset B) /∴ (A \vee C) \supset B$$

At this point you are well-armed, with both rules and supplementary "miniproofs," and you should be ready for more formidable opponents. In the next section, then, we will introduce more complex proofs and discuss the strategies for deriving various sorts of conclusions.

4. Strategies for More Complex Proofs

In constructing more complex proofs, especially now that you have so many rules, you will want to use a variety of strategies. One will be simply to use the "top down" method, that is, derive as much as you can from the premises, and see whether the conclusion follows. In order for this to work, of course, you must know what follows from the premises! That is, again, you will have to know your rules well enough to be able to see where they apply. If you do know your rules, the "top down" method should work for the following argument:

h. 1. $F \equiv \sim D$ Pr.
    2. $D \supset C$ Pr.
    3. $\sim (B \vee C) \vee \sim (A \vee D)$ Pr.
    4. A Pr.   /∴ $F \vee G$

Of course, you do have to keep your wits about you, and generate only the things which are likely to be useful. What you will need to do is to have an eye for what will advance your cause; that is, look for applications of the rules which will tie in with whatever else there is in the premises. In the problem above, for instance, we have A, which should be useful, and we need to figure out what to do with it—how to combine it with our other premises. We should notice that in premise 3 we have $\sim (A \lor D)$ as a subformula; it is one of the disjuncts. Now, it should occur to you that we could get $A \lor D$ by Add., and then double negate it in preparation for the use of D.S. If you didn't see this yourself, don't despair; you will read through many problems and do many exercises before you readily see such things for yourself. But we have now seen it, so let's do it:

5. $A \lor D$	Add. 4
6. $\sim \sim (A \lor D)$	D.N. 5
7. $\sim (B \lor C)$	D.S. 3,6

At this point we have a candidate for DeMorgan's rule, and it is always a good idea to apply this rule, especially in cases where there is a negated disjunction, since this will yield a conjunction which can then be separated into its parts. Let us generate what can be generated from that formula, then:

8. $\sim B \cdot \sim C$	DeM. 7
9. $\sim B$	Simp. 8
10. $\sim C$	Simp. 8

What next? Well, if you look through what you already have, you can hardly miss it; you have $D \supset C$ in premise 2, and $\sim C$ now in step 10, so we have:

11. $\sim D$	M.T. 2,10

Does this help us? Indeed; given our B.E. rule we can now derive F from premise 1, and since our conclusion is $F \lor G$, we can complete our proof in just a few more steps.

12. $(F \supset \sim D) \cdot (\sim D \supset F)$	B.E. 1
13. $\sim D \supset F$	Simp. 12
14. F	M.P. 11,13
15. $F \lor G$	Add. 14

There are many cases in which this direct approach won't be very helpful, however, unless you keep in mind some of the little tricks we learned in the last section. In the following proof, for instance, unless you know how to apply DeM. in the right way, you won't get very far.

i. 1. $(A \lor B) \supset (C \cdot D)$ Pr.
   2. $\sim D$ Pr. $/\therefore$ $\sim A$

Here, notice first the overall structure of the argument: you are given a conditional as a premise and the negation of *part* of the consequent as another premise, and you want to derive the negation of part of the antecedent. It looks as if M.T. will be involved somehow, but obviously we cannot apply it directly. What we would need for M.T. would be the negation of the whole consequent, $\sim (C \cdot D)$. But here our mini-proofs come in handy: we already know how to derive a negated conjunction, given the negation of one conjunct. Surely you still remember:

   3. $\sim C \lor \sim D$ Add. 2
   4. $\sim (C \cdot D)$ Dem. 3

Now we can just apply M.T. to get:

   5. $\sim (A \lor B)$ M.T. 1,4

From here on it is clear sailing; you know enough by now always to apply DeM. to a negated disjunction, so you get:

   6. $\sim A \cdot \sim B$ Dem. 5   and thus:
   7. $\sim A$ Simp. 6

Notice that here our strategy was two or threefold: we had to analyze the structure of the premises and conclusion, which tipped us off to a probable application of M.T. Given that, we applied our mini-proofs, knowing that since we had $\sim D$ we could derive $\sim (C \cdot D)$, and that, when we had $\sim (A \lor B)$, we could get $\sim A$. Once we got to step 5, we just derived what we could, which gave us the conclusion.

In many cases, however, even this combination of strategies will not be enough, and you will have to resort to the method of working backwards, or from the bottom up. You have already seen this method at work in the last unit; here it will be even more useful. Let us apply it to the following problem:

j. 1. $(A \cdot B) \lor (C \cdot \sim D)$ Pr.
   2. $A \supset \sim B$ Pr.
   3. $C \supset (D \lor F)$ Pr. $/\therefore$ F

Here we look at our conclusion, and then check back into the premises to see where it might come from and what we would need in order to get it. F appears in the third premise as one disjunct in the consequent of a conditional. This means that if we could first get C, for M.P., and then $\sim D$, for D.S., we would have our F. Now, looking further into the premises, we see that $C \cdot \sim D$, just what we need, appears as one disjunct in a disjunction. We could get what we need, then, if we could get the negation of the other disjunct, $\sim (A \cdot B)$. The question is, how do we get it? Well, we have another premise, and a reasonable presumption is that it will come from that premise; the only problem is to figure out exactly how. Here we can apply the working backwards method to this little part of the problem. We want $\sim (A \cdot B)$, and we need to ask where it could come from. We should be able to see that we could derive it by DeM. if we could get $\sim A \lor \sim B$. Now if we ask where $\sim A \lor \sim B$ comes from we should see the connection between it and our second premise; the two are mutually derivable by C.E. At this point, we can put it all together: we will need $\sim (A \cdot B)$, so that we can use D.S. on the first premise. So:

4. $\sim A \lor \sim B$	C.E. 2
5. $\sim (A \cdot B)$	DeM. 4
6. $C \cdot \sim D$	D.S. 1,5
7. C	Simp. 6
8. $\sim D$	Simp. 6
9. $D \lor F$	M.P. 3,7
10. F	D.S. 8,9

If you have forgotten the rationale for these steps, go back and reread the preceding paragraph.

The above was a rather modest application of the working backwards method; let us take a somewhat more complicated problem. Here we will number our reasoning steps, but keep in mind that we are reasoning from the conclusion back to the premises, so when we come to actually write down the proof our steps will be in reverse order.

k.	1. $(\sim A \lor \sim B) \supset (\sim C \lor D)$	Pr.		
	2. $\sim C \supset (E \cdot F)$	Pr.		
	3. $E \cdot \sim (F \lor D)$	Pr.	$/\therefore$	A

(1) Here we notice that our conclusion A appears only in the first premise, and there we have $\sim A$ appearing as part of the antecedent of a conditional. This makes it look as if M.T. will be relevant, since by M.T. we would get the *negation* of $\sim A$, which would yield our concluson A. (2) If we are going to use M.T. on premise 1 we will need the negation of the consequent,

$\sim (\sim C \vee D)$. (3) $\sim (\sim C \vee D)$ can be derived from $(\sim \sim C \cdot \sim D)$ by DeM. (4) We could derive $\sim \sim C$ from premise 2 by M.T., provided we had $\sim (E \cdot F)$. (5) We could derive $\sim D$ by DeM. from $\sim (F \vee D)$, which we can get from premise 3 by Simp. (6) Our only remaining problem is how to derive $\sim (E \cdot F)$. It is the negation of a conjunction, so we could get it if we had the negation of one conjunct. We do have E, but that is the conjunct, not its negation. However, we can get $\sim F$, and that will give us, by another application of one of the miniproofs, $\sim (E \cdot F)$. Now all we have to do is put it together:

4.	$\sim (F \vee D)$	Simp. 3
5.	$\sim F \cdot \sim D$	DeM. 4
6.	$\sim F$	Simp. 5
7.	$\sim D$	Simp. 5
8.	$\sim E \vee \sim F$	Add. 6
9.	$\sim (E \cdot F)$	DeM. 8
10.	$\sim \sim C$	M.T. 2,9
11.	$\sim \sim C \cdot \sim D$	Conj. 10,7
12.	$\sim (\sim C \vee D)$	DeM. 11
13.	$\sim (\sim A \vee \sim B)$	M.T. 1,12
14.	$\sim \sim A \cdot \sim \sim \sim B$	DeM. 13
15.	$\sim \sim A$	Simp. 14
16.	A	D.N. 15

Again, if you do not understand why we took the steps we did, go back and reread the preceding paragraph.

Part of the method of working backwards is to be aware of the ways in which you can derive various kinds of formulas. Unless you can get it directly—for instance, by M.P. or Simp.—the best way to get a *biconditional* is to get the two conditionals separately, conjoin them, and then use B.E. To derive a *disjunction*, you have several methods available. From your basic rules you can use Addition or Dilemma, and from the replacement rules you have Conditional Exchange and DeMorgan's (if the disjuncts are negated), and even Distribution (if the disjuncts are related conjuncts). In general, a *conjunction* will be derived by getting each of the conjuncts separately and then conjoining them. A *conditional* may be obtained from H.S. in our basic rules, C.E. from the replacement rules, possibly B.E. (if the conditional you want is one-half of the biconditional), or from Exportation or Contraposition (if you have the equivalent conditionals). *Negations* can come from M.T. or D.N., or by a combination of one form of DeM. and Simp.

In order to be efficient in constructing proofs, you need to keep all these things in mind, and you need to be able to juggle, as well, the use of mini-proofs and the methods of working from the top down and from the bottom up. In general, you need to be able to analyze the structure of premises and conclusion to see what applications of the rules are likely; and you need to see how premises and conclusion "match up," in order to figure out where the conclusion is likely to come from. No wonder students find this rather overwhelming at first! But again, perseverence pays off, at least if you learn your rules and are diligent in your exercises. It is absolutely essential that you do the exercises, including both the preliminary exercises 1 through 3 and the proof constructions in 4 and 5 at the end of the unit. The more problems you work, and see worked out, the more little tricks you will pick up, and the more moves you will notice. Practice is a necessary condition for doing at all well in this unit. Answers are provided to some of the proofs, but remember that there are many different ways of doing them, especially now that you have so many different rules; thus, if your proof is different, that does not necessarily mean it is wrong. The answers are meant to provide only examples of how the proofs might be carried out.

At this stage, if you are very frustrated, you may wonder what the point is of this whole process, what good constructing proofs will ever do you. There are at least two answers to this: for one thing, you have now learned some very basic rules, which are frequently used in ordinary language (or at least should be). DeMorgan's rules are very common and very important, as are Exportation, H.S. and so on. When you know the rules, when you are aware of them, you are more likely to use them correctly. Second, the proof process itself, constructing extended chains of reasoning, will develop your reasoning powers and stretch your mental muscles. As you practice thinking out the steps you need, planning strategy, working out ahead of time what you will need to get the conclusion, and how best to aim for it, you are really learning problem solving, something that will be extremely useful in all sorts of everyday situations.

There are certain problems involving conditionals which are extremely complex to prove, given only the rules we have so far. You might, just for the fun of it, try to construct a proof for the following:

$$(A \lor B) \supset \sim (C \lor D), (\sim C \cdot E) \supset (F \cdot \sim O), (F \lor \sim H) \supset (J \cdot \sim K)$$
$$/\therefore (A \cdot E) \supset \sim K \cdot$$

Unless you are an exceptional student, your skill and patience are likely to run out. It is valid, however, and there is a much more direct and natural way to show this than by the rules we have so far. This natural method for providing conditionals, not unreasonably, is called *Conditional Proof*, and the next unit will be devoted to a discussion of this and one other closely related rule, *Indirect Proof*. These two rules will make a great many proofs a great deal easier, and will also allow us to carry out proofs that are not possible with only the basic rules and replacement rules. They will thus round out our system of Sentential Logic.

SUMMARY OF REPLACEMENT RULES

DOUBLE NEGATION (D.N.)

$$p :: \sim \sim p$$

DUPLICATION (Dup.)

$$p :: (p \lor p)$$
$$p :: (p \cdot p)$$

COMMUTATION (COMM.)

$$(p \lor q) :: (q \lor p)$$
$$(p \cdot q) :: (q \cdot p)$$

ASSOCIATION (ASSOC.)

$$((p \lor q) \lor r) :: (p \lor (q \lor r))$$
$$((p \cdot q) \cdot r) :: (p \cdot (q \cdot r))$$

CONTRAPOSITION (CONTRAP.)

$$(p \supset q) :: (\sim q \supset \sim p)$$

DEMORGAN'S (DEM.)

$$\sim (p \lor q) :: (\sim p \cdot \sim q)$$
$$\sim (p \cdot q) :: (\sim p \lor \sim q)$$

BICONDITIONAL EXCHANGE (B.E.)

$$(p \equiv q) :: ((p \supset q) \cdot (q \supset p))$$

CONDITIONAL EXCHANGE (C.E.)

$$(p \supset q) :: (\sim p \lor q)$$

DISTRIBUTION (DIST.)

$$(p \cdot (q \lor r)) :: ((p \cdot q) \lor (p \cdot r))$$
$$(p \lor (q \cdot r)) :: ((p \lor q) \cdot (p \lor r))$$

EXPORTATION (EXP.)

$$((p \cdot q) \supset r) :: (p \supset (q \supset r))$$

EXERCISES

*1. Which of the following are correct applications of the rule cited? For those which are not, say why not, and what would be needed in order for it to be a correct application.

a. $\sim (\sim A \supset B)$ $/\therefore A \supset B$ D.N.

b. $(\sim A \supset \sim B) \lor (\sim C \cdot \sim D)$ $/\therefore (\sim A \supset \sim B) \lor (\sim D \cdot \sim C)$ Com.

c. $\sim A \lor ((B \cdot C) \lor ((A \cdot C) \lor (A \cdot B)))$
    $/\therefore \sim A \lor (((B \cdot C) \lor (A \cdot C)) \lor (A \cdot B))$ Assoc.

d. $\sim A \lor (\sim (B \lor C) \lor \sim (C \lor D)) /\therefore A \lor \sim ((B \lor C) \lor (C \lor D))$ DeM.

e. $((A \cdot B) \cdot (C \supset D)) \supset (B \cdot E)$ $/\therefore (A \cdot B) \supset ((C \supset D) \supset (B \cdot E))$ Exp.

f. $(D \lor F) \supset (\sim A \supset \sim B)$ $/\therefore (D \lor F) \supset (\sim B \supset \sim A)$ Com.

g. $(\sim (A \lor B) \lor \sim (A \lor B)) \lor (A \lor \sim B) /\therefore \sim (A \lor B) \lor (A \lor \sim B)$ Dup.

h. $A \equiv ((B \equiv C) \equiv (A \equiv C))$
    $/\therefore A \equiv (((B \equiv C) \equiv (A \supset C)) \cdot ((C \supset A) \equiv (B \equiv C)))$ B.E.

i. $\sim (B \supset C) \lor \sim \sim ((A \lor B) \lor (B \lor C))$
    $/\therefore \sim (B \supset C) \lor \sim (\sim (A \lor B) \cdot \sim (B \lor C))$ DeM.

j. $\sim (\sim (\sim B \supset \sim C) \supset \sim \sim (B \supset C)) / \therefore \sim (\sim (B \supset C) \supset (\sim B \supset \sim C))$ Contrap.

k. $\sim A \supset \sim (B \lor (A \supset C))$ $/\therefore \sim A \supset (B \supset (A \supset C))$ C.E.

l. $(A \equiv B) \equiv (C \equiv (A \equiv C))$
    $/\therefore (A \equiv B) \equiv ((C \supset (A \equiv C)) \cdot ((A \equiv C) \supset C))$ B.E.

m. $(\sim A \lor \sim B) \lor (\sim C \cdot (D \lor E))$
     $/\therefore ((\sim A \lor \sim B) \lor \sim C) \cdot ((\sim A \lor \sim B) \lor (D \lor E))$ Dist.

n. $\sim A \supset \sim (\sim B \lor \sim C)$ $/\therefore \sim A \supset (B \lor C)$ D.N.

o. $(A \cdot B) \supset (C \supset (D \cdot E))$ $/\therefore (A \supset B) \supset (C \supset (D \cdot E))$ Exp.

p. $\sim A \lor ((\sim \sim B \cdot C) \lor (D \cdot E))$ $/\therefore \sim A \lor ((\sim B \cdot C) \supset (D \cdot E))$ C.E.

q. $(\sim (C \cdot D) \cdot (A \lor (C \cdot D))) \lor (\sim (C \cdot D) \cdot A)$
    $/\therefore \sim (C \cdot D) \cdot ((A \lor (C \cdot D)) \lor A)$ Dist.

r. $A \supset ((B \cdot \sim E) \supset (\sim B \lor \sim E))$
    $/\therefore A \supset (\sim (B \cdot \sim E) \lor (\sim B \lor \sim E))$ C.E.

s. $(\sim A \supset \sim B) \supset (\sim B \supset \sim A)$ $/\therefore (\sim B \supset \sim A) \supset (\sim A \supset \sim B)$ Contrap.

t. $\sim A \lor (\sim A \lor C)$ $/\therefore \sim A \lor (\sim A \cdot \sim C)$ DeM.

*2.   Cite the rule which was used in the following valid inferences.

a. $\sim \sim A \supset \sim \sim B / \therefore \sim B \supset \sim A$

b. $(A \cdot (B \cdot C)) \supset \sim D / \therefore A \supset ((B \cdot C) \supset \sim D)$

c. $\sim \sim A \supset \sim \sim B / \therefore \sim \sim A \supset B$

d. $\sim A \supset \sim \sim B / \therefore \sim \sim A \lor \sim \sim B$

e. $(A \cdot (B \cdot C)) \supset \sim D / \therefore ((A \cdot B) \cdot C) \supset \sim D$

f. $\sim \sim A \lor \sim \sim B / \therefore \sim (\sim A \cdot \sim B)$

g. $\sim (B \cdot A) \lor (\sim B \cdot \sim C) / \therefore (\sim (B \cdot A) \lor \sim B) \cdot (\sim (B \cdot A) \lor \sim C)$

h. $(\sim A \cdot \sim B) \supset (\sim \sim C \cdot \sim D) / \therefore (\sim A \cdot \sim B) \supset \sim (\sim C \lor D)$

i. $(\sim A \cdot \sim B) \supset (\sim \sim C \lor \sim \sim C) / \therefore (\sim A \cdot \sim B) \supset \sim \sim C$

j. $\sim (C \cdot \sim C) \lor \sim (A \cdot \sim B) / \therefore (C \cdot \sim C) \supset \sim (A \cdot \sim B)$

k. $(\sim A \lor \sim (B \cdot C)) \cdot (\sim A \lor C) / \therefore \sim A \lor (\sim (B \cdot C) \cdot C)$

l. $\sim A \lor \sim \sim (B \cdot C) / \therefore \sim A \lor \sim (\sim B \lor \sim C)$

m. $\sim (A \supset \sim B) \supset \sim (\sim B \supset \sim A) / \therefore \sim (A \supset \sim B) \supset \sim (A \supset B)$

n. $\sim (A \supset \sim B) \supset (\sim \sim B \supset \sim A) / \therefore \sim (A \supset \sim B) \supset (B \supset \sim A)$

o. $(\sim (A \lor B) \cdot \sim (A \lor B)) \lor \sim A / \therefore \sim (A \lor B) \lor \sim A$

p. $(\sim B \cdot \sim C) \supset (\sim (C \cdot D) \supset \sim A) / \therefore (\sim B \cdot \sim C) \supset (\sim \sim (C \cdot D) \lor \sim A)$

q. $(A \cdot B) \equiv ((B \cdot C) \supset (B \equiv C)) / \therefore (A \cdot B) \equiv ((C \cdot B) \supset (B \equiv C))$

r. $\sim \sim (A \lor B) \lor (\sim A \lor \sim (B \cdot C)) / \therefore \sim \sim (A \lor B) \lor \sim (A \cdot (B \cdot C))$

s. $(A \equiv B) \supset (((C \equiv D) \supset (D \equiv C)) \cdot ((D \equiv C) \supset (C \equiv D)))$
    $/\therefore (A \equiv B) \supset ((C \equiv D) \equiv (D \equiv C))$

t.　$((\sim A \lor \sim A) \lor (\sim B \lor \sim C)) \supset \sim (A \lor B)$
　　$/\therefore (((\sim A \lor \sim A) \lor \sim B) \lor \sim C) \supset \sim (A \lor B)$

u.　$(((A \equiv B) \supset C) \cdot ((C \equiv B) \supset A)) \supset (A \equiv C)$
　　$/\therefore ((A \equiv B) \supset C) \supset (((C \equiv B) \supset A) \supset (A \equiv C))$

v.　$(\sim B \lor \sim C) \supset (\sim C \lor ((\sim C \lor \sim B) \lor (\sim C \lor \sim B)))$
　　$/\therefore (\sim B \lor \sim C) \supset (\sim C \lor (\sim C \lor \sim B))$

w.　$\sim A \supset (\sim (B \lor \sim C) \supset \sim (\sim C \lor B)) /\therefore \sim A \supset ((\sim C \lor B) \supset (B \lor \sim C))$

x.　$\sim (B \lor \sim C) \cdot ((\sim A \supset C) \lor (\sim A \supset B))$
　　$/\therefore (\sim (B \lor \sim C) \cdot (\sim A \supset C)) \lor (\sim (B \lor \sim C) \cdot (\sim A \supset B))$

y.　$(\sim (B \lor \sim C) \cdot (\sim A \supset C)) \lor \sim (A \supset C)$
　　$/\therefore (\sim (B \lor \sim C) \cdot (\sim A \supset C)) \lor \sim (\sim A \lor C)$

*3.　Justify each step which is not a premise in the following derivations.

a.　1. $\sim A \equiv \sim (B \equiv C)$　　　　　　　　　　　Pr.
　　2. $\sim (D \lor C) \lor \sim B$　　　　　　　　　　Pr.
　　3. $\sim B \supset \sim (E \supset F)$　　　　　　　　　Pr.
　　4. $\sim (E \lor H)$　　　　　　　　　　　　　Pr.
　　5. $\sim E \cdot \sim H$
　　6. $\sim H$
　　7. $\sim E$
　　8. $\sim E \lor F$
　　9. $E \supset F$
　　10. $(E \supset F) \supset B$
　　11. B
　　12. $\sim \sim B$
　　13. $\sim (D \lor C)$
　　14. $\sim D \cdot \sim C$
　　15. $\sim C$
　　16. $\sim \sim B \cdot \sim C$
　　17. $\sim (\sim B \lor C)$
　　18. $\sim (B \supset C)$
　　19. $\sim (B \supset C) \lor \sim (C \supset B)$
　　20. $\sim ((B \supset C) \cdot (C \supset B))$
　　21. $\sim (B \equiv C)$
　　22. $(\sim A \supset \sim (B \equiv C)) \cdot (\sim (B \equiv C) \supset \sim A)$
　　23. $\sim (B \equiv C) \supset \sim A$
　　24. $\sim A$
　　25. $\sim A \cdot \sim H$
　　26. $\sim (A \lor H)$

b.　1. $(A \lor B) \supset (C \cdot D)$　　　　　　　　　　Pr.
　　2. $C \supset (E \cdot F)$　　　　　　　　　　　Pr.
　　3. $\sim (A \lor B) \lor (C \cdot D)$
　　4. $(\sim (A \lor B) \lor C) \cdot (\sim (A \lor B) \lor D)$
　　5. $\sim (A \lor B) \lor C$
　　6. $C \lor \sim (A \lor B)$
　　7. $C \lor (\sim A \cdot \sim B)$

    8. $(C \vee \sim A) \cdot (C \vee \sim B)$
    9. $C \vee \sim A$
  10. $\sim C \vee (E \cdot F)$
  11. $(\sim C \vee E) \cdot (\sim C \vee F)$
  12. $\sim C \vee F$
  13. $\sim A \vee C$
  14. $A \supset C$
  15. $C \supset F$
  16. $A \supset F$

c.   1. $\sim (((D \cdot F) \vee (F \cdot C)) \vee ((C \vee B) \cdot (A \vee B)))$ Pr.
    2. $\sim (C \vee B) \supset (\sim A \supset (F \supset C))$ Pr.
    3. $(A \vee B) \vee (D \cdot F)$ Pr.
    4. $\sim ((D \cdot F) \vee (F \cdot C)) \cdot \sim ((C \vee B) \cdot (A \vee B))$
    5. $\sim ((D \cdot F) \vee (F \cdot C)) \cdot (\sim (C \vee B) \vee \sim (A \vee B))$
    6. $\sim ((F \cdot D) \vee (F \cdot C)) \cdot (\sim (C \vee B) \vee \sim (A \vee B))$
    7. $\sim (F \cdot (D \vee C)) \cdot (\sim (C \vee B) \vee \sim (A \vee B))$
    8. $(\sim F \vee \sim (D \vee C)) \cdot (\sim (C \vee B) \vee \sim (A \vee B))$
    9. $\sim (C \vee B) \vee \sim (A \vee B)$
  10. $(D \cdot F) \vee (A \vee B)$
  11. $\sim ((D \cdot F) \vee (F \cdot C))$
  12. $\sim (D \cdot F) \cdot \sim (F \cdot C)$
  13. $\sim (D \cdot F)$
  14. $A \vee B$
  15. $\sim \sim (A \vee B)$
  16. $\sim (C \vee B)$
  17. $\sim (C \vee B) \supset ((\sim A \cdot F) \supset C)$
  18. $\sim (C \vee B) \supset (\sim C \supset \sim (\sim A \cdot F))$
  19. $\sim (C \vee B) \supset (\sim C \supset (\sim \sim A \vee \sim F))$
  20. $\sim (C \vee B) \supset (\sim C \supset (\sim A \supset \sim F))$
  21. $\sim C \supset (\sim A \supset \sim F)$
  22. $(\sim C \overset{.}{\cdot} \sim A) \supset \sim F$
  23. $\sim (C \vee A) \supset \sim F$
  24. $F \supset (C \vee A)$

d.   1. $(A \supset (B \vee C)) \vee (A \supset E)$ Pr.
    2. $(A \vee (B \vee C)) \cdot (A \vee D)$ Pr.
    3. $\sim (((B \vee C) \cdot D) \vee ((B \vee C) \cdot A))$ Pr.
    4. $\sim ((B \vee C) \vee (B \vee C))$ Pr.
    5. $A \vee ((B \vee C) \cdot D)$
    6. $\sim ((B \vee C) \cdot D) \cdot \sim ((B \vee C) \cdot A)$
    7. $\sim (B \vee C)$
    8. $((B \vee C) \cdot D) \vee A$
    9. $\sim ((B \vee C) \cdot D)$
  10. A
  11. $A \cdot \sim (B \vee C)$
  12. $\sim \sim A \cdot \sim (B \vee C)$
  13. $\sim (\sim A \vee (B \vee C))$
  14. $\sim (A \supset (B \vee C))$

15. $A \supset E$

16. E

4. Construct proofs for the following, using both the basic rules and the replacement rules. Answers are provided for starred exercises.

a. $\sim B /\therefore \sim (A \cdot B)$

b. $\sim (A \vee B) /\therefore \sim B$

c. $\sim (A \supset B) /\therefore \sim B$

d. $\sim A, \sim B /\therefore A \equiv B$

*e. $\sim A \supset \sim B, B /\therefore A$

*f. $A \supset B, A \supset C /\therefore A \supset (B \cdot C)$

*g. $A \supset B, C \supset B /\therefore (A \vee C) \supset B$

*h. $\sim (A \cdot B), A /\therefore \sim B$

*i. $\sim (A \cdot B) /\therefore B \supset \sim A$

j. $(A \vee B) \supset C /\therefore A \supset C$

k. $A, B /\therefore \sim (A \supset \sim B)$

*l. $(A \supset B) \vee (A \supset C) /\therefore A \supset (B \vee C)$

m. $A \supset C /\therefore (A \cdot B) \supset C$

n. $\sim ((A \vee B) \vee (C \vee D)) /\therefore \sim D$

o. $A, \sim B /\therefore \sim (A \equiv B)$

p. $A \supset \sim A /\therefore \sim A$

*q. $\sim A \supset A /\therefore A$

5. Construct proofs for the following more complex arguments.

a. $A \supset B, B \supset \sim C, C \vee D, \sim D /\therefore \sim A$

*b. $(A \vee B) \supset \sim (C \vee D), (A \cdot E) \vee \sim F, F /\therefore \sim C$

c. $\sim (C \vee D), D \equiv (E \vee F), \sim A \supset (C \vee F) /\therefore A$

d. $S \vee P, P \supset (G \cdot R), \sim G, P \equiv T /\therefore S \cdot \sim T$

*e. $(A \equiv B) \supset C, \sim (C \vee A) /\therefore B$

f. $(P \cdot G) \supset R, (R \cdot S) \supset T, P \cdot S, G \vee R /\therefore R \vee T$

*g. $X \equiv \sim Y, (Y \vee Z) \supset T, \sim (T \vee W) /\therefore P \supset X$

*h. $(F \cdot \sim G) \vee (T \cdot \sim W), W \cdot H, \sim (F \supset G) \supset (H \supset \sim S) /\therefore \sim S$

i. $(A \vee B) \supset (C \vee D), A \supset \sim C, \sim (F \cdot \sim A), F /\therefore D$

j. $B \supset (C \supset E), E \supset \sim (J \vee H), \sim S, J \vee S /\therefore B \supset \sim C$

k. $A \supset \sim B, \sim C \supset B, \sim A \supset \sim C /\therefore A \equiv C$

l. $\sim (C \cdot D), \sim (D \supset E) /\therefore \sim C \cdot \sim E$

m. $(A \supset B) \supset \sim (C \supset D), \sim (A \vee F) /\therefore \sim (D \vee F)$

n. $\sim B \supset \sim (S \cdot T), \sim S \equiv (P \vee O), \sim (P \vee (\sim T \vee O)), \sim A \supset P /\therefore A \cdot B$

o. $F \supset (\sim G \supset \sim F), \sim (H \supset G), \sim (H \cdot W), A \supset (S \cdot W) /\therefore \sim (A \vee F)$

p. $P \equiv \sim Q, Q \supset R, T \equiv \sim (Q \cdot \sim R), \sim (T \vee W) \vee \sim Q \mathbin{/\therefore} P$

q. $P \vee Q, Q \supset R, (R \supset (S \supset \sim T), (\sim S \supset U) \cdot (U \supset V) \mathbin{/\therefore} (\sim P \cdot \sim V) \supset \sim T$

r. $G \supset (H \cdot I), J \supset (H \cdot K), ((L \supset \sim G) \cdot M) \supset N, (M \supset N) \supset (L \cdot J) \mathbin{/\therefore} I \vee K$

s. $(P \cdot S) \supset (T \vee W), \sim T \equiv \sim (M \cdot O), \sim (W \vee (\sim S \vee M)), \sim A \supset P \mathbin{/\therefore} A$

t. $(A \vee F) \supset \sim (B \cdot \sim G), \sim (B \supset G), \sim S \supset (\sim T \supset A)\, T \supset F \mathbin{/\therefore} S \cdot B$

*6.   Symbolize and construct proofs for the following valid arguments.

a.   If the mind and brain are identical, then the brain is a physical entity if and only if the mind is a physical entity. If the mind is a physical entity then thoughts are material entities. Thoughts are not material, but the brain is a physical entity. Therefore, the mind and the brain are not identical.

b.   I will find a job when I graduate only if I am well-prepared, and I will be well-prepared only if I can read and write extremely well and have a good technical education. I will read and write extremely well if and only if I take a lot of Humanities courses, but if I take a lot of Humanities courses I will not take many technical courses, and if I don't take many technical courses then I won't have a good technical education. Therefore, I won't find a job when I graduate.

c.   If the cat is ill, either she was fighting or ate too many mice. She was fighting only if she was attacked, and she was attacked only if either the large Siamese or the small Beagle was out. The large Siamese was out only if it was sunny, and the small Beagle was out only if it was warm. It was neither warm nor sunny, but the cat is ill. Therefore, she ate too many mice.

d.   If I drink too much coffee then I can't sleep well and I don't study properly. If I don't drink enough coffee I can't stay awake and I don't study at all. Either I drink too much coffee or not enough. Therefore, either I don't study at all or I don't study properly.

e.   If the Monetarists are right, then there is an increase in inflation if and only if the money supply increases too fast. If the Keynesians are right then there is an increase in inflation if and only if there is a decrease in unemployment. If the Libertarians are right, there is an increase in inflation if and only if the federal government spends more than it takes in. The money supply increases too fast only if taxes are too low, and the federal government spends more than it takes in only if taxes are too low. There is no decrease in unemployment and taxes are not too low, but there is an increase in inflation. Therefore, neither the Monetarists, the Keynesians, nor the Libertarians are right.

Conditional Proof and Indirect Proof

A. INTRODUCTION

In this unit we will be rounding out our rule system for sentential logic with two very powerful rules which are designed for very special purposes, and which have a structure quite different from the rules you have learned so far. These rules are Conditional Proof (C.P.), which, as its name indicates, is used to prove conditionals, and Indirect Proof (I.P.), which is used to prove negations. These rules differ radically from the ones you have learned so far in that they do not have premises in the usual sense. How, you may reasonably ask, can one infer something when one has no premises from which to infer it? The answer is that, in both these rules, we make *assumptions* and then see what follows from the assumptions. In Conditional Proof we assume the antecedent of the conditional we wish to prove, and see whether the consequent follows. If it does, we may infer the conditional itself. In Indirect Proof we assume the *opposite* of what we want to prove, and see whether this leads to a contradiction. If it does, then that assumption, the opposite of the desired conclusion, must be *rejected*, and so we may infer the conclusion itself.

Both these rules are extremely powerful and can be used in a wide variety of proof constructions, and they will considerably simplify the proof process. Once you have learned these rules, you will find that your proofs go much more quickly and are much easier and more "natural" to construct. Using these rules, it will be easier to figure out strategies and to see where you are going; in particular, the working backwards method will be greatly enhanced. Both rules are used frequently in everyday life, and both have a long history in formal logic and mathematics; the rule of I.P., for instance, has been used explicitly for millennia, under the name *Reductio ad absurdum* (reducing an assumption to an absurdity).

Because these rules require no previously given premises, they can be used in a special kind of proof: proofs without premises, which are used to derive theorems. A *theorem* is simply a formula which can be proved without any initially given premises. Theorems are "obvious truths" such as $((p \cdot q) \supset p)$ (if both p and q, then p), and in fact, as we shall see at the end of the unit, the set of theorems turns out to be exactly the same as the set of tautologies. We will now need a new definition of "proof," since, included among the steps in our proof, we will have assumptions as well as premises and derived lines. What you will need to master in this unit is listed below.

B. UNIT 9 OBJECTIVES

1. Learn the rules of Conditional Proof (C.P.) and Indirect Proof (I.P.) in their symbolic forms, and understand the explanations which accompany them.
2. Learn the *restrictions* which apply to the rules of C.P. and I.P.
3. Be able to use the rule of C.P. in constructing proofs.
4. Be able to use the rule of I.P. in constructing proofs.
5. Learn the definition of "theorem," and be able to construct proofs of theorems.
6. Learn the new definition of "proof" and some facts about the relationship between the proof method and the "semantic," or truth table method.
7. Be able to demonstrate that an argument is invalid.

C. UNIT 9 TOPICS

1. Conditional Proof

A conditional is a formula which asserts that *if* one thing (say A) is the case, *then* another (say B) is the case. As we saw in Unit 8, one way to prove a conditional is to derive the corresponding disjunction and then use C.E., but this is not a particularly direct or intuitive way to validate if-then statements. A far simpler and more direct method is just to *assume* that the first thing, *A*, is the case, and then see whether the second, *B*, follows from that assumption. If it does, then we are justified in concluding that *if A, then B*, that is ($A \supset B$). This is, in a nutshell, the rule of Conditional Proof: to prove a conditional ($A \supset B$), simply *assume A temporarily* and see whether, given *A*, *B* follows. If it does, then you are permitted to infer ($A \supset B$). There is nothing especially mysterious about the rule of C.P. (although it may take some practice to catch on to how it works); it simply reflects the meaning of the conditional: "if-then."

Let us now take a rather simple example to demonstrate the application of the rule of Conditional Proof.

a. 1. $(A \lor B) \supset (C \cdot D)$ Pr.   /∴ $A \supset C$
    2. ┌→ A Assp. (C.P.)
    3. │ $A \lor B$ Add. 2
    4. │ $C \cdot D$ M.P. 1,3
    5. └ C Simp. 4
    6. $A \supset C$ C.P. 2–5

Notice that we have assumed *A*, since this is the antecedent of the conditional we are trying to prove, and given *A*, we have been able to deduce the consequent *C*. Thus at step 6 we are permitted to infer ($A \supset C$), because we have shown in steps 2 through 5 that indeed *if A* is the case, then so is *C*. In using C.P. remember that you must always *assume the antecedent of the conditional you want to prove.*

Note also that we have here *set in* our assumption and the steps which followed from it, and marked them off with an arrow and a vertical line. As we will emphasize later, assumptions are not on a par with the given premises; premises are given for the duration of the proof, while *assumptions are only temporary, and are made for very specific purposes.* (Here, for instance, we assumed *A* to see whether *C* would follow). To emphasize this difference between premises and assumptions, we will always set in the assumption and what follows from it, and mark it off with an arrow and a vertical line. Note also that we must always *justify* our assumptions, like every other step in a proof. We will justify an assumption simply by noting that it is an assumption (which we will abbreviate Assp.), and indicating whether the assumption is for C.P. or I.P.

The rule of C.P. will help you enormously in constructing proofs of conditionals, especially where premises and conclusion are rather complex. Instead of having to manipulate rather long and complicated formulas, using our most complex rules, you are permitted to assume just the antecedent, a simpler formula, which then makes it possible to break the proof down into easier bits. At the end of the last unit you were given a rather complex example, and it was suggested that you try to carry out the proof with only your basic rules and replacement rules. If you tried (and succeeded), you discovered that it required almost 30 steps, and that those steps required the frequent use of one of your more complex rules, Distribution. Furthermore, there was no very good way of planning strategy, no natural way to "reason through" the proof; without C.P. on problems like this, it is often just a matter of perseverance and blind luck. If you happen to hit on the right combination, well and good; if not, you may go round in circles indefinitely. The rule of Conditional Proof is by far the more rational approach to problems like this, and since logic is supposed to be a rational discipline, C.P. is the appropriate strategy. Let us use this rule, then, in constructing a proof of that rather refractory problem from the last unit.

b.   1. $(A \lor B) \supset \sim (C \lor D)$ Pr.
     2. $(\sim C \cdot E) \supset (F \cdot \sim O)$ Pr.
     3. $(F \lor H) \supset (J \cdot \sim K)$ Pr.   /∴ $(A \cdot E) \supset \sim K$

Here the antecedent of the conclusion is $(A \cdot E)$, and since *our assumption for C.P. will always be the antecedent*, at step 4 we may assume $(A \cdot E)$. *Given* that assumption, the consequent $\sim K$ follows quite directly (assuming you learned the material in the last two units), so we can infer the conclusion, $(A \cdot E) \supset \sim K$, at step 17.

4.	$A \cdot E$	Assp. (C.P.)
5.	A	Simp. 4
6.	E	Simp. 4
7.	$A \lor B$	Add. 5
8.	$\sim (C \lor D)$	M.P. 1,7
9.	$\sim C \cdot \sim D$	DeM. 8
10.	$\sim C$	Simp. 9
11.	$\sim C \cdot E$	Conj. 6,10
12.	$F \cdot \sim O$	M.P. 2,11
13.	F	Simp. 12
14.	$F \lor H$	Add. 13
15.	$J \cdot \sim K$	M.P. 3,14
16.	$\sim K$	Simp. 15
17. $(A \cdot E) \supset \sim K$		C.P. 4–16

Note that this proof is just half as long as the one that did not make use of Conditional Proof. Note also that, given the assumption $(A \cdot E)$, we are able to use all our most basic rules—Simp., Add., M.P., etc.—instead of the rather complicated replacement rules we would have to use otherwise. Not only are proofs generally *shorter* using C.P., but they are almost always *easier* as well.

Now that you have seen some examples of the way in which Conditional Proof works, it is time to state the rule itself.

$$\begin{array}{l} \rightarrow p \\ \quad \cdot \\ \quad \cdot \\ \rule{2cm}{0.4pt} q \\ p \supset q \end{array}$$

If, given the assumption p we are able to derive q, then we are allowed to infer $(p \supset q)$, citing all the steps from p to q inclusive.

In using C.P. it is important to remember that the conclusion of the rule must always be a conditional; this is why it is called "Conditional Proof." You may then go on to use the conditional in other inferences—it could be a premise for H.S., for instance—but the *immediate* conclusion of the rule will be a horseshoe formula. Also, as noted, you must always assume the antecedent of the conditional you are trying to prove, and then derive the consequent. It is essential, of course, that you be able to figure out what the antecedent and consequent of the formula are; this is a matter of being able to analyze structure. Some examples are given below. To prove:

$((A \supset B) \supset C)$	assume $(A \supset B)$,	derive C
$(A \supset (B \supset C))$	assume A,	derive $(B \supset C)$
$((A \supset B) \supset C) \supset (D \supset F)$	assume $((A \supset B) \supset C)$,	derive $(D \supset F)$
$((A \cdot B) \supset C) \supset (A \supset (B \supset C))$	assume $((A \cdot B) \supset C)$,	derive $(A \supset (B \supset C))$
$(A \supset B) \supset (C \supset (A \supset (B \supset C)))$	assume $(A \supset B)$,	derive $(C \supset (A \supset (B \supset C)))$

As noted earlier, we will *set in* the assumption and all of the steps which lead to the consequent we are trying to derive, and will mark them off by an arrow and a vertical line. *We will call the sequence of steps set in and marked off by this vertical line a subproof,* since it is really a little *proof within a proof* (just as a subformula is a formula within a formula); it *demonstrates* (for C.P.) that if the antecedent is true then the consequent is true. *The first step of a subproof for C.P. will be the assumption—the antecedent of the conditional to be proved—and the last step of the subproof will be the consequent.*

Another important and closely related concept is the *scope of an assumption*, which is roughly its extent or for how long it is operative, for how long we are assuming it. *The scope of an assumption includes all (and only) the steps of the subproof,* and is indicated by the arrow and vertical line. This concept will be especially important when we talk about *discharging* assumptions.

The arrow and vertical line, which we will call the *scope marker*, *must be included* in any application of C.P. or I.P., the rules in which we make assumptions. It is essential in such proofs to know exactly how far the scope of the assumption extends, and it is the scope marker which gives us this information. These scope markers will also be extremely useful as visual aids, and will help you plot strategy and keep your assumptions straight, especially in proofs with several uses of C.P. and I.P.

Finally, in the step in which we apply C.P.—in which we infer the conditional we wanted to prove—we will cite *all* the steps included in the subproof. This is because no *one* step suffices to show that the antecedent implies the consequent; it is all the steps taken together which do this, and so we cite them all.

It is perfectly possible to use more than one application of C.P. within a single proof, just as you may use more than one application of M.P. or Add. There are some restrictions on this procedure, but these will be discussed in more detail in a later section; here it will be sufficient to note that it is possible, and to give one rather simple example.

c. 1. $A \supset (B \supset C)$ Pr /∴ $(\sim B \supset \sim A) \supset (A \supset C)$
    2. $\sim B \supset \sim A$ Assp. (C.P.)
    3. A Assp. (C.P.)
    4. $A \supset B$ Contrap. 2
    5. B M.P. 3,4
    6. $B \supset C$ M.P. 1,3
    7. C M.P. 5,6
    8. $A \supset C$ C.P. 3–7
    9. $(\sim B \supset \sim A) \supset (A \supset C)$ C.P. 2–8

At step 2, we assumed $(\sim B \supset \sim A)$ because that was the *antecedent* of the conclusion we wanted to prove. Given that assumption we wanted to derive the *consequent* of our conclusion, $(A \supset C)$. Since the consequent is also a conditional, we can use C.P. to prove it, which means assuming *its* antecedent, which is just A, and trying to derive its consequent, which is C. Thus, we assumed A at step 3. Given these assumptions, we were first able to derive C, which showed that if A, then C. Thus we were entitled to infer $(A \supset C)$ at step 8. Then, since we did derive the consequent of the conclusion of our problem from its antecedent, we were entitled at step 9 to infer the entire conclusion. We will see more such problems, with assumptions inside assumptions, in later sections.

2. Indirect Proof

The rule of Indirect Proof (I.P.) has a structure similar to that of C.P.—we also make an assumption—but its intent is a little different. In

I.P., which is sometimes called the rule of *Reductio ad absurdum* (reducing to an absurdity), we try to show that the assumption we make has absurd consequences, leads to a contradiction. Why would we want to do this? Well, sometimes is it very difficult to prove your conclusion *directly*, just by applying the rules of inference you have learned so far. In these cases, the best way to proceed is often to assume the *opposite* of what you want to prove, and show that it results in a contradiction. If you want to prove ~ *A*, for instance, and can see no direct method of proceeding from premises to conclusion, you can try an *indirect* proof by assuming just the opposite, that is, *A*, and seeing whether this assumption leads to a contradiction. If it does, your assumption, *A*, must be *wrong*, since it leads to absurd consequences; and so *you are permitted to infer the negation of what you assumed*, namely, ~ *A*, which is the conclusion you wanted.

An example of a proof using the rule of I.P. is worked out below.

d. 1. $N \supset O$ Pr.
    2. $(N \cdot O) \supset P$ Pr.
    3. $P \supset\; \sim O$ Pr. $/\therefore\; \sim N$
    4. $\quad N$ Assp. (I.P.)
    5. $\quad O$ M.P. 1,4
    6. $\quad N \cdot O$ Conj. 4,5
    7. $\quad P$ M.P. 2,6
    8. $\quad \sim O$ M.P. 3,7
    9. $\quad O \cdot \sim O$ Conj. 5,8
  10. $\quad \sim N$ I.P. 4–9

Here the assumption N is shown in steps 4 through 9 to lead to a contradiction; it results in both O and ~ O. Thus N cannot be the case, and we are permitted to infer ~ N by I.P. Notice that here, as in C.P., we cite *all* of the steps 4 through 9, which show that N results in a contradiction.

We may state the rule of I.P. schematically as follows:

$$\begin{array}{l} p \\ \cdot \\ \cdot \\ \cdot \\ q \cdot \sim q \\ \hline \sim p \end{array}$$

If, given an assumption p we are able to derive a contradiction $q \cdot \sim q$, then we may infer the negation of our assumption, ~ p, citing all the steps from p to $q \cdot \sim q$ inclusive.

Just as in C.P., there is a certain pattern in the use of I.P., and certain conditions which must be fulfilled. First, *the conclusion of I.P. is always a negation*, the negation of what was assumed, just as the conclusion of C.P. is always a conditional. Second, when using I.P., *always assume the opposite of what you want to prove*. If you want to prove the negation $\sim p$ simply assume p, and if you want to prove an unnegated formula p, assume $\sim p$. In the latter case, if you assume $\sim p$ and derive a contradiction, you will be able to infer $\sim \sim p$ by I.P., and p will follow by D.N. *The first step in the subproof for I.P., then, will be the opposite of the conclusion you wish to derive.* Third, *the last step in the subproof for I.P. is a contradiction.*

A *contradiction*, for purposes of I.P., is simply *any formula conjoined to its negation.* $(A \cdot \sim A)$ is a contradiction, but so is $(A \equiv B) \cdot \sim (A \equiv B)$ and so is $(A \lor (B \cdot C)) \cdot \sim (A \lor (B \cdot C))$. The contradiction need *not* be just a single letter and its negation. *But beware:* there are certain combinations which may *look* like contradictions but which are not, such as $(A \supset B) \cdot (A \supset \sim B)$. Here only *part* of the first formula is negated in the second, but to be a contradiction, the *whole* of the first conjunct must be negated in the second conjunct. A "proper" contradiction would be $(A \supset B) \cdot \sim (A \supset B)$, where the second conjunct is exactly the negation of the first. Finally, the contradiction you derive may include the assumption itself. If you assume A, and manage to get $\sim A$, the conjunction $A \cdot \sim A$ is a perfectly good contradiction, and will allow you to conclude $\sim A$ by I.P. An example in which this situation occurs is given below.

e. 1. $A \supset \sim (B \lor C)$ Pr.
    2. $(\sim B \lor \sim D) \supset F$ Pr.
    3. $F \supset \sim A$ Pr. /∴ $\sim A$
    4. ┌─► A Assp. (I.P.)
    5. │ $\sim (B \lor C)$ M.P. 1,4
    6. │ $\sim B \cdot \sim C$ DeM. 5
    7. │ $\sim B$ Simp. 6
    8. │ $\sim B \lor \sim D$ Add. 7
    9. │ F M.P. 2,8
   10. │ $\sim A$ M.P. 3,9
   11. └─ $A \cdot \sim A$ Conj. 4,10
   12. $\sim A$ I.P. 4–11

Notice here that we are *not* finished with the proof at step 10, even though we have derived $\sim A$. The reason is that *we have not yet discharged our assumption*, that is, we are still *within* the subproof, and no proof is complete until all assumptions have been discharged. More will be said about this in the next section.

Another example of the use of I.P. is the following, in which the conclusion is a negated compound.

f. 1. $A \supset (\sim F \supset D)$ Pr.
    2. $D \supset ((B \lor \sim C) \supset F)$ Pr.

3.	$\sim (E \supset F)$	Pr. $/\therefore \sim (A \cdot B)$
4.	$A \cdot B$	Assp. (I.P.)
5.	A	Simp. 4
6.	B	Simp. 4
7.	$\sim F \supset D$	M.P. 1,5
8.	$\sim (\sim E \vee F)$	C.E. 3
9.	$\sim \sim E \cdot \sim F$	DeM. 8
10.	$\sim F$	Simp. 9
11.	D	M.P. 7,10
12.	$(B \vee \sim C) \supset F$	M.P. 2,11
13.	$B \vee \sim C$	Add. 6
14.	F	M.P. 12,13
15.	$F \cdot \sim F$	Conj. 14,10
16.	$\sim (A \cdot B)$	I.P. 4–15

Notice here that our conclusion is the negation of the entire conjunction, not just of one of the conjuncts. To conclude $\sim A$, or $\sim A \cdot \sim B$, would be incorrect, since it is the *conjunction* $(A \cdot B)$ which led to the contradiction and which must therefore be rejected.

The rules of C.P. and I.P. may be used at any point in a proof, as may any of our other rules. You may need to prove a negation or a conditional in the *middle* of a proof, and if so, it is perfectly correct to make use of the rules in this way. An assumption for C.P. or I.P. may be introduced at *any* point in a proof (except for the last step, since then, by definition, it would not be the last step, as we shall see in the next section). Furthermore, we may introduce assumptions within assumptions, use C.P. inside an assumption for I.P., and vice versa. There is no limit (theoretically) to the number of assumptions we may make in a proof, *provided* we discharge them all, as we will see in the next section.

At this point we need to revise our definition of proof. In Unit 7 we defined a proof as a sequence of justified steps which resulted in the conclusion, and a justified step was defined as either a premise or a step which was derived from previous steps by application of one of the given rules of inference. At this point, however, we can also make *assumptions*, and our definition must be modified to accommodate these new steps. The modification is very simple, and is exactly what you would expect: we will just add to our definition of "justified step" the assumptions which we can now make. Thus, our new definition must read: *a justified step is either a premise, an assumption, or a step that follows from previous steps according to one of the given rules of inference.* A proof is still a derivation (a sequence of justified steps) in which the last step is the desired conclusion, but we could also define it now as *a sequence of steps, each of which must be either a premise, an assumption, or a step which follows from previous steps according to one of the given rules of inference, and such that the last step in the sequence is the desired conclusion.*

3. Discharging Assumptions; Restrictions on C.P. and I.P.

As mentioned earlier, an assumption for C.P. or I.P. is not on the same footing as a premise. A premise is given "for keeps," at least within a certain problem, *while an assumption is only temporary.* An assumption is made for a particular purpose: in C.P. to see whether the consequent follows from the antecedent, and in I.P. to see whether a contradiction follows, and once it has served its purpose, once the consequent (for C.P.) or the contradiction (for I.P.) has been derived, it must be rendered inoperative, discharged. The notion of discharging an assumption fits hand in glove with the concept of the *scope* of an assumption, the set of steps from the assumption through the consequent or contradiction. *To discharge an assumption is simply to cut off the scope of that assumption, to end the subproof.* Once we reach the last step in the subproof, we have reached our goal, we know what happens given the assumption; at this point, since the assumption is no longer needed, we discharge it.

Note that the *conclusion* of C.P. or I.P., the conditional or negation, is *not* included within the scope of the assumption, does not occur within the subproof. This means that it is not derived *from* that assumption, but can be asserted independently of it, even though the point of the assumption was to prove the conditional. Although this sounds somewhat paradoxical, the explanation is straightforward. We might say that the *conclusion* of C.P. or I.P. *sums up* the results of the subproof, "describes" in a way (a highly metaphorical way) what has gone on in the subproof, reflects a kind of "overview" of the subproof. Thus, it does not occur within the subproof itself, but rather, falls outside it. We look over the completed subproof and then write down *outside* the subproof of the results of our investigation.

We will say, then, that *the assumption is discharged at the step in which we apply C.P. or I.P.* In example f, for instance, where we used Indirect Proof, the assumption is discharged at step 16, where we apply I.P. to conclude $\sim (A \cdot B)$, and in our first C.P. example, the assumption is discharged at step 6, where we used C.P. to infer $(A \supset C)$.

There are certain restrictions related to the notion of discharging an assumption. In the first place, *every assumption made in a proof must eventually be discharged.* The reason is that in a proof the goal is to derive the conclusion from the *given* premises and no others. If we left assumptions undischarged we would in effect be adding further premises, and would not be giving an answer to the problem posed. Second, and just as important, *once an assumption has been discharged you may not use that assumption or any step that falls within the scope of that assumption again.*

All those steps derived within the scope of the assumption should be thought of as only temporary steps on the way to proving the conditional or negation. Once they have done their job of getting to the consequent (for C.P.) or the contradiction (for I.P.) they are out of the picture, just as is the assumption itself. This makes good sense, since they were only derived under that assumption, and once the assumption has been discharged, we are no longer operating under its scope, or influence. Thus, we have no right to those steps which were derived only given that assumption. So again, once the assumption has been discharged, neither it nor any step which falls within its scope may be used again. Another way to put this restriction is that *once the assumption has been discharged, no step in its subproof may be used again.* An instance of this kind of error is made in the "proof" below.

g. 1. $(A \lor B) \supset C$ Pr.

     2. $(F \lor G) \supset H$ Pr. /∴ $C \cdot H$

     3. ┌→A Assp. (C.P.)

     4. │ $A \lor B$ Add. 3

     5. │ C M.P. 1,4

     6. $A \supset C$ C.P. 3–5

     7. ┌→F Assp. (C.P.)

     8. │ $F \lor G$ Add. 7

     9. │ H M.P. 2,8

   10. $F \supset H$ C.P. 7–9

   11. $C \cdot H$ Conj. 5,9   **X Error**

Here the error comes in step 11, where we have used steps from previously discharged assumptions. The proper use of the scope markers and set-in steps should help you avoid this kind of mistake; once a hypothesis has been discharged, this should be fairly clear just from the way the proof is written. You might even draw a big X through the whole subproof once it has been discharged, to make sure you won't be tempted to use anything inside it again. A third restriction must be imposed for cases in which one assumption is made *within* the scope of another; this restriction will be discussed in the next section.

4. Using C.P. and I.P.

The use of C.P. and I.P., as noted earlier, can greatly simplify the process of constructing proofs. This will only be true, however, if you know how to use them effectively, and to their best advantage. In this section we will go over several points of strategy for using these rules, which should help you understand some of the finer points of their application. We will also discuss the third restriction on the use of the rules.

In using these rules, it is of the utmost importance that you *make the proper assumption*; for C.P. this means the antecedent of the conditional you want to prove, and for I.P. it means the opposite of what you want to prove. Theoretically you can make any assumption you want provided it is discharged in the end, but unless you keep clearly in mind where you are going, and make your assumptions accordingly, you will run into a lot of dead ends. *You should never make an assumption unless you know exactly what it is for—what you are aiming at.*

As mentioned earlier, you may sometimes have to use C.P. or I.P. in the middle of a proof, rather than just discharging the assumption at the last step. You might have to derive a negation, for instance, to use in M.T. or D.S., and you might need a conditional for use in H.S. or M.P. The following is an example of this sort of proof:

h.	1. $(A \supset B) \supset (\sim C \supset D)$	Pr.
	2. $\sim (C \lor E)$	Pr.
	3. $(A \lor F) \supset B$	Pr. /∴ D
	4. A	Assp. (C.P.)
	5. $A \lor F$	Add. 4
	6. B	M.P. 3,5
	7. $A \supset B$	C.P. 4–6
	8. $\sim C \supset D$	M.P. 1,7
	9. $\sim C \cdot \sim E$	DeM. 2
	10. $\sim C$	Simp. 9
	11. D	M.P. 8,10

Here we need to derive $(A \supset B)$ in order to use M.P. with premise 1. We can do this by assuming A at step 4 and deriving B at step 6. We then go on to complete the proof using the formula we have derived by C.P.

In the following problem there are a number of assumptions, one inside the other. Notice also that an I.P. subproof occurs within a C.P. subproof. This is perfectly correct; there is no limit to the number or variety of assumptions we may make in a proof, or to the "degree" of the proof (the number of assumptions inside other assumptions). However, there is a third restriction which applies to those cases in which one assumption is made inside the scope of another, namely, that *assumptions made inside the scope of other assumptions must be discharged in the reverse order in which they were made.* If you have a second assumption inside the scope of a first, and a third inside the scope of the second, the third must be discharged first, the second next, and the first assumption discharged last. Note that this is the procedure followed in problem i. What this amounts to is that our subproofs must be *nested*, one entirely within the other. *At no time should the scope marker lines cross.* If you find patterns like the ones below, you know you have done something wrong.

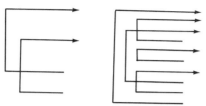

This is one reason it is so important to use the scope markers; they will tell you if you have your assumptions crossed. The following proof will illustrate what happens in these more complex cases, where you have several assumptions, one inside the other. Here we first assume $(A \cdot B)$, since this is the antecedent of what we want to prove, and then we try to derive $(A \supset (F \supset \sim E))$, the consequent. This is itself a conditional, with A as the antecedent, so at step 5 we assume A and try to derive $(F \supset \sim E)$. To get $(F \supset \sim E)$, we assume F (for C.P.), and then try to derive $\sim E$. For this we will use I.P., so at step 7 we assume E and try to derive a contradiction. Once all the assumptions have been made, the rest of the proof goes smoothly.

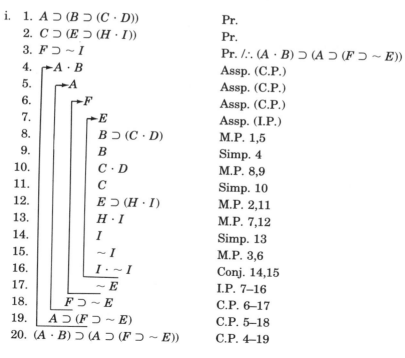

i. 1. $A \supset (B \supset (C \cdot D))$ Pr.
    2. $C \supset (E \supset (H \cdot I))$ Pr.
    3. $F \supset \sim I$ Pr. /∴ $(A \cdot B) \supset (A \supset (F \supset \sim E))$
    4. $A \cdot B$ Assp. (C.P.)
    5. A Assp. (C.P.)
    6. F Assp. (C.P.)
    7. E Assp. (I.P.)
    8. $B \supset (C \cdot D)$ M.P. 1,5
    9. B Simp. 4
    10. $C \cdot D$ M.P. 8,9
    11. C Simp. 10
    12. $E \supset (H \cdot I)$ M.P. 2,11
    13. $H \cdot I$ M.P. 7,12
    14. I Simp. 13
    15. $\sim I$ M.P. 3,6
    16. $I \cdot \sim I$ Conj. 14,15
    17. $\sim E$ I.P. 7–16
    18. $F \supset \sim E$ C.P. 6–17
    19. $A \supset (F \supset \sim E)$ C.P. 5–18
    20. $(A \cdot B) \supset (A \supset (F \supset \sim E))$ C.P. 4–19

The proof above may look somewhat complicated, but *without* the use of C.P. and I.P. it would probably be twice as long and would require more complex rules such as Distribution. In general, C.P. and I.P. will greatly simplify your work, not only shortening proofs but also eliminating most applications of the more difficult rules. Another advantage of C.P. and I.P. is that they make it easy to plan your proof strategy, since you should always be making assumptions with a particular aim in mind and know exactly what you are looking for. You may also work from the bottom up; the strategy of working backwards is particularly effective once you have C.P. and I.P. The arrows and vertical lines—scope markers—help to keep straight what assumptions go with what conclusions, and allow us to see exactly where we are going at each stage of the proof. Using C.P. and I.P. clarifies the proof and makes it much easier to find an effective strategy for solving problems. Now that you have these rules you should, in general, have much less trouble figuring out how to "get started" on a proof; to further assist you, some strategy hints are given below.

There are some cases in which the direct method, working from the top down, will work, but with more complex problems your best bet will usually be the method of working backwards. That is, *think about the structure of the conclusion, and what you need to do to derive a conclusion of that form*. For *conditionals*, most often the best approach is to use Conditional Proof; it will generally clarify and simplify your proof structure. For *biconditionals* the best method is often to use *two* applications of C.P., derive the two conditionals, conjoin them, and then use the rule of Biconditional Exchange. For *conjunctions*, in general, you will just get each conjunct separately, by whatever method works best in that particular case, and then conjoin them. For *disjunctions* you may use Addition or Dilemma, and now that you have C.P. you can often use this in conjunction with the C.E. rule. If you want to derive a formula $A \lor B$, for instance, you can assume the negation of one disjunct, $\sim A$, derive B, infer $\sim A \supset B$ by C.P., use C.E. to get $\sim \sim A \lor B$, and then use D.N. to get $A \lor B$. To derive a *negation* or a *single letter*, if there is no obvious direct method available, the best approach is to use Indirect Proof: assume the opposite of what you want to prove and derive a contradiction from it. This is an extremely powerful rule and can, in fact, be used to derive *any* formula. It is usually helpful if you get stuck on a proof; if you just can't see what to do or how to get started, try I.P. In the following problem, for instance, which looks deceptively simple, the direct approach doesn't work very well. If we use I.P., however, the solution is almost immediate.

j. 1. $S \lor T$ Pr.
    2. $S \supset T$ Pr. $/\therefore T$

3.	$\sim T$	Assp. (I.P.)
4.	$\sim S$	M.T. 2,3
5.	S	D.S. 1,3
6.	$S \cdot \sim S$	Conj. 4,5
7.	$\sim \sim T$	I.P. 3–6
8.	T	D.N. 7

These are the last rules you will be given for sentential logic; together with the other rules they form a very powerful deductive system—powerful enough, as we shall see in Section 7 that we can derive any truth-functionally valid argument, and any tautology, by using just these rules. At this point your job is to learn to use these rules effectively, and this requires doing as many problems as possible. There are a great many exercises at the end of the unit, and you should do all the preliminary questions and as many of the proofs as you can. A summary of the rules of C.P. and I.P. is also given at the end of the unit for handy reference.

5. Proofs of Theorems

As noted in the introduction to this unit, there are certain formulas, "obvious truths," such as $((p \cdot q) \supset p)$, which can be proved without any initially given premises. These formulas are called *theorems*, and they play a very important role in logic because they are the formulas which are absolutely reliable: they are always derivable, and they are always true, no matter what.

We may define a *theorem* as *a statement form that can be proved without any given premises*. This means that the proof contains only assumptions and steps derived from previous steps. It is interesting that the set of theorems is exactly the same as the set of tautologies; that is, every theorem is a tautology, and every tautology is a theorem, so can be proved without premises.[1]

In proving theorems we use exactly the same methods and rules as in the other proofs; the only difference is that here we have no premises to begin with, so in order to get started we need to use one of the rules in which we can make *assumptions*, namely, C.P. or I.P. *The proof of a theorem will always require the use of C.P. or I.P.* since, without premises, this is the only way we can begin the proof. *The first step of the proof for a theorem will always be an assumption.* You may use the same kind of

---

[1] We use statement forms in this section rather than statements, but an analogous definition could be given for statements. We would then have that all statement theorems are tautologous statements, and vice versa.

proof strategy as you have used before; if the theorem is a conditional, for instance, the best method is to use C.P. The proof of the theorem below is extremely simple, requiring only three steps. We simply assume the antecedent of the formula, derive its consequent in one step, by Simp. and then apply C.P.:

k. 1. $\quad p \cdot q$ Assp. (C.P.)
    2. $\quad p$ Simp. 1
    3. $(p \cdot q) \supset p$ C.P. 1,2

Other theorems, of course, will require more extensive proofs, but there is nothing different in principle from the kinds of proofs we use for theorems and the ones we use in arguments. We simply have to start off with *assumptions* in the proofs of theorems. This is illustrated below in the proof of $(p \supset (q \supset r)) \supset (q \supset (p \supset r))$.

l. 1. $\quad p \supset (q \supset r)$ Assp. (C.P.)
    2. $\quad q$ Assp. (C.P.)
    3. $\quad p$ Assp. (C.P.)
    4. $\quad q \supset r$ M.P. 1,3
    5. $\quad r$ M.P. 2,4
    6. $\quad p \supset r$ C.P. 3–5
    7. $\quad q \supset (p \supset r)$ C.P. 2–6
    8. $(p \supset (q \supset r)) \supset (q \supset (p \supset r))$ C.P. 1–7

Another example is given below, in which we make use of both C.P. and I.P. to prove $((p \lor q) \supset \sim r) \supset \sim (r \cdot p)$. There are alternative ways, in which only one assumption is made, to do both these problems. You might try to work out the proofs in that way.

m. 1. $\quad (p \lor q) \supset \sim r$ Assp. (C.P.)
    2. $\quad r \cdot p$ Assp. (I.P.)
    3. $\quad p$ Simp. 2
    4. $\quad p \lor q$ Add. 3
    5. $\quad \sim r$ M.P. 1,4
    6. $\quad r$ Simp. 2
    7. $\quad r \cdot \sim r$ Conj. 6,5
    8. $\quad \sim (r \cdot p)$ I.P. 2–7
    9. $((p \lor q) \supset \sim r) \supset \sim (r \cdot p)$ C.P. 1–8

An example of a theorem in which the best approach is to use the rule of I.P. is the following: $\sim ((p \equiv q) \cdot (p \cdot \sim q))$. Here we will assume the opposite, the *un*negated formula, and derive a contradiction. At that point we will be entitled to infer, by I.P., the negation of the formula which we assumed, that is, the theorem itself.

n.   1. $\sim(p \equiv q) \cdot (p \cdot \sim q)$ Assp. (I.P.)
       2. $p \equiv q)$ Simp. 1
       3. $p \cdot \sim q$ Simp. 1
       4. p Simp. 3
       5. $\sim q$ Simp. 3
       6. $(p \supset q) \cdot (q \supset p)$ B.E. 2
       7. $p \supset q$ Simp. 6
       8. q M.P. 4,7
       9. $q \cdot \sim q$ Conj. 5,8
    10. $\sim ((p \equiv q) \cdot (p \cdot \sim q))$ I.P. 1–9

We have done problems in which the conclusions are conditionals and negations; the obvious strategy in these cases is to use C.P. or I.P. As noted earlier, if the conclusion is a *biconditional*, $(p \equiv q)$, the best method is probably to use two applications of C.P. to get $(p \supset q)$ and $(q \supset p)$, and then conjoin them and use B.E. If the conclusion is a *conjunction*, you must derive each conjunct separately and then use Conj. (A conjunction will be a theorem only if each of its conjuncts alone is a theorem.) If you want to prove a *disjunction*, $(p \supset q) \lor (q \supset p)$, for instance, there are two possibilities: you can simply assume its *negation*, $\sim ((p \supset q) \lor (q \supset p))$, and try to derive a contradiction, that is, use I.P.; or you can assume the negation of the left disjunct, $\sim (p \supset q)$, and try to derive the other. This will yield, in this case, $\sim (p \supset q) \supset (q \supset p)$, which, by an application of C.E. and D.N., will give you the desired result. We will work the problem both ways below, to illustrate both methods for proving disjunctions.

o.   1. $\sim ((p \supset q) \lor (q \supset p))$ Assp. (I.P.)
       2. $\sim (p \supset q) \cdot \sim (q \supset p)$ DeM. 1
       3. $\sim (p \supset q)$ Simp. 2
       4. $\sim (q \supset p)$ Simp. 2
       5. $\sim (\sim p \lor q)$ C.E. 3
       6. $\sim \sim p \cdot \sim q$ DeM. 5
       7. $\sim (\sim q \lor p)$ C.E. 4
       8. $\sim \sim q \cdot \sim p$ DeM. 7
       9. $\sim \sim q$ Simp. 8
    10. $\sim q$ Simp. 6
    11. $\sim q \cdot \sim \sim q$ Conj. 9,10
    12. $\sim \sim ((p \supset q) \lor (q \supset p))$ I.P. 1–11
    13. $(p \supset q) \lor (q \supset p)$ D.N. 12

1.	$\sim (p \supset q)$	Assp. (C.P.)
2.	$\sim (\sim p \vee q)$	C.E. 1
3.	$\sim \sim p \cdot \sim q$	DeM. 2
4.	$\sim q$	Simp. 3
5.	$\sim q \vee p$	Add. 4
6.	$q \supset p$	C.E. 5
7.	$\sim (p \supset q) \supset (q \supset p)$	C.P. 1–6
8.	$\sim \sim (p \supset q) \vee (q \supset p)$	C.E. 7
9.	$(p \supset q) \vee (q \supset p)$	D.N. 8

6. Invalidity

We must also say something at this point about *invalidity*. So far we have only used the proof method to show that an argument is *valid*: that the conclusion *does* follow from the premises. As it turns out, this is *all* we can do with proofs; the method of proofs is only good for demonstrating *positive* results, for concluding that the argument is *valid*. *It can never be used to show that an argument is invalid.* You can never conclude that an argument is invalid, for instance, just because you are unable to find a proof for it, as you no doubt have discovered. There can be all sorts of reasons why you are unable to find a proof—too much noise in the dorms, too much pasta and beer for supper, too many distracting thoughts about current loves (or the lack of them). And this is true not just for beginning logic students; there have been many cases in the history of logic and mathematics in which theorems were *conjectured*—were thought to be true—but for which no proof was found for centuries. Even the best logicians and mathematicians are, on occasion, unable to find proofs for valid arguments. Thus, no matter who you are, failure to *find* a proof does not mean that no proof exists; it may just be very well hidden.

How, then, do we show that an argument is invalid? One proposal that sounds reasonable, which may have occurred to you, is to try to derive the *negation* of the conclusion. Surely if you could do that, the argument would be invalid; that is, you would be *unable* to derive the conclusion. Right? Unfortunately, wrong. Deriving the negation of the conclusion does *not* mean that you cannot derive the conclusion itself. If the premises turn out to be inconsistent (contradictory), you will be able to derive *both* the conclusion and its negation (and anything else you want, for that matter)! So this method won't work either.

To show that an argument is invalid, you cannot use the proof method at all; you must revert to the method of counterexample. *The only way to demonstate that an argument is invalid is to construct a counterexample*, that is, an instance with true premises and a false conclusion. At this point, since our arguments are rather complex, this generally means *using the short truth table method*. You have already gone over this in Unit 5, but to refresh your memory, one example is given here, and there are a few exercises at the end of the unit. We can show that the argument below is invalid in the following way:

$$A \supset (B \supset \sim C), B \cdot (D \supset \sim E), (E \lor F) \supset (C \lor H) /\therefore A \supset H$$

To make the conclusion false, we must make A *true* and H *false*. In order for the second premise to be true we know, at least, that B *must be true*. If both A and B are true, then for the first premise to be true, C *must be false*. But if C and H are both false, the only way for the third premise to be true is to *make both E and F false*. We now have the conclusion false, with the first and third premises true. All we need for a counter-example is to make the second conjunct of the second premise true. But this is already done: we had to make E false, so $\sim E$ is true, so $D \supset \sim E$ is true. Therefore, all the premises are true and the conclusion is false, which means that there is a counterexample; so the argument form, and hence the argument, is invalid.

7. Truth and Proof

Finally, our last word on the topic of proofs, and in fact on sentential logic in general, concerns the relationship between the two methods you have learned for testing arguments for validity: the truth table method and the method of proofs. As we mentioned in Unit 7, the two methods are theoretically independent of each other. However, there are some very interesting relationships between them which should be mentioned to round out your understanding of sentential logic.

The most important fact about the two methods—truth tables and proofs—is that for classical logic they are exactly equivalent, which means that they give exactly the same results. If an argument is valid according to the truth table method—sometimes called the *semantic* method—then it is possible to construct a proof for it; this means that the system is *complete*. Also, whatever argument we can prove will also be semantically valid, which means that the system is *consistent*. And as noted in Section 5, every theorem is a tautology and every tautology is a theorem. This equivalence between the two methods holds for all classical, two-valued logic through relational predicate logic with identity. (In predicate logic we can't use truth tables *per se*, but there is an analogous semantic procedure using the notion of possible truth value assignments.)

Interestingly, this equivalence between truth and proof, the consistency and completeness of the system, does *not* hold even in simple arithmetic. In 1931, Kurt Gödel, possibly the greatest logician ever, showed that the proof method falls short of the semantic method, which means that for *any* proof system for arithmetic, there will be true arthmetical statements that cannot be proved in that system. Arithmetic is irredeemably incomplete. Even worse, perhaps, Gödel proved that it is impossible ever to show that all the statements *provable* in arithmetical systems are *true*. In other words, we can never show that our arithmetical system is consistent. We may assume that arithmetic contains no false statements, but we will never be able to demonstrate this fact.

These results, which would be more fully explored in an advanced course in logic, have been the source of a great deal of philosophical controversy about the nature of mathematics and mathematical truth. If mathematics is not certain, not provably consistent, then is it just another empirical science? And if being true in mathematics does not mean the same as being provable, as many had thought, then what could it possibly mean? This issue is still being debated, and the implications of Gödel's results for the concepts of certainty and truth in mathematics are still not clear.

SUMMARY OF RULES OF CONDITIONAL PROOF AND INDIRECT PROOF

A. CONDITIONAL PROOF (C.P.)

If, given the assumption p we are able to derive q, then we are allowed to infer $(p \supset q)$, citing all the steps from p to q inclusive.

B. INDIRECT PROOF (I.P.)

If, given an assumption p we are able to derive a contradiction $q \cdot \sim q$, then we may infer the negation of our assumption, $\sim p$, citing all the steps from p to $q \cdot \sim q$ inclusive.

C. RESTRICTIONS ON THE USE OF C.P. AND I.P.

1. Every assumption made in a proof must eventually be discharged.
2. Once an assumption has been discharged, neither it nor any step which falls within its scope may be used in the proof again.

3. Assumptions inside the scope of other assumptions must be discharged in the reverse order in which they were made; that is, no two scope markers may cross.

D. GENERAL INSTRUCTIONS FOR USING C.P. AND I.P.

1. For both C.P. and I.P., an assumption may be introduced at any point in the proof, provided it is justified as such, that is, provided we label it as an assumption.

2. In using C.P., we assume the antecedent of the conditional to be proved and then derive the consequent. In using I.P., we assume the opposite of what we want to prove and then derive a contradiction. All the steps from the assumption to the consequent (for C.P.) or the contradiction (for I.P.) are said to be *within the scope* of the assumption.

3. The sequence of steps within the scope of an assumption is called a *subproof*.

4. We indicate the scope of an assumption, and set off the subproof, by an arrow (pointing to the assumption) and a vertical line which runs to the left of the subproof and includes every step in the subproof. This arrow and vertical line is called the *scope marker* for the assumption. We also *set in* or *indent* every step in the subproof.

5. There is no limit to the number of assumptions we may introduce in a given proof, and we may make one assumption inside the scope of another.

6. The scope of the assumption ends immediately prior to the step in which we infer the conditional or negation. We say that *the assumption is discharged* at this point. Thus neither the conditional nor the negation, the result of applying C.P. or I.P., falls within the scope of the assumption. We indicate that the assumption has been discharged by cutting off the vertical line (the scope marker) at this point.

DEFINITIONS

1. A **theorem** is a statement form that can be proved without any given premises.

2. A **justified step** is either a premise, an assumption, or a step which follows from previous steps according to one of the given rules of inference.

3. A **proof** is a sequence of justified lines in which the last step is the desired conclusion, that is, it is a sequence of steps, each one of which is either a premise, an assumption, or a step which follows from previous steps according to one of the given rules of inference, and in which the last step is the desired conclusion.

EXERCISES

*1.   Given the following as conclusions, what would be our first assumption if we were using C.P.? If we were to use more than one application of C.P., what would be the other assumptions (in order of appearance)?

a. $(((A \lor B) \supset C) \supset (A \lor C))$

b. $((A \supset B) \supset ((A \supset C) \supset A))$

c. $(((A \supset (C \supset A)) \supset (B \supset C)) \supset (A \supset C))$

d. $((A \supset B) \supset (A \supset ((B \supset C) \supset (A \supset C))))$

e. $(((A \supset B) \supset A) \supset A)$

f. $((A \supset A) \supset ((B \supset (C \supset B)) \supset (B \supset C)))$

g. $(A \supset (B \supset (C \supset ((A \lor B) \supset (B \lor C)))))$

h. $((((A \supset B) \supset C) \supset (A \lor B)) \supset (B \lor C))$

i. $(((A \supset A) \supset (B \supset (A \supset (A \supset B)))) \supset (A \supset (B \supset A)))$

j. $(((A \supset (B \supset A)) \supset B) \supset ((B \supset A) \supset (((C \supset B) \supset B) \supset (A \supset C))))$

*2.   Which of the following qualify as contradictions, and so could be the last step in a subproof for I.P.?

a. $(A \lor B) \cdot (\sim A \lor \sim B)$

b. $A \supset (B \cdot \sim B)$

c. $(\sim B \cdot A) \cdot \sim (\sim B \cdot A)$

d. $(A \supset B) \cdot (\sim A \supset B)$

e. $(\sim A \supset \sim B) \cdot \sim (\sim A \supset \sim B)$

f. $(A \supset B) \cdot (\sim A \supset \sim B)$

g. $(A \lor \sim B) \cdot \sim (\sim A \lor B)$

h. $(\sim A \lor \sim B) \cdot \sim (\sim A \lor \sim B)$

i. $((\sim A \equiv B) \equiv \sim C) \cdot \sim ((A \equiv B) \equiv \sim C)$

j. $((\sim A \supset A) \supset \sim A) \cdot \sim ((\sim A \supset A) \supset \sim A)$

k. $((\sim A \supset A) \supset \sim A) \cdot (\sim (\sim A \supset A) \supset \sim A)$

*3.   Answer the following questions.

a.   Could $\sim A \supset \sim B$ be the conclusion of the rule of I.P.? C.P.?

b.   Could $\sim (A \supset B)$ be the conclusion of the rule of I.P.? C.P.? M.T.? M.P.?

c.   Could $\sim A \lor B$ be the conclusion of the rule of I.P.? M.T.? List all the rules you can think of for which it could be the conclusion.

d.   Could $\sim A \lor B$ be the premise for Add.? D.S.? H.S.? What rules could it be a premise for?

e.   What is always the last step in a subproof for C.P.? I.P.?

4.   Construct proofs for the following, using the rules of C.P. and I.P., plus the rules from Units 7 and 8.

a. $(\sim A \lor \sim B) \supset \sim C /\therefore C \supset A$

*b. $(D \cdot E) \supset \sim F, F \lor (G \cdot W), D \supset E /\therefore D \supset G$

c. $(A \supset B) \supset C, A \supset \sim (E \lor F), E \lor B /\therefore A \supset C$

d. $P \supset Q, (P \cdot Q) \supset R, P \supset (R \supset S), (R \cdot S), \supset T /\therefore P \supset T$

e. $W \supset X, (W \supset Y) \supset (Z \lor X), (W \cdot X) \supset Y, \sim Z /\therefore X$

*f. $(A \lor B) \supset \sim C, D \supset (\sim F \cdot \sim G) /\therefore (A \lor D) \supset \sim (C \cdot F)$

g. $A \supset (B \supset C), (C \cdot D) \supset E, F \supset \sim (D \supset E) /\therefore A \supset (B \supset \sim F)$

*h. $(A \lor B) \supset (A \cdot B) /\therefore A \equiv B$

*i. $A \lor B, \sim A \lor \sim B /\therefore \sim (A \equiv B)$

j. $P \supset S, S \supset \sim (B \cdot D), \sim B \supset T, \sim (D \supset T) /\therefore \sim P$

k. $A \supset (B \lor C), C \supset (D \equiv F), D \cdot \sim F, B \supset F /\therefore \sim A$

l. $(A \cdot B) \lor (\sim A \cdot \sim B) /\therefore A \equiv B$

m. $A \supset (B \lor C), E \supset (C \lor P), \sim C /\therefore \sim (B \lor P) \supset \sim (A \lor E)$

n. $C \supset (D \supset \sim C), C \equiv D /\therefore \sim C \cdot \sim D$

o. $A \equiv \sim (B \lor C), B \equiv (D \cdot \sim E), \sim (E \cdot A) /\therefore A \supset \sim D$

p. $F \equiv \sim (Z \cdot Y), \sim (G \lor Z) \supset \sim H, \sim (F \cdot H) \lor Y /\therefore F \supset (H \supset G)$

*q. $(A \lor B) \supset \sim (F \cdot D), \sim (A \cdot \sim D), \sim F \supset \sim (C \cdot D) /\therefore \sim (A \cdot C)$

5. Construct proofs for the following more challenging problems.

a. $A \equiv (\sim B \lor (C \cdot D)), \sim (A \equiv (D \cdot F)), F \lor (C \equiv G) /\therefore A \supset (B \supset G)$

b. $\sim (A \lor B) \lor \sim (E \lor F), (C \lor D) \supset F, G \equiv (E \cdot H) /\therefore A \supset \sim (C \lor G)$

c. $((P \lor T) \equiv R) \supset (Z \lor W), R \supset (T \lor Z), \sim (T \lor W), Z \equiv T /\therefore P$

d. $\sim F \supset (R \lor S), (F \lor R) \supset (V \cdot T), V \equiv \sim (P \lor (G \lor H)) /\therefore P \supset S$

*e. $(X \cdot Y) \lor \sim (Z \lor W), (Z \cdot X) \supset (Y \supset \sim B), C \supset \sim C, \sim (B \lor C) \supset \sim Y$
   $/\therefore \sim Z$

f. $D \supset \sim (T \cdot Z), S \lor \sim (D \supset \sim Z), A \equiv \sim (S \cdot T), W \supset \sim W, E \supset (T \lor W)$
   $/\therefore D \supset (A \supset \sim E)$

g. $(A \cdot B) \supset \sim (C \cdot D), \sim C \supset (E \lor \sim F), \sim (F \supset E), \sim (D \lor \sim A) \supset (P \equiv Q),$
   $P \cdot \sim Q /\therefore A \supset \sim B$

6. Construct proofs for the following theorems.

*a. $(\sim p \supset (\sim q \supset \sim r)) \supset (r \supset (p \lor q))$

b. $(p \supset (p \cdot q)) \lor (q \supset (p \cdot q))$

c. $(p \supset (q \supset (r \cdot s))) \supset ((p \supset q) \supset (p \supset s))$

*d. $\sim ((p \equiv \sim q) \cdot \sim (p \lor q))$

e. $(p \equiv \sim q) \equiv \sim (p \equiv q)$

f. $((p \lor q) \supset (p \cdot q)) \equiv (p \equiv q)$

g. $((p \lor q) \supset (r \cdot s)) \supset (\sim s \supset \sim p)$

h. $p \supset (\sim p \supset q)$

7.   Show that the following are invalid.

*a. $(A \lor B) \supset (C \lor \sim D), A \equiv \sim D, C \equiv (E \cdot F) / \therefore A \supset F$

 b. $(X \cdot Y) \lor (\sim A \cdot \sim Y), X \supset (Y \lor \sim B), / \therefore X \supset Y$

*c. $A \supset (B \lor C), \sim (B \lor D), B \equiv (C \cdot \sim F) / \therefore \sim A$

 d. $(G \cdot H) \equiv (\sim A \lor \sim B), \sim (G \supset \sim H), \sim A \supset (F \lor \sim Z), \sim B \supset (\sim F \lor Z)$
      $/ \therefore \sim (F \cdot Z)$

*e. $(A \lor \sim B) \supset (C \supset \sim D), \sim C \equiv (E \cdot \sim F), \sim (E \lor \sim H), D \lor (F \cdot H)$
      $/ \therefore F \supset B$

*8.   Symbolize the following arguments and determine whether they are valid or invalid. If valid, construct a proof; if invalid, give a truth-functional counterexample.

a.   If we rely primarily on nuclear power or coal, then either there will be a nuclear accident or an increase in air pollution. We will rely primarily on neither nuclear power nor coal, but will develop solar energy. Therefore, there will not be an increase in air pollution.

b.   If the Bible is literally true, then both God and the Devil exist, and the story of Adam and Eve is correct. If the story of Adam and Eve is correct, then God is wrathful and not kind. If God exists, then he is omniscient and kind. Therefore the Bible is not literally true.

c.   If the Bible is literally true, then both God and the Devil exist. If God exists there is goodness in the world. If the Devil exists there is evil in the world. There is both goodness and evil in the world. Therefore the Bible is literally true.

d.   There will be nuclear war if and only if there is a proliferation of nuclear weapons and unrest in the developing nations. Nuclear weapons will proliferate if and only if there is an increase in the use of nuclear power and nuclear safeguards are inadequate. There will be unrest in the developing nations if economic conditions do not improve. There will be an increase in the use of nuclear power and economic conditions will not improve. Therefore, nuclear war will be avoided only if there are adequate nuclear safeguards.

e.   If the Monetarists are right, then there is an increase in inflation if and only if the money supply increases too fast. If the Keynesians are right, then there is an increase in inflation if and only if there is a decrease in unemployment. If the Libertarians are right, there is an increase in inflation only if the federal government spends more than it takes in. The government spends more than it takes in only if taxes are too low. There is no decrease in unemployment and taxes are not too low, but there is inflation. Therefore, neither the Monetarists, the Keynesians, nor the Libertarians are right.

PART TWO: MONADIC PREDICATE LOGIC

UNIT 10
Singular Sentences

A. INTRODUCTION

There are certain arguments which are intuitively valid, but which cannot be *shown* to be valid by the methods of sentential logic. If we know, for instance, that all rhododendron leaves are poisonous, and we have correctly identified this as a rhododendron leaf, then we can correctly infer that this leaf is poisonous (and presumably refrain from chewing on it). But this argument, simple and obvious as it is, *cannot* be proved using only the resources of sentential logic. Nor, if we used just sentential logic (and not our heads) could we prove that, if all cats are mammals, and all mammals are vertebrates, then all cats are vertebrates. It is clear that in the second case Hypothetical Syllogism has *something* to do with the argument, since it has the form all A are B, and all B are C, therefore all A are C. But we *cannot* use the rule of H.S. here because the premises and conclusion simply do not have the form of conditionals. They are what we have been calling *simple* sentences in sentential logic, sentences which we have taken as unanalyzed wholes. In the first example, the rule of Modus Ponens seems to be involved, since the form is basically all A are B, x is an A, therefore x is a B. But again, these are sententially simple sentences, and thus are not amenable to treatment by propositional logic.

Modus Ponens just does not apply here. Both arguments would have to be symbolized simply as $p, q /\therefore r$, since they each just involve three different noncompound sentences. And obviously $p, q /\therefore r$ is *not* valid according to the canons of propositional logic.

What, then, needs to be done in these cases? Well, in the second example we might notice that the subject of the first premise is "cats" and the predicate "mammals," that the subject of the second premise is "mammals" and the predicate "vertebrates," and that the subject of the conclusion is again "cats" while the predicate is "vertebrates." Hypothetical Syllogism seems to apply once we take into account the *subjects* and *predicates* of the simple sentences. This indicates that what we need to do in such cases is to undertake an analysis of the *internal* structures of the sentences, rather than just being content to take them as unanalyzed wholes. This is the first thing you will learn to do in quantifier logic, which is also called *predicate logic*. One of the basic differences between sentential logic and quantifier logic, or predicate logic, is that in sentential logic we take the sententially simple sentence as an unanalyzed whole, while in predicate logic we analyze these simple sentences into their component parts.

Another thing you may have noticed about the examples above is the occurrence of the word "all." In the second argument, for instance, the premises and conclusion have the form "All P are Q." The words "all" and "some" are called "quantifiers" (hence the term "quantifier logic"), and they play an extremely important role in the sorts of inferences we will be studying in the rest of the book. The second basic difference between sentential and quantifier logic, then, is that in quantifier logic we will be using and analyzing these basic quantifier concepts and the relations between them.

Traditional, Aristotelian logic, which dominated the logic scene for over 2200 years (up to the beginning of the twentieth century), dealt almost exclusively with arguments made up of sentences like "All A are B" and "Some A are B." Such propositions are called "categorical propositions," because they state the relationship between two categories, or as we would now say, classes. They have been extremely important in the history of logic, partly because they do play a large role in many of the inferences we make, and partly because of the enormous influence of Aristotle. We will be studying categorical propositions for both of these reasons—their historical importance and the role they play in our language—but we will place them within the broader context of quantifier logic.

In this unit your job will be to learn about the most fundamental elements of predicate logic, in particular, singular sentences and propositional functions, and the concepts associated with them. You will learn to analyze singular sentences into their components, will refresh your memory on subjects and predicates, and will learn about three different kinds of subject–predicate propositions. In the next unit you will learn some of the basic facts about quantifiers, and in later units we will cover categorical propositions and inferences involving quantifiers. Toward the end of the book we will introduce an even more powerful system involving the logic of relations.

B. UNIT 10 OBJECTIVES

1. Be able to state the definitions of, and give examples of, each of the following concepts: name, individual constant, singular sentence, individual variable, and propositional function.
2. Be able to identify and symbolize the propositional function (or functions) of any given singular sentence, and be able to symbolize the singular sentence.
3. Learn the three senses of "is," the three kinds of subject–predicate propositions, and the major differences between them.

C. UNIT 10 TOPICS

1. Singular Sentences and Propositional Functions

Most sentences in English (though not all, as we shall see in Unit 11) can be analyzed into their subject and predicate components. Loosely speaking, the subject is what the sentence is talking about, or refers to, and the predicate is what is being asserted about the subject. You are undoubtedly familiar with this kind of analysis, and are able to identify correctly the subjects and predicates of simple sentences. In "The Eiffel Tower is in France," for example, "The Eiffel Tower" is the subject, and "is in France" is the predicate. In the sentence "Cats like to chase fireflies," "Cats" is the subject and "like to chase fireflies" is the predicate.

We may divide subject–predicate sentences into two groups: those in which the subject refers to an individual such as John or the Eiffel Tower, and those in which the subject refers to an entire class such as cats. We will call the first type of sentence a *singular sentence*, since it refers to a single individual; and the second type a *categorical sentence*, since it refers to an entire class, or category. The sentences in the preceding paragraph are examples, respectively, of singular and categorical sentences. We will have a great deal to say about categorical sentences in Unit 12, but in this unit we will focus on singular sentences because of their central role in predicate logic.

Simple singular sentences and propositional functions are the most fundamental, elementary units of predicate logic; they are the building blocks out of which more complex formulas are constructed, and are thus analogous to the simple letters—constants and variables—of sentential logic. Because they are so fundamental, it is important that you understand them thoroughly from the outset, so this unit will be devoted to a fairly detailed examination of these and related topics. At times it may seem as if we are taking an unnecessarily circuitous route to what are, after all, very simple formulas, but in the end the careful details will pay off in a more thorough understanding of predicate logic.

We may begin by noting again that a *singular sentence* is one that predicates something, asserts something, about a particular named individual, such as Ronald Reagan or the moon. (An individual, for logical purposes, is not necessarily a human being; it is just any particular single thing. It could be the book you are reading, the chair you are sitting in, your best friend, your cat, your bicycle, or even a particular point in space and time.) "Reagan is a conservative" would be an example of a singular sentence. We may then define a *singular sentence* very simply as a *sentence that contains a name*. Names, of course, are expressions such as "Dallas," "the moon," "Leo Tolstoy," and "*War and Peace*," which are conventionally used to refer to particular things. They are often, though not always, capitalized in English.

Some examples of singular sentences, with the names italicized, are below. Although these are all simple sentences with just one name, singular sentences may be compound and, as we shall see in Unit 17, may contain more than one name. Also, a name may appear at any point in the sentence, though to simplify the discussion, in all the sentences below the name is at the beginning.

1. *Leo Tolstoy* was Russian.
2. *War and Peace* is a novel.
3. *The moon* has an elliptical orbit.
4. *Dallas* is a large city.
5. *Angela* is happy.
6. *Angela* is wealthy.
7. *Bob* is happy.
8. *Bob* is wealthy.

Sentences such as "The best movie in town is a western" or "The cat that lives next door is Siamese" also predicate something of an individual, but are not considered to be singular sentences because they contain no names. The referring expressions—"the best movie in town" and "the cat that lives next door"—are what are called *definite descriptions*, which simply *describe* an entity in enough detail to be able to identify it uniquely.

Definite descriptions have the form "the so-and-so." Sentences containing definite descriptions require for their proper symbolization the use of identity, which will not be introduced until Unit 19, so we will defer their discussion until that unit. It is important to realize, however, that singular sentences are not the only ones that assert something about individuals, even though they are the only ones we will be discussing at this point.

Singular sentences, again, are those containing names, and a simple singular sentence will thus consist of two parts: the name, which refers to the individual, and the predicate, which asserts something about the named individual. To symbolize singular sentences, then, we will need two kinds of symbols: symbols representing names and symbols representing predicates.

Our symbolizations for names will be very simple: we will use what are called *individual constants*, which are simply lowercase letters that serve to abbreviate the name. We will formally define *individual constant* as *a lowercase letter, generally the first letter of the name, that is used as an abbreviation for the name*. Thus we would use *l* for "Leo Tolstoy," *w* for "*War and Peace*," *m* for "the moon," and *d* for "Dallas."

To symbolize predicates—what is said about the individuals—we will use *propositional functions*, and to explain this concept we need to introduce another very important item, the *individual variable*. You are probably already familiar with individual variables from elementary algebra; we will be using them in the same way here, except that they will stand for *any* individuals, rather than just numbers. By definition, *an individual variable is a lowercase letter (we will use x, y, z . . .) that serves as a placeholder for, or takes as a substitution instance, any name or individual constant*. In algebra, the formula $x + y = y + x$ uses individual variables, and a substitution instance of it, using the individual numbers 3 and 4, would be $3 + 4 = 4 + 3$.

Once we have individual variables, the notion of a propositional function is easy to define. Remember that a singular sentence consists of the name plus the predicate, so if we *remove the name* from the singular sentence, we will be left with the predicate. The predicate of the singular sentence "Angela is happy," for instance, is just "is happy." The *propositional function*, which will represent the predicate, is very like the predicate itself, but *has an individual variable in the place where the subject was*. The propositional function for "Angela is happy," then, is simply "x is happy," where the individual variable x indicates where the subject of the sentence would be placed.

We will formally define a *propositional function* as *the result of replacing some or all of the names in a singular sentence with individual variables*. All of the following are propositional functions, and represent, in order, the predicates of the singular sentences listed previously. Any variable can be used in a propositional function.

1. x was Russian.
2. y is a novel.
3. z has an elliptical orbit.
4. y is a large city.
5. x is happy.
6. x is wealthy.
7. x is happy.
8. x is wealthy.

Note that propositional functions 5 and 7 are identical, as are 6 and 8. This indicates that *sentences* 5 and 7 (and 6 and 8) are *saying the same thing* about the two individuals, that the predicates are the same.

Note also that propositional functions are just like singular sentences *except that they contain variables instead of names*. This distinction is extremely important, however; because functions contain variables, they do not say anything about particular individuals, and so *cannot be either true or false. They are not sentences*, but merely represent a *part* of a sentence, the predicate. We may say, however, that propositional functions, though not true or false in themselves, are *true of* or *false of* individuals. Function 3, for instance, is true of the Earth, but false of our sun, and function 1 is true of Leo Tolstoy but false of Charles Dickens. If a propositional function is true (false) of an individual, then, naturally, the sentence obtained by replacing the variable with the name of that individual will be true (false). A sentence obtained from a propositional function by replacing the variable with a name will be considered a *substitution instance of that function*. Thus, "*War and Peace* is a novel" is a substitution instance of "x is a novel."

The sentences we have been using so far have only one name, so their propositional functions have only one variable. Such functions are called "monadic" or "one-place" propositional functions. As noted earlier, however, there are also sentences containing more than one name, such as "John loves Mary," and the associated propositional functions will then have more than one variable, as in "x loves y." Such functions are called "polyadic" or "many-place" functions. We will discuss such sentences and functions in Units 17 and 18, but for the next few units we will concentrate on monadic functions.

2. Symbolizing Singular Sentences

Once you understand the concepts introduced in Section 1, symbolizing propositional functions and singular sentences will be very easy. You will first learn to symbolize propositional functions, and then use these as a step toward symbolizing singular sentences.

If you are given a simple, monadic propositional function, such as "*x* is happy," the symbolization is almost automatic. All you have to do is *pick a capital letter*, preferably one which reminds you of the meaning of the predicate, then *write down that letter followed by the individual variable*. The most natural symbolization for "*x* is Russian," for instance, would be simply *Rx*. Appropriate symbolizations for the other monadic functions mentioned in the preceding section would be the following:

y is a novel.	*Ny*
z has an elliptical orbit.	*Ez*
y is a large city.	*Cy*
x is happy.	*Hx*
x is wealthy.	*Wx*

Naturally, the variable that occurs in the unabbreviated function will be the same one that appears in the symbolization. Note also that no matter where the variable occurs in the unabbreviated function, it will always be written to the right of the capital letter. We will normally indicate abbreviations for propositional functions with the triple bar, as we did for abbreviations of English sentences in sentential logic. Thus, we will say "*Wx* ≡ *x* is wealthy," "*By* ≡ *y* is a baseball player," and so on. Finally, since the capital letters here are very much like abbreviations for predicates—what is said about an individual—we will call them predicate letters. Thus the predicate letters for the abbreviations above, in order, are *R*, *N*, *E*, *C*, *H*, and *W*.

In symbolizing simple singular sentences we will first *identify the propositional function* of which the sentence is an instance, then *abbreviate the function*, and finally, *abbreviate the sentence by replacing the variable with an appropriate individual constant*. Let us take the simple sentence "Mary is a scientist." We get the propositional function by removing the name and replacing it with a variable. The propositional function here would thus be "*x* is a scientist." An appropriate abbreviation for this function would be *Sx*. The proper individual constant, the abbreviation for the proper name, would be *m*, and in replacing the variable with the constant we get the symbolization for the singular sentence: *Sm*. This may seem like a roundabout way to get at the symbolization, which is, after all, very simple but it is important to state the function explicitly.

It should now be clear how we can symbolize the sentences of Section 1, since their propositional functions were symbolized previously.

1. Leo Tolstoy was Russian.	*Rl*
2. *War and Peace* is a novel.	*Nw*
3. The moon has an elliptical orbit.	*Em*

4. Dallas is a large city.	Cd
5. Angela is happy.	Ha
6. Angela is wealthy.	Wa
7. Bob is happy.	Hb
8. Bob is wealthy.	Wb

The functions and sentences we have been discussing so far have all been truth-functionally simple, but, of course, we will be using compounds as well. Compound singular sentences and propositional functions are essentially those that contain one or more of our five operators, and will be symbolized using the techniques of Unit 4.

An example of a compound propositional function would be "x went to the party but did not see either Betty or Marge." To symbolize this we need to make explicit its compound structure, to "expand" it so that each of its simple components is explicitly contained in the compound. The expanded function would be "x went to the party but x did not see Betty and x did not see Marge." It is then clear that the simple functions are "x went to the party" (Px), "x saw Betty" (Bx), and "x saw Marge" (Mx), and the symbolization for the compound function would be $Px \cdot \sim Bx \cdot \sim Mx$. Note that as in Unit 4, *negated expressions must be considered to be compound*. To symbolize "x did not see Betty," we abbreviate the function "x did see Betty" as Bx, and then put the negation sign in front of it.

Further examples of compound functions, with their symbolizations, are given below. The simple functions which are their components are given in parentheses, with their abbreviations.

1. y is neither fish nor fowl. ($Fy \equiv$ $\sim Fy \cdot \sim Ly$
y is fish; $Ly \equiv y$ is fowl)

2. If z lives in a glass house, then $Gz \supset \sim Tz$
z should not throw stones. ($Gz \equiv$
z lives in a glass house; $Tz \equiv z$
should throw stones)

3. If x proposed to both Betty and $((Bx \cdot Mx) \supset (Lx \lor Hx)) \cdot \sim Mx$
Marge, then x will either leave
town or hide, but x did not pro-
pose to Marge. ($Bx \equiv x$ proposed
to Betty; $Mx \equiv x$ proposed to
Marge; $Lx \equiv x$ will leave town;
$Hx \equiv x$ will hide)

In symbolizing compound singular sentences, those that contain sentential operators, the idea is the same as with compound functions: identify the simple components, symbolize them, and then put the whole thing together. We will here go over just two examples. The symbolizations for the propositional functions and names are given directly under the sentence, followed by the symbolization for the sentence itself.

> If John plays either basketball or football, and Bill plays both, then the coach will be happy.
>
> (Bx = x plays basketball; Fx = x plays football; Hx = x is happy;
>
> j = John; b = Bill; c = the coach)
>
> $((Bj \lor Fj) \cdot (Bb \cdot Fb)) \supset Hc$
>
> If John and Mary are going to the movie, then unless John cooks supper, neither one will have anything to eat.
>
> (Mx = x is going to the movie; Cx = x cooks supper;
>
> Ex = x has something to eat; j = John; m = Mary)
>
> $(Mj \cdot Mm) \supset (\sim Cj \supset \sim (Ej \lor Em))$

The only thing you need to watch out for is not to get the letters mixed up; when you have both predicate letters and individual constants this is sometimes easy to do. There will be more examples of compound singular sentences in the exercises.

At this point you should have a fairly thorough understanding of singular sentences, and we can now discuss two other types of sentences, the categorical sentence and the identity sentence. We will not be able to symbolize these other sentences until later, but it is important at this point to be aware of their existence, since they are often confused with singular sentences.

3. Three Kinds of Subject–Predicate Sentences; Three Senses of "Is"

Most sentences in English, as noted earlier, can be analyzed according to their subjects and predicates, and subjects may be either individuals or classes. This gives us a convenient, if informal, way of classifying subject–predicate (S–P) sentences in English, and that will be the topic of this section.

We will consider the *subject expression* of a sentence to be the *noun* (*common or proper*) *plus all modifiers*; it is the expression that specifies the thing being talked about. Thus, the subject expression of "Left-handed tennis players have good backhands" is "left-handed tennis players" rather than "tennis players." We are saying something not about *all* tennis players, but about the more specific group of *left-handed* tennis players.

In the following sentences, the subject expressions have been *italicized.* "*The man in the white hat* is the good guy." "*Susan* is a surgeon." "*White whales with grudges* are dangerous." "*Any man of ambition, intelligence, and compassion, who has a bright smile,* can be elected president."

The predicate of a sentence, as indicated, is what is said about the subject, and *the predicate expression will be simply the sentence minus its subject expression and any quantifiers* ("all" and "some"). In general, the predicate expression of a sentence will simply be a propositional function without the variables. In the sentences below, the predicate expressions have been *italicized*. "Susan *never uses nylon thread in her surgery*." "The Eiffel Tower *is in Paris and has been modeled with toothpicks*." "Whales *are sea-going mammals, and are the largest animals known*." "Elementary school teachers *need patience and understanding*." "Any chimpanzee *can learn to communicate*."

In English, the copula "is" is used to join subject and predicate expressions, as in the sentence "John is a painter." Although this very common word may seem unambiguous, as it turns out there are actually three different senses of "is," and three corresponding types of subject–predicate sentences.[1] In the following three sentences, for instance, although the grammatical structure and the meaning of "is" superficially look exactly the same, in fact they are very different.

1. Moby Dick is a whale.
2. A whale is a fish.
3. Moby Dick is the devil.

The differences between these sentences are primarily a matter of what kind of subjects and predicates they have, whether individuals or classes. In the third sentence, *both subject and predicate expressions refer to individuals* (at least in standard demonology), and the function of the "is" is to state an identity between them, to assert that Moby Dick and the Devil are one and the same. This sense of "is" we will thus call *the "is" of identity*, and the sentence using this sense of "is" will be an *identity sentence*.[2] Since an identity sentence has two names we will not be able to symbolize it until Unit 19, when we discuss relational predicate logic with identity, but the symbolization will be very simple: we will simply put an identity sign between the two names.

In the second sentence, *subject and predicate expressions both refer to classes* rather than individuals, and here, quite obviously, the "is" is *not* stating an identity between them. It is not asserting that the class of whales and the class of mammals are the *same*, but rather that the class of whales *is included in* the class of mammals, is a *part of* the class of mammals. For this reason we will call this sense of "is" *the "is" of inclusion*. A sentence employing this sense of "is" will be called a *categorical sentence* since it states a relation between two classes, or categories. The symbol-

[1] There is a fourth sense, the "is" of existence, as in "There is a God" (or "There is a fourth sense of 'is'"), but these are not subject–predicate sentences. We are talking here only about the various senses of "is" that join subjects and predicates.

[2] An identity sentence is a special case of a singular sentence. It is certainly a kind of singular sentence since it contains at least one name.

ization for categorical sentences is considerably more complex than that for singular sentences, despite the superficial grammatical similarity, and we will discuss this in Unit 12.

In the first sentence *the subject expression refers to an individual, and the predicate expression refers to a class*, and the sentence asserts that the individual is a member of the class of whales. The "is" here is obviously not that of identity, nor is it that of inclusion, since inclusion is a relation that holds between two classes. Here we have a third sense of "is," which we will call the *"is" of predication*, since its function is to predicate something—the property of being a whale—of an individual. (As we shall explain later, having a certain property and being a member of a certain class are, for logical purposes, interchangeable.) This kind of sentence, since it contains a name, is a *monadic singular sentence*.

These three senses of "is" are very different, so different that many languages have three different words for them. It is a quirk of English that the same word is used for each.

It should be noted that this classification into three kinds of *S–P* propositions is not exact, since it is often possible to analyze a sentence in more than one way, and the choice of subject is not always uniquely determined. In "Cats love Mary," for instance, we might take cats as subject and consider the predicate to be "love Mary," or we might take *Mary* as subject and consider the predicate *being loved by cats*. In the first case we would have a categorical sentence and in the second case a singular sentence. However, for our limited purposes the distinction is useful, and it is important to be aware of the logical differences between the three kinds of sentences.

A summary of the basic facts about the three senses of "is" and the three different kinds of subject–predicate propositions is given below.

KIND OF SENTENCE	SUBJECT AND PREDICATE EXPRESSIONS	FUNCTION OF "IS"	KIND OF "IS"
Simple monadic singular	Subject expression refers to an individual; predicate expression refers to a class.	Predicates a property of an individual	Predication
Categorical	Subject expression and predicate expression both refer to classes.	States inclusion relationship between classes	Inclusion
Identity	Subject expression and predicate expression refer to individuals.	Asserts identity of two individuals	Identity

Several examples of the various kinds of sentences are given below; you should then be able to do Exercise 3, at the end of the unit.

SIMPLE SINGULAR SENTENCES

John is one of the Smith clan.

Ronald Reagan ran for president in 1980.

Bob loves to fish.

Mercury is a planet in our solar system.

Andrew was late for the elegant dinner party.

π has a long history.

SIMPLE CATEGORICAL SENTENCES

Whales are mammals.

People who watch a lot of television get brain lesions.

Presidential candidates are ambitious.

Unscrupulous people sometimes become bank presidents.

Anyone who gets fired will be depressed.

People who live in glass houses shouldn't throw stones.

SIMPLE IDENTITY SENTENCES

Dr. Jekyll is Mr. Hyde.

Mark Twain is Samuel Clemens.

Marlene Dietrich is "The Blue Angel."

Clark Kent is Superman.

DEFINITIONS

1. A **name** is an expression such as "George Washington" which is conventionally used to refer to a particular individual thing.
2. An **individual constant** is a lowercase letter, generally the first letter of the name, which is used as an abbreviation for the name.
3. A **singular sentence** is any sentence that contains a name.
4. An **individual variable** is a lowercase letter from the end of the alphabet which serves as a placeholder for, or takes as a substitution instance, any name or individual constant.
5. A **propositional function** is the result of replacing some or all of the names in a singular sentence with individual variables.
6. A **predicate letter** is a capital letter, preferably one that reminds us of the meaning of the predicate, which is used in conjunction with variables to symbolize propositional functions.

7. A **categorical proposition** is a proposition in which subject expression and predicate expression both refer to classes, and which states an inclusion (or exclusion) relation between the two classes.

8. An **identity sentence** is a sentence in which subject expression and predicate expression both refer to individuals, and which asserts that the two individuals are identical, are one and the same.

EXERCISES

1. For each of the sentences below, pick out and symbolize the (monadic) propositional functions and then symbolize the singular sentence. For "John likes everyone," for example, you should have (1) x likes everyone, (2) Lx (3) Lj. For those sentences which are sententially compound, isolate and abbreviate *all* the simple functions and then symbolize the compound singular sentence. Remember that negated sentences are compound.

*a. Rocky is a master safecracker. Mr

b. The Chase Manhattan Bank has been buglarized.

*c. Rocky's apartment contained huge diamond rings.

d. The New York Police Department could not find the burglars.

*e. The FBI solved the mystery.

f. Andrew is an FBI agent, and he saw several suspicious-looking characters. $Fa \cdot Sa$

*g. If Andrew does not get enough sleep, he misses a lot, but if he is alert he is very sharp. $(Sa \supset Ma) \cdot Aa$

h. Andrew neither runs nor swims, but he loves to ice-skate, though he falls down a lot. $(Ra \vee Sa) \cdot (La \cdot Fa)$

*i. Andrew loves police work, but neither Jane nor Bob is happy with his job, and he will stay in police work if and only if his supervisor gives him a raise and a commendation.

j. It is not true that if Andrew does not get a raise then he will quit, but if he gets married he will request less hazardous duty.

2. Given the following abbreviations, symbolize the following compound singular sentences.

a = Amy; b = Bob; k = Kathy; p = Pete; $Px \equiv x$ likes popcorn; $Sx \equiv x$ can swim; $Dx \equiv x$ will drown; $Tx \equiv x$ likes television; $Lx \equiv x$ likes logic; $Fx \equiv x$ falls into the water; $Rx \equiv$ someone rescues x; $Cx \equiv x$ likes to climb trees; $Gx \equiv x$ is good at logic; $Wx \equiv x$ watches a lot of television; $Hx \equiv x$ is happy.

*a. Amy, Bob, and Kathy like popcorn. $Pa \cdot Pb \cdot Rk$

b. Bob likes television but not logic, and Amy likes both.

*c. Pete can swim, but Kathy can't, and she will drown if she falls into the water and no one rescues her.

d. Kathy likes television and popcorn, but Pete likes neither.

*e. Amy and Kathy like to climb trees, but neither Bob nor Pete does.

f. Bob will be good at logic if and only if he likes it and doesn't watch a lot of television. $Gb \equiv Lb \cdot \sim Wb$

*g. Amy will be happy if someone rescues Kathy and Kathy doesn't drown.

h. Kathy will drown if and only if she can't swim and falls into the water and no one rescues her, but she likes to climb trees and will not fall into the water.

*i. Bob will be happy only if Amy likes television and logic and he does not fall into the water.

j. Amy, Bob, Kathy, and Pete will be happy, provided none of them falls into the water and drowns, and all watch a lot of television.

3. For each of the following sentences, indicate whether it is a singular, a categorical, or an identity sentence.

*a. Amy has a tree house. S

b. "The Good Book" is the Bible. I

*c. Tree houses are for the birds. C

d. Mephistopheles is Satan. I

*e. Devils are evil. C

f. Superman can fly. S

*g. Clark Kent is Superman. I

h. Clark Kent is wonderful. S

*i. Superman catches criminals.

j. Criminals hate Superman.

*k. Fish gotta swim.

l. Moby Dick is not a fish.

*m. Jimmy Carter was defeated in 1980.

n. Presidents have power.

*o. Truman was a powerful president.

p. God is all-powerful.

*q. Allah is God.

r. The Gods are angry tonight.

*s. Jupiter is a plant.

t. Planets have elliptical orbits.

UNIT 11
Quantifiers

A. INTRODUCTION

As noted in Unit 10, the quantifiers "all" and "some" constitute one of the basic elements of quantifier logic. In this unit you will be learning their precise meanings, the relations between them, and how to symbolize simple quantified sentences. We have in English any number of quantifiers besides "all" and "some"; "many," "most," "a few," and "almost all" are all quantifiers in the sense that they indicate quantity, but they can play no role in our logic because their meanings are vague and the relations between them are imprecise. The meanings of "all" and "some," by contrast, are definite and fixed. ("Some" may seem like the epitome of a vague, imprecise concept but, as we shall see, in logic it has a very definite meaning.) The relationships between them are so determinate that we will even have rules of inference telling us how we may infer sentences containing one quantifier from those containing the other.

The role of a quantifier is to tell us of how many things the function is true. It is always placed *in front of* the propositional function, and when prefixed to the function in this way it yields a sentence. Remember from Unit 10 that one way to get a sentence from a propositional function is to replace the variable with a constant, which gives us a singular sentence.

The other way, which we will discuss in this unit, is to prefix the function with a quantifier. "x is an astronaut," for instance, is not a complete sentence and has no truth value. If we prefix the quantifier "some" to the function, however, we get the true sentence. "Some x is an astronaut," or more colloquially, "There are some astronauts." *We will call the result of prefixing a quantifier to a propositional function a general sentence.*

We will have two kinds of general sentences in quantifier logic. The *existential* sentence, in which the quantifier "some" is used, will say that there is *something* of which the propositional function is true. An example of an existential proposition would be "There are footprints on the moon." The *universal* sentence, in which the quantifier "all" is used, says that the propositional function is true of *everything*. An example of a universal sentence would be "Everything has a purpose."

In this unit we first examine the differences between singular sentences, general sentences, and propositional functions; we then discuss the meaning and function of the quantifiers, and introduce the symbolism we will be using; finally, we discuss the relations between the two quantifiers. More specific objectives are listed below.

B. UNIT 11 OBJECTIVES

1. Be able to distinguish between singular sentences, simple quantified sentences, and propositional functions.
2. Learn the meanings and symbolism for the universal and existential quantifiers.
3. Learn the definitions for free and bound variables, and for the scope of a quantifier.
4. Learn the English phrases for, and relations between, negated quantifiers.
5. Be able to identify and symbolize simple quantified sentences and their negations.

C. UNIT 11 TOPICS

1. Simple General Sentences

There are certain English sentences which, though they look superficially like subject–predicate propositions, cannot really be interpreted in this way, and which have given rise to very difficult philosophical problems. Consider the sentence "Something trampled the strawberries," for instance. Although this might be interpreted by the uninitiated as a singular sentence, with "trampled the strawberries" as predicate and "something" as subject, this would be a mistake. The predicate analysis is correct, but we *cannot* consider "something" to be the subject expression of a singular sentence. The subject expression of a singular sentence must be a *name*, a term which refers to a particular individual, and "something" does not fill the bill. In fact, we generally use the term "something" precisely because we do *not* know and cannot name the particular individual. The problem is even clearer if we consider sentences with negative terms, such as "Nothing will help." Here we could not possibly interpret the word "nothing" as the *subject* of the sentence, the *individual* about which the sentence is speaking. It would be ludicrous, for instance, to reply to such a sentence, "Well, thank goodness; I'm glad *something* will help," as if "nothing" was the name of the thing that would help! "Something" and "nothing" are simply *not names*, and sentences using these terms cannot be interpreted as singular sentences.

Nor can they be interpreted as propositional functions, despite the fact that they contain no names. A propositional function, if you remember, is *not* a complete sentence and cannot be considered to be either true or false. The sentence "Something trampled the strawberries," on the other hand, is certainly a complete, meaningful sentence, and you can verify it by taking a tour of the strawberry patch (even if you can't identify the culprit). It would also be incorrect to interpret these sentences as categorical propositions, since "something" does not refer to any particular class, and the sentence cannot therefore be said to be stating a relationship between two classes. Thus, these sentences fit none of the categories we have discussed so far.

How, then, can a sentence such as "Something trampled the strawberries" be interpreted? It does have a propositional function, a predicate, "*x* trampled the strawberries," but it does not have a subject; it does not say who did the trampling. What it does have is the word "something," an expression which indicates simply that the propositional function is true of at least one thing, even though that thing is not identified. Such an expression is called an *existential quantifier*, and a sentence like the above, which consists of an existential quantifier followed by a simple propositional function, will be called a *simple existential sentence*. A sentence consisting of a *universal* quantifier, such as "everything," followed by a simple propositional function, will be called a *simple universal sentence*. An example of a simple universal sentence would be "Everything has a purpose," where the universal quantifier "everything" indicates that the propositional function "*x* has a purpose" is true of every individual. A simple *general* sentence is either a simple existential or a simple universal sentence; sentences of the sort we have been considering will be interpreted as simple general sentences. Negative sentences, such as "Nothing will help," will be interpreted as *negated general sentences* ("It is not the case that there is something that will help"), thereby avoiding the problems we would have if we treated the sentence as singular, and the term "nothing" as a name.

In later units we will discuss general sentences with compound functions, but in this unit we will confine ourselves to simple general sentences. It is important to distinguish between propositional functions, singular sentences, and simple general sentences, and the following paragraphs will summarize the differences.

A *propositional function*, again, is the result of replacing the name in a singular sentence with an individual variable, so it is a predicate expression plus an individual variable indicating where a name could be placed to yield a singular sentence. It is not a complete sentence, and is neither true nor false. An instance of a propositional function is "*x* trampled the strawberries."

A *singular sentence* is a substitution instance of a propositional function, in which a name has been substituted for the variable. It asserts that the propositional function is true of the named individual. It is, of course, a complete sentence, and is either true or false. An instance of a singular sentence would be "Jennifer trampled the strawberries."

A *simple quantified (or general) sentence* is the result of prefixing a quantifier to a simple propositional function. An existential sentence says that the function is true of *something*, while a universal sentence says that the function is true of *everything*. These are complete sentences, and are either true or false. An instance of a simple existential proposition is "Something trampled the strawberries," and an instance of a simple universal proposition is "Everything is composed of quarks."

Exercise 1, at the end of the unit, will test your ability to distinguish between these three kinds of expressions, and at this point you should complete that exercise.

2. Universal and Existential Quantifiers

A *universal quantifier is an expression which, when prefixed to a propositional function, yields a universal sentence. A universal sentence is one which asserts that the function is true of everything.* Examples of universal propositions would be "Everything has a purpose," "Everything is beautiful in its own way," and more "restricted" universal statements such as "All politicians are ambitious" and "Every student wants to learn something." Sentences which contains words like "all," "every," "each," "each and every," "any" (in some contexts), "everything," "anything," or any word which means the same, will be translated as universal propositions.

The symbol, or abbreviation, for the universal quantifier will be a variable (the same one that occurs in the function to be quantified) enclosed in parentheses, such as (x), (y), or (z). This symbol may be read as *"For every x"* (or *y* or *z*). A formula such as $(x)Fx$ would then be read literally as "For every *x*, *x* is an *F*," or more colloquially, as "Everything is an *F*." We could symbolize the sentence "Everything has a purpose" by abbreviating the propositional function "*x* has a purpose" as *Px*, and then prefixing to the function the universal quantifier (x). This gives us the formula $(x)Px$ as the proper symbolization, which could be read literally as "For every *x*, *x* has a purpose." The quantifier itself is the expression, such as "For every *x*," that is prefixed to the propositional function. Alternative readings for the symbolized universal quantifier are "For all *x*" and "Every *x* is such that"; any equivalent phrase will do as well.

Any English sentence which asserts simply that everything, without qualification, has a certain property can be symbolized by means of what we will call a *simple universal formula. A simple universal formula is just a simple (noncompound) propositional function, such as Fx, preceded by a universal quantifier.* In this unit we are concerned only with such simple quantified expressions; in later units we discuss quantifiers which extend over compound functions. *To symbolize a simple universal sentence, you just identify and abbreviate the propositional function, and then prefix to it the universal quantifier.* In the sentence "Everything exerts a gravitational pull on the Earth," for instance, the propositional function is "*x* exerts a gravitational pull on the Earth," which could be abbreviated as *Ex*. To say that *everything* does that, all we need to do is put the universal quantifier (x) in front, so the sentence can be completely symbolized as $(x)Ex$. The sentence would be read literally as "For every *x*, *x* exerts a gravitational pull on the Earth." Other examples of simple universal sentences are given below, with their propositional functions immediately beneath them; the symbolizations are given on the right.

1. Anything is better than nuclear war. $(x)Bx$
   x is better than nuclear war. Bx
2. Everything is unique. $(y)Uy$
   y is unique. Uy
3. Everything was created by God. $(z)Gz$
   z was created by God. Gz

Notice that we have sometimes used '*y*' or '*z*' as the quantified variable, rather than '*x*'. For quantified expressions it makes no difference what letter we use; $(x)Fx$ means exactly the same as $(y)Fy$: namely, that for any individual thing, that individual has the property *F*. We do have to make sure, however, to use the *same* variable in the quantifier as appears in the function; normally it will not make sense to say "For every *x*, *y* is an *F*."

An existential quantifier is a phrase which, when prefixed to a propositional function, yields an existential statement. An existential statement is a statement which asserts that the propositional function is true of at least one thing, although it is not specified what. Analogously to the universal quantifier, we can read the existential quantifier as *"for some x,"* although there are other readings which are sometimes more appropriate, which we will discuss shortly. *The existential quantifier, which conveys the idea of "some," has a very precise meaning in logic.* Unlike its counterpart in ordinary English, which is extremely vague on the matter of quantity, the logical "some" is exact: it means only *"at least one."* It does *leave open* the possibility that there is more than one, but unlike the ordinary English word, it does *not* have the connotation of more than one. It means only that *there is at least one*, and what it is really doing is asserting *existence* (hence the term "existential quantifier"). Thus any English phrase which conveys the notion of existence, as well as those which use the phrase "some," will be rendered by the existential quantifier. Furthermore, since "at least one" leaves open the possibility of more than one, we can translate any quantity which falls short of "all" by using the existential quantifier. Such phrases include "many," "a few," even "most," and while their full sense is not captured by the existential quantifier, at least some part of their *logical* sense can be indicated.

The symbol we will use for the existential quantifier is a backwards E followed by the appropriate variable, and the whole enclosed in parentheses, such as $(\exists x), (\exists y)$ or $(\exists z)$. Thus the expression "$(\exists x) x$ is heavier than gold" would be read "For some x, x is heavier than gold." More standard renderings in English would be "Something is heavier than gold" or "There is something that is heavier than gold." The sentence could be fully abbreviated as $(\exists x)Hx$. Some examples of *simple existential statements*, with their propositional functions and abbreviations, are given below.

1. There are black holes in the universe. $(\exists y)By$
   y is a black hole. By
2. There is at least one dollar in my pocket. $(\exists x)Dx$
   x is a dollar in my pocket. Dx
3. There are many translations of the Bible. $(\exists y)Ty$
   y is a translation of the Bible. Ty
4. Some things are mysterious. $(\exists z)Mz$
   z is mysterious. Mz

As the examples above suggest, there are other phrases besides "For some x" which will do as a reading for the existential quantifier $(\exists x)$: "There is an x such that," "There is some x such that," "There is at least one x such that," "Some x exists such that," and so on. Any of these, prefixed to a propositional function, will yield an existential sentence. "There is some x such that x is mysterious," for instance, is a perfectly meaningful English sentence, though it is not in the form we would normally use.

You should get in the habit, especially at first, of reading the quantifier formulas literally, using one of the phrases for the quantifier, and then reading out the propositional function. This will make the transition between the normal English sentence and the logical formula more apparent, and will be an aid in symbolization.

At this point you should be able to identify and symbolize simple universal and existential sentences, and so should be able to complete Exercise 2 at the end of the unit.

3. Free and Bound Variables, and Related Concepts

Now that you have been introduced to the two quantifiers, there are several other concepts closely related to them which you should learn, though they will play no role until the next unit. The first of these is the *scope* of a quantifier. A variable which falls within the scope of a quantifier is governed by that quantifier, or refers back to that quantifier. The *scope of a quantifier* is defined in the same way as the scope of a negation sign: it *is the first complete formula that follows the quantifier.* (A formula in quantifier logic will be an expression built up of simple propositional functions, which will be analogous to the single letters in sentential logic.) In the formula $(x)Fx$ the Fx falls within the scope of the quantifier, since it is the first complete formula following the quantifier. In the more complex formula $(x)Fx \supset Gy$, the Fx is within the scope of (x), but the Gy is not, since the scope extends *only* as far as the first formula, which is Fx. On the other hand, in the formula $(x)(Fx \supset Gy)$, Gy is in the scope of (x) because the first complete formula following the quantifier is the whole thing enclosed in parentheses.

Other important concepts in quantifier logic are those of *free* and *bound* variables. Given the definition of "scope" they are very easy to define. A *bound variable is simply one which falls within the scope of its own quantifier,* that is, a quantifier using the same variable. A *free variable is one which does not fall within the scope of its own quantifier.* Obviously, a variable is free if and only if it is not bound. Thus in the formula $(x)(Fx \supset Gy)$, the x in Fx is bound, but the y in Gy is free since, although it does fall within the scope of a quantifier, it is the wrong quantifier. In the formula $(x)Fx \supset Gx$, the x in Fx is bound, but the x in Gx is free, since it does not fall within the scope of the quantifier at all. In $(x)(Fx \supset Gx)$, all occurrences of x are bound.

Another useful concept, which will be discussed more fully in Unit 16, is the *domain of discourse* of a quantified statement, or the *range of bound variables* of a quantifier. This is, roughly, the set of objects for which the statement is defined, about which the statement could be speaking. In the sentence "Something is eating the bean sprouts," for example, the domain of discourse would have to be a set which included bean sprout eaters, such as the set of all animals. It wouldn't make sense to say that the domain of discourse was the set of all rocks, or all oceans, or all houses. In many cases the domain is simply the set of all existing things, as in the sentence "Everything is subject to electromagnetic fields." This set of all existing things is the *universal domain*, and in the following units, unless otherwise specified, this is the domain we will be using. In this unit, however, to simplify the symbolizations, we will frequently restrict the domain of discourse to human beings. In any sentence that uses terms like "any*one*," "some*one*," "every*one*," and so on, you can assume that the domain of discourse is people.

Now that you have the basic quantifier concepts in hand, it is time to turn to the more challenging topic of negated quantifier statements.

4. Negated Quantifiers

As noted earlier, a negative sentence, such as "Nobody is here," cannot be interpreted as a singular sentence whose subject is "nobody." Such a reading just does not make sense, and would lead one into meaningless inquiries about, for instance, the color of "nobody's" eyes. (Lewis Carroll, in the *Alice* books, makes a number of jokes based on this kind of misunderstanding; he was, in fact, a logician.) This kind of sentence can also lead into various philosophical perplexities; if we say, for instance, that nothing exists, and interpret this as a singular sentence, then it seems that the sentence implies that something exists after all, namely *nothing*, even though the sentence *says* that *nothing* exists (which is precisely what *does* exist, and so on). Many sleepless nights were avoided by logicians' discovery that propositions of that sort, using terms such as "none," or "nothing," could be interpreted as *negated existential* propositions—existential sentences with a negation sign in front. To say that nothing will help, for instance, is simply to say that *there does not exist* anything that will help. If we symbolize "Something will help" as $(\exists x)Hx$, then "Nothing will help" can be symbolized as its negation, $\sim (\exists x)Hx$. This could be read literally as "It is not the case that there is some x such that x will help," not a very elegant English sentence, but one that does mean the same as "Nothing will help." There are a number of these expressions in English which can similarly be considered as negated existentials. Such phrases as "none," "nothing," "there are no," "there aren't any," and "no one" all indicate the *denial of existence*, and so can be interpreted as the negation of existentials, symbolized by using the expression $\sim (\exists x)$.

The following are examples of negated existential sentences, along with their symbolizations. The propositional functions are given immediately below the sentence, and then the literal reading of the symbolized expression, to assist you in making the transition from English to symbols.

1. Nothing cleans like Lysol. $\sim (\exists x)Cx$
   x cleans like Lysol. Cx
   There is no x (or, there does not exist an x) such that x
   cleans like Lysol. $\sim (\exists x)Cx$
2. There are no happy wanderers. $\sim (\exists x)Hx$
   x is a happy wanderer. Hx
   There is no x such that x is a happy wanderer. $\sim (\exists x)Hx$
3. Devils do not exist. $\sim (\exists x)Dx$
   x is a devil. Dx
   There is no x such that x is a devil. $\sim (\exists x)Dx$

Interestingly enough, those negated expressions can equivalently be considered as *universal propositions*. To say that nothing will help, for instance, is the same as saying that, for *every* x, x will not help, that is, $(x) \sim Hx$. Symbolically, we have an equivalence between $\sim (\exists x)Hx$ and $(x) \sim Hx$. That is, *the negation of the existential proposition is equivalent to the universal proposition with the propositional function negated*. Some further examples may help convince you of this.

"Nothing is clear," $\sim (\exists x)Cx$ means "Everything is unclear," $(x) \sim Cx$

"Nobody has money," $\sim (\exists x)Mx$ means "Everybody is without money," $(x) \sim Mx$

"There are no flying saucers," $\sim (\exists x)Fx$ means (roughly) "Whatever it is, it isn't a flying saucer," $(x) \sim Fx$.

In general, where ϕx is any propositional function, we have an equivalence between $\sim (\exists x)\phi x$ and $(x) \sim \phi x$.

Negated universal propositions are those which *deny* that *all* things have a certain property, as in "Not everything has mass." In English the phrases which will signal a negated universal proposition are "not all," "not every," "not everything," and similar expressions. Such sentences will be symbolized as universal formulas preceded by a negation sign. If Mx abbreviates "x has mass," for instance, then "Not everything has mass" can be symbolized as $\sim (x)Mx$. Other examples of negated universal sentences, with their symbolizations, follow.

1. Not everything is beautiful. $\sim (x)Bx$
   x is beautiful. Bx
   It is not the case that for every x, x is beautiful. $\sim (x)Bx$
2. Not all things are made of atoms. $\sim (x)Ax$
   x is made of atoms. Ax
   It is not the case for every x, x is made of atoms. $\sim (x)Ax$

There is also an equivalence between negated universals and existentials. If we say that *not everything is beautiful*, this means that *there are some things which are not beautiful*. Here we have an equivalence between $\sim (x)Bx$ and $(\exists x) \sim Bx$. In general, *the negation of a universal proposition will be equivalent to an existential proposition with the function negated*, that is, $\sim (x)\phi x$ is equivalent to $(\exists x) \sim \phi x$. Again, some examples may be helpful.

"Not everything is easy," $\sim (x)Ex$	means "Some things are not easy," $(\exists x) \sim Ex$
"Not everyone is here," $\sim (x)Hx$	means "Someone is not here," $(\exists x) \sim Hx$
"Not everything can be understood," $\sim (x)Ux$	means "Some things cannot be understood," $(\exists x) \sim Ux$

If we again let ϕx stand for *any* propositional function (simple or complex) then we can state the following two general equivalences, which will be used as replacement rules later on: $\sim (\exists x)\phi x \equiv (x) \sim \phi x$ and $\sim (x)\phi x \equiv (\exists x) \sim \phi x$. Note that if we substitute for the ϕx a negated propositional function on the left, and use D.N. on the right, we get two more forms: $\sim (\exists x) \sim \phi x \equiv (x)\phi x$ and $\sim (x) \sim \phi x \equiv (\exists x)\phi x$. Since these forms are also useful, you should be convinced that they are correct. Some examples may again be helpful.

"There is nothing which does not have a purpose," $\sim (\exists x) \sim Px$, means the same as "Everything has a purpose," $(x)Px$. "Nobody was unjust," $\sim (\exists x) \sim Jx$, means the same as "Everybody was just," $(x)Jx$. For the last form, we can say that "Not everything is unclear," $\sim (x) \sim Cx$, means the same as "Something is clear," $(\exists x)Cx$, and "Not everyone was unsure," $\sim (x) \sim Sx$, means "Someone was sure," $(\exists x)Sx$.

If we put the four symbolic forms of the equivalences together, there is a neat pattern which emerges. See if you can figure it out for yourself before going on.

QUANTIFIER NEGATION (Q.N.) EQUIVALENCES

$$\sim (\exists x)\phi x \equiv (x) \sim \phi x$$
$$\sim (x)\phi x \equiv (\exists x) \sim \phi x$$
$$\sim (\exists x) \sim \phi x \equiv (x)\phi x$$
$$\sim (x) \sim \phi x \equiv (\exists x)\phi x$$

If we consider a quantified sentence as made up of a quantifier plus a propositional function, then we can negate the quantifier, the function, both, or neither. Notice that in the above equivalences, from one side to another, *everything changes*. If the quantifier is negated on the left, it is unnegated on the right, and vice versa; if the propositional function is negated on one side, it is unnegated on the other; and the *quantifier* changes from universal to existential and from existential to universal.

If this pattern of equivalences seems vaguely familiar to you, there may be a good reason, which is closely connected with DeMorgan's laws. If we think of the truth conditions for the quantifiers we see a *strong analogy between the universal quantifier and conjunction, and between the existential quantifier and disjunction.* A universal statement $(x)Fx$ will be true just in case *each* instance is true, which means Fa is true, Fb is true, Fc is true, and so on. That is, just in case the *conjunction* $Fa \cdot Fb \cdot Fc$ . . . is true. Similarly, the existential statement is true just in case *some instance or other* is true, which means *either* Fa is true, *or* Fb is true, *or* Fc is true . . . etc. That is, just in case $Fa \lor Fb \lor Fc \lor$ . . . is true. Thus, if we negate a universal statement we are doing something like negating a conjunction, and we should come up with the analogue of the disjunction, which is, in fact, the existential proposition. That is, negating $(x)Fx$ is similar to negating $(Fa \cdot Fb \cdot Fc$ . . .), so that $\sim (x)Fx$ is very close in meaning to $\sim (Fa \cdot Fb \cdot Fc$ . . .). This latter form will be equivalent, by a generalization of DeMorgan's rules, to $(\sim Fa \lor \sim Fb \lor \sim Fc \lor$ . . .). And since the disjunction is very like an existential statement, this disjunction will be very close in meaning to the existential with the function negated, that is, $(\exists x) \sim Fx$. The details for the analogous negated existential will be left to you, but keep in mind that just as the negation of a conjunction is always equivalent to a disjunction, and vice versa, so *the negation of a universal proposition will always be equivalent to an existential*, and *the negation of an existential statement will always be equivalent to a universal*.

Given these equivalences between negated universals and existentials, and negated existentials and universals, there will be more than one correct symbolization for negated quantifier sentences, just as there is for "neither" statements and "not both" statements. "Nothing is working out," for instance, can be equally well symbolized as $\sim (\exists x)Wx$ or as $(x) \sim Wx$, and "Not everything is a disaster" can be symbolized either as $\sim (x)Dx$ or as $(\exists x) \sim Dx$. Exercises 3 and 4, at the end of the unit, will give you practice in carrying out these symbolizations and in understanding quantified formulas.

DEFINITIONS

1. A **universal statement** is a statement which asserts that the propositional function is true of everything.
2. An **existential statement** is a statement which asserts that the propositional function is true of at least one thing.

3. A **universal quantifier** is an expression (such as "For every x") which, when prefixed to a propositional function, yields a universal statement.

4. An **existential quantifier** is an expression (such as "For some x") which, when prefixed to a propositional function, yields an existential statement.

5. A **general sentence** is either a universal or existential statement.

6. The **scope of a quantifier** is the first complete formula following the quantifier.

7. A **bound variable** is a variable that falls within the scope of its own quantifier.

8. A **free variable** is a variable that does not fall within the scope of its own quantifier, that is, one that is not bound.

9. The **domain of discourse** of a general sentence is the set of objects over which the bound variables range.

10. The **universal domain** is the set of all existing objects.

EXERCISES

1. For each of the following expressions, indicate whether it is a singular sentence, a propositional function, or a simple existential or universal sentence.

*a. John has three cats.   _Sing. S._   f. Mary is in this class. _Sing. S._

 b. Something happened here.   *g. x took this class and did well.

*c. Kilroy was here.   h. Everything makes me laugh.

 d. x was here.   *i. John did well in this class.

*e. There is a "Mary" in class. _Simp. Ex._ j. Some things make me cry. _Simp. Ex._

2. Identify each of the following sentences as universal or existential, pick out and abbreviate their propositional functions, and symbolize them.

*a. Some things are clear.   f. Anything goes.

 b. Everything has mass.   *g. Something is wrong.

*c. Flying saucers exist.   h. All's right with the world.

 d. There is a Loch Ness monster.   *i. There are evil beings.

*e. Everything has a price.   j. Anything makes John happy.

3. Symbolize the following sentences, using the abbreviations indicated for the propositional functions.

$Lx \equiv x$ is a free lunch; $Ax \equiv x$ is an angel; $Ex \equiv x$ is evil; $Ux \equiv x$ is a unicorn; $Bx \equiv x$ is beautiful; $Gx \equiv x$ is good; $Dx \equiv x$ is a devil; $Jx \equiv x$ is enjoyable; $Cx \equiv x$ is certain; $Ux \equiv x$ is usual.

*a. There is no such thing as a free lunch. $(x)\sim Lx$ or $\sim(\exists x)Lx$

 b. Angels do exist. $(\exists x)Ax$

 _or_

 $\sim(x)\sim Ax$

*c. Not everything is evil.

d. Some things are evil.

*e. There are no unicorns.

f. Everything is beautiful.

*g. There is nothing which is not good. $\sim (\exists x) \sim Gx.$ or $(x)Gx$

h. Devils do not exist.

*i. Not everything is without beauty.

j. There are some things which are not enjoyable. $(\exists x) \sim Jx.$
$\sim (x)Jx.$

*k. Everything is uncertain.

l. There is nothing which is not certain.

*m. Certitude exists.

n. Nothing is certain.

*o. There is no thing which is not unusual. $\sim (\exists x) \cup x$ or $(x) \sim Ux.$

4. Write out the *English* sentence which corresponds to the following symbolizations. Use the abbreviations above.

*a. $\sim (\exists x)Ax$

b. $(x) \sim Ex$

*c. $\sim (x)Bx$

d. $(\exists x) \sim Cx$

*e. $\sim (\exists x)Bx$

f. $(\exists x) \sim Ux$

*g. $(\exists x)Dx$

h. $(x) \sim Jx$

*i. $\sim (x) \sim Ax$

j. $\sim (\exists x) \sim Jx$

UNIT 12
Categorical Propositions

A. INTRODUCTION

As noted in Unit 10, categorical propositions play a rather large role in our logic and in our language. Most of the sentences we use are of subject–predicate form, and of these a substantial portion are categorical propositions, those in which subject and predicate are both classes, or categories. In this unit you will learn about the four types of categorical propositions, the relations between them, how to symbolize them in quantifier logic, and how to diagram them using a special sort of Venn diagram.

Categorical propositions occupy a central position in quantifier logic; they are the simplest of a very large class of propositions which will fit into these categorical forms rather like substitution instances. Most of the more complex propositions you will be learning to symbolize later are just more elaborate versions of the four basic categorical propositions. Thus, it is extremely important that you understand this material thoroughly; in fact, if you don't, it is safe to say that you do not understand quantifier logic. For this reason this will be a rather long unit with many subdivisions. But each section in itself is fairly simple, and if you take it bit by bit you should have an excellent grasp of the basics of quantifier logic by the end of the unit. Most of the rest of quantifier logic will be variations on the same theme.

B. UNIT 12 OBJECTIVES

1. Learn the definition of "categorical proposition" and the four types of categorical propositions.
2. Understand the relationships between sets, properties, and individuals.
3. Using Venn diagrams, be able to diagram all four types of categorical propositions.
4. Know the symbolizations in quantifier logic for the four types of categorical propositions.
5. Learn the negation equivalences that hold between categorical propositions.
6. Be able to derive the categorical negation equivalences from the simple quantifier negation equivalences.
7. Be able to symbolize English sentences that have the form of categorical propositions or their negations.

C. UNIT 12 TOPICS

1. The Four Categorical Propositions

In this unit we discuss the *categorical proposition*, in which both subject and predicate are classes. In the categorical proposition a relationship is stated between the subject class and the predicate class, and because of the nature of this relation, it turns out that there can be only four kinds of categorical sentences. The basic relationship between two classes is that of *inclusion*, or *containment*, and we have both the positive relation, *inclusion*, and the negative relation, *exclusion*. In addition, we can talk about either *total* or *partial* inclusion or exclusion, which gives us four types of relation between classes, which are exemplified by the four types of categorical proposition.

We say that the subject class (S) is *totally included* in the predicate class (P) (or is a *subset of* the predicate class) if every member of S is also a member of P. A sentence asserting such total inclusion is a *universal affirmative* sentence: universal because it says something about *each* member of S, and affirmative because it says something positive. We will refer to such propositions as A *propositions*. An example of an A proposition is "All cats are mammals," which asserts something positive (that it *is* a mammal) about each cat.

We say that S is *partially included* in P if *some* members of S are also members of P; in other words, if there is some *overlap* between the classes S and P. (We may also say that the two classes intersect.) A sentence asserting partial inclusion is a *particular affirmative* sentence, and we will refer to such sentences as *I propositions*. An example of an I proposition is "Some cats are spotted."

The negative categorical propositions are those that assert total or partial *exclusion* of the subject class S from the predicate class P. We say that S is *totally excluded* from P if no member of S is a member of P, that there is *no overlap* between S and P. (We may also say that S and P are *disjoint*.) A sentence asserting total exclusion is a *universal negative*: it is saying something negative, that it is *not* included in P, about *every* member of S. We will refer to such sentences as *E propositions*. An example of an E proposition is "No cats are amphibious."

Finally, the class S is said to be *partially excluded* from P if some members of S fall outside of P. Sentences asserting partial exclusion are *particular negative* sentences; we will refer to them as *O propositions*. An example of an O proposition is "Some cats are not quiet," which asserts that a part of the cat class falls outside of, or is excluded from, the class of quiet things. The four types of categorical sentence are summarized below.

NAME	TYPE OF SENTENCE	RELATION BETWEEN CLASSES	ENGLISH FORM
A	Universal affirmative	Total inclusion	All S are P
I	Particular affirmative	Partial inclusion	Some S are P
E	Universal negative	Total exclusion	No S are P
O	Particular negative	Partial exclusion	Some S are not P

The choice of the letters A, I, E, and O is not arbitrary. A and I are the first two vowels of the Latin "*affirmo*," which means "I affirm," so naturally they represent the affirmative sentences. E and O are the two vowels of the Latin "*nego*," which means "I deny," so it is natural that they should represent the negative propositions. Notice also that the *first* vowel in each case represents the *universal* sentence, while the *second* vowel represents the *particular* sentence. This should help you remember which letter goes with which sentence.

You can generally tell whether an English sentence is of the form A, I, E, or O by the occurrence of words such as "all" and "some," and by the presence or absence of negations. "All squares are rectangles," for example, is clearly an A proposition, because it is *affirmative*, that is, there are no negations involved; and it is *universal*, that is, it asserts something about the entire subject class. The sentence "Some people like to drive trucks" is an I proposition, since it is again affirmative but asserts something only about *some* people, rather than all. Universal propositions will usually be indicated by the use of words such as "all," "every," "anybody," "everything," and so on, while particular sentences will contain words such as "some," or "something." Since particular sentences will be existential, they will also be indicated by words that mention existence, such as "there are." An example of such a sentence would be "There are birds in the trees" (an I proposition).

Words such as "no," "none," or "nothing" will indicate E propositions; they are clearly negative, and the force of "none" is universal. To say "None of the players was injured," for instance, is to say something *negative* (they were *not* injured) about *all* the players; thus the sentence is *universal negative*, an E proposition. One rather tricky phrase is "there are no"; students often want to interpret this as an existential proposition (particular) because it begins with the words "there are." This locution, however, indicates an *E* proposition, which will turn out to be a *universal* sentence rather than a particular. To say "There are no children in the room," for instance, is *not* to say *there are* children; it is equivalent to saying simply "No children are in the room," which is an E proposition. Sentences that contain the phrase "some are not" or "there are some that are not" will be O propositions. "Some of the leaves have not yet fallen," for instance, is an O sentence, since it is *negative* but *particular*, and asserts something about some of the subject class (leaves), but not all.

In some sentences, words such as "all" and "some" are absent, and here you just have to use your common sense about what is meant. If a sign informs you, for instance, that "Children are present" it would be ludicrous to interpret this as a universal proposition; it clearly means that *some* children are present, not all. In other cases, however, such a sentence is best interpreted as universal, as in "Whales are mammals."

Categorical propositions are those that state a relationship between classes. To fully understand their logical form, then, you must be able to identify these classes, represented by the subject and predicate expressions. This often takes a little rephrasing. Some sentences are straightforward; in the sentence "All cats are mammals," for instance, it is clear that the subject class is cats and the predicate class is mammals. But what about the sentence "Some cats have fleas"? Here the subject class is again cats, but it wouldn't make sense to say the predicate *class* was *have fleas. Have fleas* is simply not a class. The class must be taken to be *things that have fleas*, and the sentence could be rephrased as "Some cats are things that have fleas." In general, if the verb of the sentence is something other than a form of "to be," you will have to make this kind of reinterpretation, in which the predicate class is taken to be the *class of things that have the property* mentioned in the predicate of the sentence. To take one more example, in the sentence "Apricot trees grow in my yard," the subject class is "apricot trees," and the predicate class is "things that grow in my yard." In cases where you have words such as "some*body*," "no*body*," or "every*one*," you should interpret the subject to be people—human beings.

At this point you should do Exercise 1 at the end of the unit, to see whether you are able to recognize the various types of categorical proposition and are able to pick out the subject class and predicate class.

2. Individuals, Sets, and Properties

In order to understand exactly what a categorical proposition says, and why it is symbolized the way it is in quantifier logic, it is necessary to understand the relationship between individuals, classes, and properties. Remember that an individual is a single, particular thing, while a set or class is a collection of things. A property of an individual is simply a characteristic, or attribute, of that individual. We say that an individual is *a member of* a set or class; James Carter, for instance, is a member of the class of former U.S. presidents. We represent classes by capital letters—we might use P to represent the class of former U.S. presidents—and we use lowercase letters for individuals, as we have seen in Unit 10. In order to say that an individual is a member of a certain class we place the Greek letter epsilon, ϵ, between the individual letter and the class letter. Thus, we could say that Carter is a member of the class of former U.S. presidents by writing $c \in P$. To say that an individual has a certain property, of course, we use a singular sentence. If Px means "x is a former U.S. president," then we can say that Carter is a former U.S. president by writing Pc.

For logical purposes, saying that an individual has a certain property means exactly the same thing as saying that the individual is a member of a certain class. If we say that Carter is a former U.S. president—that he has the property of being a former U.S. president—this means simply that he is a member of the class of former U.S. presidents. Thus the two expressions Pc and $c \in P$ mean exactly the same, and are interchangeable for logical purposes. In general, $Fx \equiv x \in F$. This identification will help you understand both the Venn diagrams in the next section and the symbolization for categorical propositions in the following sections.

3. Venn Diagrams

We may represent the four types of categorical propositions by a special kind of diagram, the *Venn diagram*, which uses interlocking circles to represent the subject and predicate classes. One word of caution: you may have learned a type of diagramming for *sets* somewhere else, which also went under the name "Venn diagrams." This is something different: here we are diagramming *propositions*, not sets (though sets are involved) and it is especially important that you read the instructions carefully.

In the type of Venn diagrams we will be using, there are *two interlocking circles* which represent the subject class (*S*) and the predicate class (*P*). (The subject class is to the left, and the predicate class is to the right.) The two interlocking circles divide the space into four separate regions, numbered below.

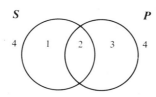

Region 1 represents the class of things that are *S* but not *P*; 2 represents things that are both *S* and *P*; 3 represents things that are *P* but not *S*; and 4 represents the class of things that are neither *S* nor *P*. We will diagram propositions by either *placing an x in a region to indicate that it is nonempty, that there is some object in that class*, or by *shading the region to indicate that it is empty, that there are no members of that part of the class*.

The two interlocking circles in themselves, without shading or x's, say nothing; they are just the shell with which we start. The actual diagrams for the four categorical propositions are given below, with accompanying explanations.

I PROPOSITION

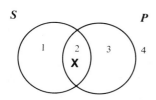

The I proposition says that *there are some S's that are P's*; that the *S* class is *partly included* in the *P* class. In other words, there is some overlap between the two. This means that *there exists* some object that is both *S* and *P*, which is indicated nicely by placing an *x* in the overlap region (region 2), the region that is both *S* and *P*, so that the *x* is in *both* circles.

O PROPOSITION

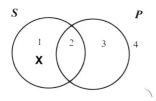

The O proposition says that *there is something which is S but not P*, that is, that the *S* class falls partly outside the *P* class. This can be indicated simply by placing an *x* in the region that is *S* but not *P* (region 1).

E PROPOSITION

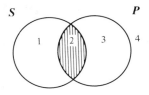

The E proposition says that *no S's are P's*, that there is *no overlap* between the S class and the P class. This means that the two classes have no members in common, that they are disjoint; that the section that is both S and P is empty. Thus, we can diagram it by *shading out the intersecting region* (region 2).

A PROPOSITION

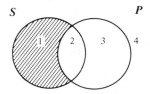

The A proposition says that *all S's are P's*, that whatever is in the S class must also fall into the P class. This means that *there can be no S that is not also a P*, nothing in region 1, which is S but not P. Thus, this region is shaded out.

One very important thing to notice about these diagrams is that the I and the O, the particular propositions, actually state that *there is something* of a certain sort; that is, they are *existential propositions*. The A and the E, by contrast, do *not* allow us to infer that there is anything of a certain kind, not even that there are S's or P's. All they do is to *rule out* certain cases: things that are both S and P in the E proposition, and things that are S but not P in the A. Furthermore, *since the shading indicates nonexistence, the A and the E are really the negations of existential propositions*. They are thus equivalent to universals, which ties in neatly with what you learned in the last unit about quantifiers. Notice in particular that since an x indicates existence in a certain region, and shading indicates nonexistence in that region, *the E proposition is exactly the opposite, or negation of, the I, and the O proposition is the negation of the A*. The I proposition says that *there is* something in region 2, and the E proposition says just the opposite, that there is *nothing* there; similarly, the O proposition says *there is* something in region 1, while the A proposition *denies* this. This coincides with our ordinary understanding: "Some S's are P's," the I proposition, is the opposite of "No S's are P's," the E proposition, and "All S's are P's," the A, is the opposite of "Some S's are not P's," the O. You will be hearing a good deal more about these "negation relations" between categorical propositions in later sections.

If you can identify the form of English sentences and have learned the diagrams for the A, I, E, and O propositions, diagramming English sentences should be relatively easy. First, pick out the subject class and predicate class, and draw two interlocking circles, using the left one for subject and the right one for predicate. Label the two circles, being sure that you have the appropriate class terms for subject and predicate. In the sentence "Some automobiles use diesel fuel," for instance, it would not do to label the predicate circle "diesel fuel"; the proper term is "things that use diesel fuel." Finally, identify the form of the sentence and fill in the diagram accordingly. To illustrate this procedure, we will take one example of each type of sentence.

The sentence "Some apples are green" is an I proposition, and is very easy to diagram. The subject class is apples and the predicate class is things that are green, or green things. The sentence says that there is some *overlap* between the two classes, that some of the apple class is also in the class of green things. This is represented by placing an X in region 2, which indicates then that there is some object which is both an apple and a green thing.

Some apples are green.

Apples   *Green things*

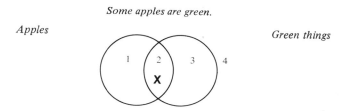

The sentence "All cacti have spines" is an A proposition; the subject is cacti and the predicate is things that have spines. The universal affirmative proposition can sometimes be tricky, but remember that to say *all cacti have spines* is equivalent to saying that *there are no cacti that do not have spines*, so we shade out the portion of the circle which represents cacti without spines, namely, the far left portion.

All cacti have spines.

Cacti   *Things that have spines*

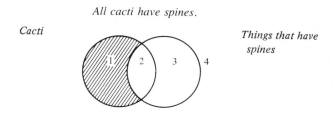

When diagramming negative sentences, you should take as classes things that are specified by *affirmative* phrases rather than negative phrases, and then represent the negation in the sentence by diagramming it as an O or an E. In the sentence "Some people don't like peanuts," for instance, the subject class is clearly people, and the predicate *class* should be taken to be things that *like* peanuts, rather than things that don't like peanuts. The sentence is then an O proposition, since it states that some

of the subject class, people, is *not* included in the predicate class, things
that *do* like peanuts. The sentence would be diagrammed as follows:

Some people don't like peanuts.

People   *Things that like*
                                                                   peanuts

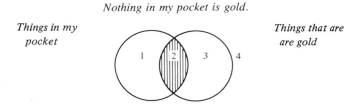

In the sentence "Nothing in my pocket is gold," an E proposition,
the subject class is things in my pocket, and the predicate class is things
that are gold. The sentence would be diagrammed as follows:

Nothing in my pocket is gold.

Things in my   *Things that are*
pocket   *are gold*

Here we shade out the center section because the sentence says that there
is *no overlap* between things in my pocket and gold things. At this point
you should do Exercise 2 at the end of the unit.

4. Symbolizing Categorical Propositions

If you have understood the Venn diagrams for categorical propositions, it should be clear why these sentences are symbolized the way they
are in quantifier logic. In fact, the symbolizations can almost be read right
off the diagrams. The I and the O propositions, for instance, are drawn
in the Venn diagrams with an X, which indicates existence, so they should
be, and are, symbolized with the *existential* quantifier. The A and E sentences are *shaded* in the diagrams, which indicates *non*-existence, so they
could be symbolized by means of a *negated existential* quantifier, or, equivalently, a *universal* quantifier. Let us look at the symbolizations in more
detail.

The I proposition says that there is some *overlap* between the subject
and predicate classes; in other words, *there is an x such that x is both S
and P.* We will use the existential quantifier ($\exists x$) to represent "There is,"
so we have ($\exists x$)(x is both S and P), and the obvious symbolization for the
propositional function is ($Sx \cdot Px$). Thus, the final symbolization for the
I proposition will be ($\exists x$)($Sx \cdot Px$).

The Venn diagrams show pictorially that $x \in S$ and $x \in P$, since x appears inside both circles, which represent the subject class and the predicate class. This means that Sx and Px, that is, $(Sx \cdot Px)$. All we need to do is add the existential quantifier and we have our symbolization. *In general, an I proposition will be symbolized by using an existential quantifier followed by a conjunction of two propositional functions, which indicates that the x is in both classes, that it has both properties.*

The O proposition is very similar to the I, differing only in the addition of a negation sign. The O sentence says that some S's are not P's, that *there are S's that are not P's*, which means that *there is some x that is an S but not a P*. This can be partially symbolized as $(\exists x)(Sx$ but not $Px)$. Clearly, this translates into $(\exists x)(Sx \cdot \sim Px)$, which will be our symbolization for the O proposition. *Again, we have an existential quantifier followed by a conjunction, but here the second conjunct is negated.*

The use of parentheses around the propositional function is essential to the correct symbolization of categorical sentences. Including parentheses extends the scope of the quantifier to the end of the formula, so that each x is bound by the quantifier, falls within its scope. This is important, because in the I proposition $(\exists x)(Sx \cdot Px)$, for instance, we want to be saying that there is some *one* object that is both S and P; the fact that Sx and Px both fall within the scope of the same quantifier is what guarantees that the x that is S is the *same x* that is P. This is because a variable that falls within the scope of a quantifier *refers back* in some sense to that quantifier, much as a pronoun such as "it" refers back to the noun. Thus, if Sx and Px are within the scope of the same quantifier, then both x's have the same referent, so we can say that it is the *same object* that has both the properties S and P.

A variable that does *not* fall within the scope of a quantifier, on the other hand, does not refer back to that quantifier and has no connection with the quantifier. Thus if we omit the parentheses we get a funny kind of expression, and one that is not categorical. If Cx means "x is a cat," and Sx means "x is striped," for instance, the formula $(\exists x)Cx \cdot Sx$ does *not* say that some cats are striped. This is because the x in Sx is not in the scope of the quantifier, is not bound, and so does not refer back to that quantifier, but stands on its own; it is a free variable. The formula would be read "There are cats, and x is striped"; it is a *conjunction* rather than an existential proposition, and has as its right conjunct a propositional function. The scope of a quantifier must include *all* the variables in order for the formula to be a quantifier formula, which means that the propositional function must be enclosed in parentheses.

We will interpret the A sentence, "All cats are mammals," as saying "For any x, if x is a cat, *then* x is a mammal." Why a conditional? Well, we are claiming only that all things *of a certain sort* have a particular property. We are saying, for instance, that all *cats* are mammals, not that everything in the universe is a mammal. What this proposition says, in other words, is that for every x, x is a mammal *provided* x is a cat. Thus we need the conditional. If we use Cx and Mx for the propositional functions, the sentence above would be symbolized as $(x)(Cx \supset Mx)$, that is, for everything, for every x, *if* x is a cat, *then* x is a mammal.

Notice again the importance of parentheses here. If we want to say that all cats are mammals, our symbolization must be $(x)(Cx \supset Mx)$, because we want to be sure for any x, that if x is a cat, then *that very same* x is a mammal. This is guaranteed by making sure that both the Cx and the Mx fall within the scope of the initial quantifier. If we used the formula $(x)Cx \supset Mx$, without the parentheses, this would say something quite different, and rather odd. It would be a conditional, and would be read "If everything is a cat, then x is a mammal," not at all what we wanted to say.

An example of an E proposition is "No dogs are simians." This sentence means to assert about *every* member of the subject class—about *every* dog—that it is not a simian, and so will be symbolized by means of the *universal quantifier*. We will interpret the sentence as saying "For any x, if x is a dog, then x is not a simian." The symbolization, then, would be $(x)(Dx \supset \sim Sx)$. *In general, an E proposition will be symbolized by using a universal quantifier, followed by a conditional whose consequent is negated.*

We can use the Venn diagrams to see why universal statements are symbolized with conditionals. Look at the diagram for the A proposition, for instance. Notice that it does *not* assert that anything exists; there are no x's in the diagram. What it does, rather, is to *rule out* the existence of certain sorts of things: things that are S but not P. Thus region 1 is shaded. What this means is that if there are any S's (things that fall into the S circle), then they will fall into region 2, which is also a part of the P circle, so they will be P's. What the Venn diagram tells us, then, is that for any x, *if* x is an S *then* it is a P, which translates neatly into our quantifier symbolism as $(x)(Sx \supset Px)$. Similarly, the E proposition also fails to assert existence, and only rules out certain possibilities—in this case, things that are both S and P. What this means is that *if* anything falls into the S class (the S circle), then it will fall into region 1 (since region 2 is shaded out); that is, if anything falls into the S class, it falls *outside* the P class. In other words, if anything is an S, then it won't be a P; that is, for any x, if x is an S then it is not a P, which gives us our symbolism for the E proposition: $(x)(Sx \supset \sim Px)$.

It is important to keep in mind that, whereas the existential propositions (the I and the O) *do* assert that something of a certain kind exists (this is why they are called "existential"), the universal propositions, the A and the E, do *not* assert existence. When we say "All S's are P's," this does *not* imply that there are any S's. This sometimes sounds rather odd in English; if someone says, for instance, "All my children are grown up," we would consider it very strange if he didn't mean to imply that he had any children. But very often in a universal proposition we do not mean to imply that anything in the subject class exists. If a sign says "All trespassers will be shot," for instance, $(x)(Tx \supset Sx)$, this certainly does *not* mean that trespassers exist there. The sign is designed, in fact, precisely to insure that there *won't* be any trespassers (one way or another). It is not unusual to have universal propositions in which no existence is asserted, and in logic, universal categorical propositions should *never* be thought of as asserting any kind of existence.

Remember again that the particular categorical propositions, the I and the O, will always be symbolized with *an existential quantifier followed by a conjunction*, and the universal propositions, the A and the E, will always be symbolized with a *universal quantifier followed by a conditional*. A summary of the four categorical propositions, including their Venn diagrams and alternative symbolizations, is given at the end of the unit, for handy reference.

5. Negated Categorical Propositions

You learned in Unit 11 that the negation of a universal proposition is always equivalent to an existential, and that the negation of an existential is always equivalent to a universal. The basic equivalences, put symbolically (where ϕx represents any function, simple or complex) are:

$$\sim (x)\phi x \equiv (\exists x) \sim \phi x \qquad \text{and} \qquad \sim (\exists x)\phi x \equiv (x) \sim \phi x$$

In this section and the next, you will learn the special quantifier negation equivalences which hold between *categorical* propositions, and see how they are derived from the basic equivalences. You should become as familiar with these equivalences as you are now, or should be, with the equivalences in propositional logic; they will be used as rules in the unit on proofs, and a thorough familiarity with them will also be helpful in your symbolizations. For ease of reference we will use C.Q.N. as an abbreviation for the four *categorical* quantifier negation equivalences, and Q.N. for the simple quantifier negation equivalences you learned earlier.

With the C.Q.N. equivalences, just as with the simple quantifier equivalences, *the negation of an existential statement is equivalent to a universal statement, and the negation of a universal is equivalent to an existential*. In particular, the negation of the I means the same as an E, and the negation of the E means the same as an I. Similarly, the negation of the A is equivalent to the O, and vice versa.

These relationships may be clarified by a few examples. Let us look first at a *negated A proposition*: "Not all women want to be housewives."

This could be symbolized simply by placing a tilde in front of the formula for the A proposition (using Wx for "x is a woman" and Hx for "x wants to be a housewife"), which would give us $\sim (x)(Wx \supset Hx)$. This sentence clearly means the same in English as "Some women do not want to be housewives," which is, of course, an O proposition that would be properly symbolized as $(\exists x)(Wx \cdot \sim Hx)$. Thus, the negated A is equivalent to an O.

An example of a *negated O proposition* would be "There is nobody who does not enjoy a good laugh," that is, "There does not exist a person who does not enjoy a good laugh." This could be symbolized as $\sim (\exists x)(Px \cdot \sim Lx)$, where Px means "x is a person" and Lx means "x enjoys a good laugh." But this sentence in English is clearly equivalent to "Everyone enjoys a good laugh," which is an A proposition that is properly symbolized as $(x)(Px \supset Lx)$. Thus, we have the equivalence between the negated O and the A propositions.

An example of a *negated E proposition* would be "It is not the case that no witness was cooperative," which could be symbolized as $\sim (x)(Wx \supset \sim Cx)$, where Wx means "x is a witness" and Cx means "x was cooperative." This is clearly equivalent to saying "Some witnesses were cooperative," an I proposition which would be symbolized as $(\exists x)(Wx \cdot Cx)$. An example of a *negated I proposition* would be "There is nothing (there does not exist anything) that is both cheap and of good quality," which could be symbolized as $\sim (\exists x)(Cx \cdot Qx)$, where Cx means "x is cheap" and Qx means "x is of good quality." This would be the same as saying "Anything that is cheap is not of good quality," an E proposition that would be symbolized as $(x)(Cx \supset \sim Qx)$. Thus the negation of the E is equivalent to the I, and the negation of the I is equivalent to the E.

The symbolic forms of these equivalences are given below; you should learn them thoroughly, and make up examples of your own to convince yourself that they are correct. ϕx has been used as the subject term, and ψx for the predicate term; later we will have more complex substitution instances of these equivalences.

CATEGORICAL QUANTIFIER NEGATION (C.Q.N.) EQUIVALENCES

$\sim$ A	$\sim (x)(\phi x \supset \psi x) \equiv (\exists x)(\phi x \cdot \sim \psi x)$	O
$\sim$ I	$\sim (\exists x)(\phi x \cdot \psi x) \equiv (x)(\phi x \supset \sim \psi x)$	E
$\sim$ E	$\sim (x)(\phi x \supset \sim \psi x) \equiv (\exists x)(\phi x \cdot \psi x)$	I
$\sim$ O	$\sim (\exists x)(\phi x \cdot \sim \psi x) \equiv (x)(\phi x \supset \psi x)$	A

6. Deriving C.Q.N. Rules from Q.N. Rules

It has already been noted that the C.Q.N. equivalences can be derived from the simpler Q.N. equivalences; they are just special instances of the simpler forms. In order to see this you need only two things: the general form of the simple Q.N. equivalences, as given in Unit 11, and your old sentential replacement rules, especially Conditional Exchange, De-Morgan's, and Double Negation. You need also to keep in mind that the replacement rules can be applied to any *part* of a formula, including the propositional function part of a quantified formula.

The negation of the A proposition, for instance, can be transformed into $(\exists x) \sim (Fx \supset Gx)$, and then, using our other equivalences, we can go on to derive the O proposition. The sequence of steps would be as follows:

$\sim$ A	$\sim (x)(Fx \supset Gx)$	
	$(\exists x) \sim (Fx \supset Gx)$	by Q.N.
	$(\exists x) \sim (\sim Fx \lor Gx)$	by C.E.
	$(\exists x)(\sim \sim Fx \cdot \sim Gx)$	by DeM.
O	$(\exists x)(Fx \cdot \sim Gx)$	by D.N.

Exercise 3, at the end of the unit, is to derive all of the other C.Q.N. equivalences. With the above as a model, this should not be difficult. Just keep in mind that ϕx can take any propositional function as a substitution instance.

7. Symbolizing English Categorical Sentences

The first thing you should do in symbolizing is to *identify the form* of the sentence. A number of examples will be given below to assist you in this enterprise. Once you have identified the form, *identify the subject and predicate* of the sentence. The last step is to *symbolize the subject and predicate*. This last step will be quite easy with the simple categorical propositions, since both the subject and predicate will be represented by simple functions, and the abbreviations for the functions will be given. In the next unit, where you will be dealing with complex subjects and predicates, which require complex functions, symbolizing the subject and predicate will take a bit more thought, but here it should be fairly simple. Again, your operating procedure should be (1) identify the form of the sentence, (2) identify the subject and predicate of the sentence, (3) symbolize the subject and predicate. When you put all these things together, you will have your symbolization.

Let us now take one example and use the three-step procedure. Suppose you are to symbolize the sentence "All plants have chlorophyll." The form of this is simply "All *S* are *P*," which indicates an A proposition. Thus, the overall structure will be $(x)(\underline{\hspace{1cm}} \supset \underline{\hspace{1cm}})$. The subject is "plants" and the predicate is "have chlorophyll" so that the subject function is "*x* is a plant" and the predicate function is "*x* has chlorophyll." We might symbolize the former as Px and the latter as Cx. If we put this all together what we get is $(x)(Px \supset Cx)$, which is the correct symbolization. This can be read literally as "For any x, if x is a plant, then x has chlorophyll."

As noted in Section 1, in English the phrases "all," "every," "any," "anything," and the like will indicate universal propositions, but there are many universal propositions that do not contain these "tip-off" words. A sentence like "Whales are mammals" is clearly intended to be universal, as is "The sloth is a lazy animal." Even sentences with an indefinite article are sometimes universal, as in "A moose is not to be trusted." You will simply have to decide in many cases whether the sentence means to be saying something about *all* of the subject class, which would make it a universal proposition, or just *part* of it, which would indicate an existential sentence. The sentence "Children are present," for instance, as we noted above, is almost surely intended to be merely existential; it is hard to imagine a case in which this would mean "All children are present."

Words such as "some," "something," "somebody," or "there are" will indicate an existential proposition, and whether it is an I or an O depends upon whether there is a negation involved. Where there is no negation it will be an I, and if there is a negation it will generally be an O. The sentence "There are some students who like advanced calculus," for instance, is clearly an I proposition, since we have the phrase "there are" and it is a positive sentence. Thus, the overall structure would be $(\exists x)\, (\underline{\hspace{1.2cm}} \cdot \underline{\hspace{1.2cm}})$. The subject is students, for which we could use Sx, and the predicate is things that like advanced calculus, for which we could use Cx. If we plug this into the above form we get our symbolization, $(\exists x)(Sx \cdot Cx)$, which would be read literally "There is some x such that x is a student and x likes advanced calculus." The sentence "Some students do not like advanced calculus" is an O proposition. The "some" indicates an existential quantifier, and there is a negation involved. Thus, the overall form would be $(\exists x)(\underline{\hspace{1.2cm}} \cdot \sim \underline{\hspace{1.2cm}})$. The subject class is again students, and *the predicate class is again things that like advanced calculus*, but here we are saying that some of the subject class is *excluded* from the predicate class, and so the second conjunct is negated. Using the same abbreviations as in the previous case, we could symbolize this sentence as $(\exists x)(Sx \cdot \sim Cx)$, which would be read "There is an x such that x is a student and x does not like advanced calculus."

In general, straightforward A, I, and O propositions will give you little difficulty; it is fairly easy to spot universal and particular affirmative sentences, and it is quite clear that sentences that contain "some are not" are O's. Where you may have some problem is in symbolizing *negated quantifier* statements, ones that contain the phrases "not all" or "none," or their equivalents. Let us take as an example "Not all plants have flowers." Here again you should first identify the *form* of the sentence, which is here indicated by the phrase "not all." This will be a *negated universal*, and so the overall structure will be $\sim (x)(\underline{\hspace{1.2cm}} \supset \underline{\hspace{1.2cm}})$.

The subject is "plants" and the predicate is "have flowers," which could be abbreviated, respectively, as Px and Fx. If we plug in the subject and predicate terms to our basic form we get the proper symbolization, $\sim (x)(Px \supset Fx)$. This would be read literally as "It is not the case that for every x, if x is a plant, then x has flowers." With negated quantifier sentences, it is important to remember that there will be more than one correct symbolization because of the quantifier equivalences. The sentence above, for instance, means the same as "Some plants do not have flowers," and could thus be symbolized as an O proposition: $(\exists x)(Px \cdot \sim Fx)$. In general, negated universal propositions, ones with the form $\sim (x)(\underline{\hspace{1cm}} \supset \underline{\hspace{1cm}})$, or equivalently $(\exists x)(\underline{\hspace{1cm}} \cdot \sim \underline{\hspace{1cm}})$, will be indicated by phrases such as "not all," "not every," "not everything," "not everybody," and so on.

The sentences with which you are likely to have the most difficulty are ones that contain words like "no," "none," "nothing," "nobody," "there aren't any," etc., which indicate an E proposition, and which can be symbolized *either as universals or as negated existentials*. "There aren't any rattlesnakes in the neighborhood," for instance, should be interpreted as an E proposition, since it means just "No rattlesnakes are in the neighborhood"; thus the general form would be $(x)(\underline{\hspace{1cm}} \supset \sim \underline{\hspace{1cm}})$. Here rattlesnakes is the subject class, and things in the neighborhood is the predicate class, and what we are saying is that the two classes are excluded from each other; whatever is in the first, whatever is a rattlesnake, is not in the second, is not in the neighborhood. Thus one symbolization would be $(x)(Rx \supset \sim Nx)$, where Rx means "x is a rattlesnake" and Nx means "x is in the neighborhood." This would be read "For any x, if x is a rattlesnake, then x is not in the neighborhood." Another way of symbolizing the E proposition, however, is as a *negated existential*, which perhaps is closer intuitively in this case to the English sentence, which says *there are no* things of a certain sort—namely, rattlesnakes in the neighborhood. The appropriate negated existential form would be $\sim (\exists x)(Rx \cdot Nx)$, read literally as "There is no x (there does not exist an x) such that x is a rattlesnake and x is in the neighborhood." Similarly, "No cats are dogs" can be correctly symbolized either as the universal $(x)(Cx \supset \sim Dx)$, or as the negated existential $\sim (\exists x)(Cx \cdot Dx)$. It is obviously important that you know your C.Q.N. equivalences here; you must know when one symbolization is equivalent to another.

It is also important to realize that $\sim (\exists x)(Fx \cdot Gx)$ is *not equivalent* to $(\exists x)(Fx \cdot \sim Gx)$, and $\sim (x)(Fx \supset Gx)$ is not equivalent to $(x)(Fx \supset \sim Gx)$. If we let Fx mean "x is a fish" and Gx mean "x lives in the Great Lakes," these nonequivalences are quite clear. $\sim (\exists x)(Fx \cdot Gx)$ means "There are no fish that live in the Great Lakes," which is obviously false (so far). $(\exists x)(Fx \cdot \sim Gx)$, however, means "There are some fish that don't live in the Great Lakes," which is true. Similarly,

$\sim (x)(Fx \supset Gx)$ means "Not all fish live in the Great Lakes" and is true, while $(x)(Fx \supset \sim Gx)$ means "No fish live in the Great Lakes" and is false. Thus the two are not equivalent.

One phrase which is a bit tricky in quantifier logic, as it was in sentential logic, is "only." This word will indicate a *universal* proposition, but in general the subject and predicate will be reversed. If we say "Only citizens are permitted to vote," for instance, we are saying that *anyone who is permitted to vote is a citizen*, that is, for any x, if x is permitted to vote, then x is a citizen, which is symbolized as $(x)(Vx \supset Cx)$. In general, a sentence of the form "*Only F's are G's*" will be symbolized as $(x)(Gx \supset Fx)$, because it means that if anything is a G, then it is an F, since F's are the *only* things that are G's; *whatever* is a G is an F. The symbolization for "Only F's are G's," in other words, is just the *converse* of the symbolization for "All F's are G's." (This is analogous to the relationship between "if" and "only if" in sentential logic.) To take another example, the sentence "Only children throw tantrums" means that *whoever* throws a tantrum is a child, that is, $(x)(Tx \supset Cx)$. It does *not* mean that *all* children throw tantrums, which would be symbolized as $(x)(Cx \supset Tx)$. An even clearer example, perhaps, is "Only those who have tickets will win the lottery." This would be symbolized as $(x)(Wx \supset Bx)$ ("For any x, if x wins the lottery, then x has a ticket.") It could *not* be symbolized as $(x)(Bx \supset Wx)$, which would say "For every x, if x has a ticket, then x will win the lottery." Obviously, these are very different propositions, and the former is true, while the latter is false (by the very nature of a lottery).

Of course, in general, as you remember from your sentential equivalences, $(p \supset q)$ means the same as $(\sim q \supset \sim p)$, so we could symbolize the tantrum sentence in another way, as $(x)(\sim Cx \supset \sim Tx)$, which would be read "For any x, if x is not a child, then x does not throw tantrums," clearly equivalent to the original version. Similarly, we could symbolize the voting sentence as $(x)(\sim Cx \supset \sim Vx)$, which would say "For any x, if x is not a citizen, then x is not permitted to vote." We could also symbolize the lottery sentence as $(x)(\sim Bx \supset \sim Wx)$, which would say "For all x, if x did not have a ticket, then x did not win the lottery." These symbolizations, with negated subject and predicate expressions, may seem more natural to you, and if so, feel free to use this version.

Sentences with double negations can also be tricky, but here you can sometimes rephrase the sentence to eliminate the negations entirely. The sentence "There was nobody who didn't get fed," for instance, is equivalent simply to "Everybody got fed," and would be symbolized as $(x)(Px \supset Fx)$. Another such example would be "Not every job was unfinished." We could symbolize this as a negated A, which would give us $\sim (x)(Jx \supset \sim Fx)$, (where Jx means "x is a job" and Fx means "x was finished"). This means the same as "Some jobs *were* finished," however, and could also be symbolized as a simple I proposition, $(\exists x)(Jx \cdot Fx)$.

One further comment: negated quantifiers often give rise to ambiguity in ordinary English. For instance, in the sentence "All politicians are not ethical," it is not clear whether what is meant is that *no* politicians are ethical—an E proposition—or only that *some* politicians are *not* ethical—an O proposition. There may be no way to resolve the question. This is precisely what is meant by an ambiguous sentence: it can mean either thing.

In summary, for symbolizing categorical propositions, you should identify the *form* of the sentence first, then pick out subject and predicate, and then symbolize subject and predicate. If you can't quite see how it goes, you should *rephrase* the English sentence until you get a sentence that can easily be symbolized. Also, you should get in the habit of *reading back* the symbolized sentence literally, as we have done on many occasions, to see whether you have captured the sense of the sentence. In many cases, this will alert you to the fact that you have done something wrong. And finally, remember that there are alternative ways of symbolizing sentences containing negations; you should learn the C.Q.N. equivalences thoroughly if you have not done so already. The summary sheet for categorical propositions on the next page may help you put it all together. And at this point, finally, you should be ready for Exercises 4, 5, 6, and 7 at the end of the unit.

SUMMARY OF CATEGORICAL PROPOSITIONS

A categorical proposition states the relationship between two classes, or categories, of things. S and P are used here to represent these classes. The universal affirmative proposition (A) says that *all* of the S class is included in the P class. The particular affirmative (I) says that *some* of the S class is included in the P class, that there is some overlap between the two classes. The universal negative (E) says that all the S class is *excluded from the P class*; that there is no overlap, or that the two classes are *disjoint*. The particular negative (O) says that some of the S class is excluded from the P class. In the Venn diagrams below, an x indicates that there is some individual in that part of the class. The shading indicates that there is nothing in that part of the class, that it is empty. The diagrams show very clearly, then, the logical relationships between these propositions; the E proposition is the negation of the I (and vice versa) and the O proposition is the negation of the A (and vice versa).

TYPE	ENGLISH SENTENCE	FORMULA	VENN DIAGRAM	EQUIVALENT FORMULA	ENGLISH	TYPE
A	All S are P	$(x)(Sx \supset Px)$		$\sim (\exists x)(Sx \cdot \sim Px)$	There are no S's which are not P's.	$\sim$ O
I	Some S is P	$(\exists x)(Sx \cdot Px)$		$\sim (x)(Sx \supset \sim Px)$	Not all S's are non-P's.	$\sim$ E
E	No S is P	$(x)(Sx \supset \sim Px,$		$\sim (\exists x)(Sx \cdot Px)$	There is no S which is P.	$\sim$ I
O	Some S is not P	$(\exists x)(Sx \cdot \sim Px)$		$\sim (x)(Sx \supset Px)$	Not all S's are P's.	$\sim$ A

EXERCISES

1. State whether the following categorical propositions are of the form A, I, E, or O. Identify the subject class and the predicate class.

*a. No man is an island. I

 b. Anybody knows that.

*c. Every dog has his day.

 d. There are some people who don't understand Picasso. O

*e. Nobody understands modern art.

 f. Some cats like turkey. E

*g. Some cats don't have tails.

 h. There are burglars coming in the window.

*i. Nobody is calling the police. I

 j. Everyone will be robbed.

*k. Some people don't care.

 l. Anyone who doesn't care is crazy.

*m. No police cars are arriving.

 n. Burglars should be thrown in jail.

*o. Some jails are not pleasant.

2. Use Venn diagrams to diagram the following sentences. Be very explicit about what is the subject term and what is the predicate term.

*a. Some corporation executives are not wealthy. $(\exists x)(C_x \cdot \sim W_x)$

 b. No apple tree can survive a southern summer.

*c. Pine trees are coniferous. $(x)(P_x \cdot C_x)$

 d. There are pine trees in Minnesota.

*e. Hickory nuts are high in fat.

 f. None of John's children knows how to swim.

*g. Some of John's children know how to ski.

 h. There are deciduous trees which do not bear fruit.

*i. Nothing that bears fruit is coniferous.

 j. Any garden is a good investment.

3. Using just Q.N. and your propositional replacement rules, derive the following C.Q.N. equivalences.

*a. $\sim (\exists x)(Fx \cdot \sim Gx) \equiv (x)(Fx \supset Gx)$ $(\sim O \equiv A)$

 b. $\sim (x)(Fx \supset \sim Gx) \equiv (\exists x)(Fx \cdot Gx)$ $(\sim E \equiv I)$

*c. $\sim (\exists x)(Fx \cdot Gx) \equiv (x)(Fx \supset \sim Gx)$ $(\sim I \equiv E)$

4.   Symbolize the following as categorical propositions, picking out appropriate abbreviations for the propositional functions. Remember that there may be more than one correct symbolization, because of the quantifier equivalences.

*a.   All dogs have feelings.

 b.   There are some dogs that don't bite.

*c.   Nothing in the room is beautiful.

 d.   Some things in the room are valuable.

*e.   Not all diamonds are valuable.

 f.   There are valuable things which are not beautiful.

*g.   There is nothing of value in the house.

 h.   There are some old books which are valuable.

*i.   Any resemblance to persons living or dead is purely coincidental.

 j.   Nothing in this consent agreement in any way implies that Hair Transplants, Unlimited has engaged in deceptive advertising.

*k.   Some candidates weren't candid.

 l.   No candidate was candid.

*m.   Candidates are hopeful.

 n.   Only candidates kiss babies.

*o.   There was no one who didn't enjoy the show.

 p.   The only things left in the house were some old tables.

*q.   Only the hardiest will survive a nuclear war.

 r.   Not everyone who didn't show up for the party dislikes the host.

*s.   There was no one who was not unhappy about the decision.

 t.   The only people who were not unhappy about the decision were those who had nothing at stake.

5.   Symbolize the following, using the abbreviations indicated.

$Bx \equiv x$ is beautiful; $Vx \equiv x$ is valuable; $Ax \equiv x$ is appreciated; $Wx \equiv x$ is in the wastebasket; $Dx \equiv x$ should be discarded; $Gx \equiv x$ is a good person; $Ex \equiv x$ is a person who defrauds the elderly; $Hx \equiv x$ gets to heaven; $Rx \equiv x$ is a person here; $Px \equiv x$ passes the exam; $Mx \equiv x$ misses the deadline; $Ox \equiv x$ gossips.

*a.   Anything beautiful is valuable.

 b.   Not all beautiful things are appreciated.

*c.   There are some valuable things in the wastebasket.

 d.   Nothing beautiful should be discarded.

*e.   Only things which should be discarded are in the wastebasket.

 f.   Good people are sometimes not beautiful.

*g.   Nobody who defrauds the elderly is good.

 h.   Only the good will get to heaven.

*i. There is nobody here who will get to heaven.

j. Not everybody here will fail the exam.

*k. Only people who are not here will miss the deadline.

l. Good people do not gossip.

*m. There is no good person who is not appreciated.

n. Not all valuable things should not be discarded.

*o. There are some beautiful things which are not unappreciated.

p. Gossips are unappreciated.

*q. The only people here are ones who will not get to heaven.

r. Not everybody here is unappreciated.

*s. There is nobody here who is not beautiful.

t. The only things in the wastebasket are things which are not beautiful.

6. Symbolize the following, which are either categorical propositions or their negations. Use the following abbreviations.

$Ax = x$ is an angel; $Bx = x$ is beautiful; $Cx = x$ is a charlatan; $Dx = x$ is a devil; $Ex = x$ is evil; $Gx = x$ is good; $Hx = x$ is honest; $Ix = x$ is innocent; $Lx = x$ is lucky; $Px = x$ is a politician; $Sx = x$ is a saint; $Yx = x$ dies young; $Ux = x$ is a unicorn.

*a. Angels are beautiful.

b. Not all charlatans are devils.

*c. Some angels are not saints.

d. The innocent are honest.

*e. Not all politicians are honest.

f. Only the good die young.

*g. Unicorns are all beautiful.

h. No devil is an angel.

*i. Not all devils are dishonest.

j. Not every saint dies young.

*k. The only things that are evil are devils.

l. Only the innocent are angels.

*m. The innocent are not all angels.

n. Not only angels are good.

*o. No devil is not dishonest.

p. Not every politican is not an innocent.

*q. Nothing beautiful is evil.

r. Only angels are not evil.

*s. The only things that are not beautiful are devils.

7.   Write down the English sentences which correspond to the following formulas. Use abbreviations from Exercise 6. (You may restrict the domain to people, where appropriate.)

*a. $\sim (x)(Hx \supset Gx)$

 b. $(\exists x)(Cx \cdot Bx)$

*c. $(x)(Sx \supset \sim Px)$ ~~No~~ saints are ~~politicians~~.

 d. $(\exists x)(Ex \cdot \sim Dx)$

*e. $\sim (\exists x)(Dx \cdot Sx)$

 f. $(x)(Bx \supset Ax)$

*g. $\sim (\exists x)(Bx \cdot \sim Lx)$

 h. $\sim (x)(Px \supset \sim Hx)$

*i. $\sim (\exists x)(Ix \cdot \sim Bx)$

 j. $\sim (x)(\sim Hx \supset \sim Bx)$

Complex Subjects and Predicates

A. INTRODUCTION

If you understood Unit 12, you should have little difficulty with this one; the sentences here are just more elaborate versions of categorical propositions. Instead of simple subjects and predicates, symbolized by simple functions such as Fx, you may have complex subjects and predicates, which will be symbolized using truth-functional compounds of simple functions. We could, of course, use simple functions for *any* subject or predicate; we could, for instance, use Ex for "x is a bright and ambitious corporate executive who is either under 35 or making over $50,000 a year." But this would not give us as much information about the structure of the sentence as we might need, and it is better for logical purposes to use simple functions for "simple" concepts, and then symbolize complex functions by means of truth-functional compounds. This is analogous to what we did in sentential logic, where we used capital letters for simple sentences and then represented the compound sentences by making use of operators. In the example above, for instance, it would be appropriate to use Bx for "x is bright," Ax for "x is ambitious," Cx for "x is a corporate executive," Ux for "x is under 35," and Ox for "x is making over $50,000 a year." Then we would symbolize the entire phrase as $((Bx \cdot Ax) \cdot Cx) \cdot (Ux \lor Ox)$, which tells us *much* more about the logical structure of the sentence. (The more detailed we are about logical structure, the more we will be able to prove about inferences, about what follows from what.)

We might then use the above phrase as a subject in a sentence, for instance, "Any bright and ambitious corporate executive who is either under 35 or is making over $50,000 a year will either be a millionaire or drop dead of a heart attack within the next 15 years." This could be symbolized as $(x)((((Bx \cdot Ax) \cdot Cx) \cdot (Ux \lor Ox)) \supset (Mx \lor Dx))$. Note, however, that although it is quite complex its overall structure is that of an A proposition: it has a universal quantifier followed by a conditional function. It is really just a complicated substitution instance of an A. This is the sort of thing you will be doing in this unit (though hardly any of the sentences will be this complex): learning to symbolize sentences which are *basically* of categorical form, but which have more complex subjects and predicates. You can generally follow the same three-step procedure outlined in the last unit for symbolizing here; the only part that will be different is that actually *symbolizing* the subjects and predicates, once you have picked them out, will be more of a challenge.

When you have complicated structures, especially those involving negation, there will often be very many equivalent formulations. Since these cannot all be included in the answers, you should learn to recognize when one formula is equivalent to another. Section 2 will give you some hints about this.

B. UNIT 13 OBJECTIVES

1. Learn to symbolize sentences which are of categorical form, but which have complex subjects and/or predicates.
2. Learn to recognize equivalent symbolizations of the English sentences.

C. UNIT 13 TOPICS

1. Complex Subjects and Predicates

If our discourse were limited to simple categorical propositions such as "All whales are mammals," or even "All flying saucers are associated with electromagnetic disturbances," communication would be a rather dull business. More important, we would not be able to make the distinctions and qualifications which enable us to convey more complex information about the world, and we would be seriously limited in the kinds of inferences we could make. If your instructor tells you, for instance, that any student who either gets an A on the final or has an A average going into the final will get at least a B in the course and will be eligible to serve as a teaching assistant, you can correctly infer that if you get an A on the final you will get at least a B in the course. However, if we used only simple categorical propositions to symbolize the premise and conclusion we could not prove it formally; we would have to symbolize it as $(x)(Fx \supset Gx)/\therefore (x)(Hx \supset Jx)$, which is pretty clearly not valid. But if we symbolize it in all its complex glory, we get something like this: $(x)((Sx \cdot (Fx \lor Ax)) \supset (Bx \cdot Ex))/\therefore (x)((Sx \cdot Fx) \supset Bx)$, which looks a good bit more promising, since we at least have common factors in the subjects and predicates. (In Unit 15 you will learn how to prove it.) We do in general use more complicated sentences, sentences with complex subjects and predicates, and your job here is to learn to symbolize them.

As noted earlier, you can follow the basic three-step procedure outlined in the last unit, of (1) identifying the form of the sentence, (2) identifying the subject and predicate of the sentence, and (3) symbolizing the subject and predicate. It should not be much more difficult to identify the form of these sentences than of the simple categorical propositions. Look for key words such as "all," "some," and "no," as before. For most sentences it should not be hard to identify the subject and predicate, but do keep in mind that this means subject *with all modifiers*, and the *whole predicate phrase*. Symbolizing the subject and predicate is the only part that will take more analysis, but even here you should not have much trouble if you go about things systematically. There are basically three sorts of complexity you will encounter, and there are standard ways of symbolizing them, which will be outlined below.

First, for complex *predicates* you will almost always symbolize "literally," that is, just *analyze the truth-functional structure of the predicate phrase* and put it down. For instance, in the sentence "Rhinos are ill-tempered and will attack if provoked," the subject phrase is clearly just "rhinos," and the predicate phrase is the conjunction "*are ill-tempered and will attack if provoked.*" The first conjunct could be symbolized as Ix (where Ix means "x is ill-tempered"). The second conjunct is in the form of a *conditional*, with "provoked" as antecedent and "will attack" as consequent. If we let Px mean "x is provoked" and Ax mean "x will attack," we can symbolize the second conjunct as $(Px \supset Ax)$ and the whole predicate as $(Ix \cdot (Px \supset Ax))$. To finish, the sentence is clearly an A proposition, and if we let Rx mean "x is a rhino," the complete symbolization would be $(x)(Rx \supset (Ix \cdot (Px \supset Ax)))$, which would be read, "For any x, if x is a rhino, then x is ill-tempered and will attack if provoked." To take a slightly more complex example, in the sentence "Elephants are not ill-tempered and will not attack unless they are provoked or frightened," "elephants" is the subject phrase (for which we can use Ex), and the predicate phrase is "*are not ill-tempered and will not attack unless provoked or frightened.*" Here the first conjunct of the predicate phrase can be symbolized as $\sim Ix$, and the second conjunct is again a conditional, but slightly more involved. It says that x will *not* attack if it is *not either* provoked or frightened. Here the antecedent is the negation of a disjunction, $\sim (Px \lor Fx)$, and the consequent is $\sim Ax$. If we put this all together, the predicate comes out as $(\sim Ix \cdot (\sim (Px \lor Fx) \supset \sim Ax))$, and the sentence as a whole can be symbolized as $(x)(Ex \supset (\sim Ix \cdot (\sim (Px \lor Fx) \supset \sim Ax)))$. This formula could be read literally as "For any x, if x is an elephant, then x is not ill-tempered, and if x is neither provoked nor frightened then x will not attack."

There are two sorts of complex subjects, *subjects with modifiers* and *compound subjects. A sentence has a compound subject if it is talking about two different classes*, as in the sentence "Cats and dogs make good pets." In such sentences you must *not* symbolize the subject with a *conjunction*, even though the word "and" appears. If we symbolized the A proposition above as $(x)((Cx \cdot Dx) \supset Px)$, for instance, it would say that if anything is *both* a cat and a dog, then it makes a good pet, which may be true, but which is obviously not the intent of the sentence. For *universal* sentences, you should interpret the subject as a *disjunction* rather than a conjunction. What the sentence above means is that anything that is *either* a cat or a dog will make a good pet, so it should be symbolized as $(x)((Cx \lor Dx) \supset Px)$. A universal sentence may, of course, have more than two classes in the subject, as in "All cats, dogs, birds, snakes, turtles, and alligators make good pets." (That the sentence is false is irrelevant for our purposes.) This could be symbolized as $(x)((Cx \lor Dx \lor Bx \lor Sx \lor Tx \lor Ax) \supset Px)$. Again, in universal sentences you should use a *disjunction* for compound subjects. The reason the English sentence may contain an "and" rather than an "or" is that such a sentence is equivalent to the conjunction of two (or more) universal sentences. "Cats and dogs make good pets," for instance, can also be symbolized as $(x)(Cx \supset Px) \cdot (x)(Dx \supset Px)$, which is equivalent to $(x)((Cx \lor Dx) \supset Px)$. This is analogous to the propositional tautology $((p \supset r) \cdot (q \supset r)) \equiv ((p \lor q) \supset r)$.

When symbolizing *existential* sentences with compound subjects we do *not* use a disjunction, but rather, *a conjunction of existential sentences*. To say "Some cats and dogs are pedigreed," for instance, means "Some cats are pedigreed and some dogs are pedigreed," and so would be symbolized as $(\exists x)(Cx \cdot Px) \cdot (\exists x)(Dx \cdot Px)$. We would *not* use $(\exists x)((Cx \lor Dx) \cdot Px)$, because this would say *only* that there was some x that was either a cat or dog, and was pedigreed. It would not imply that there are both pedigreed cats and pedigreed dogs, which is what the sentence says. We could also not use $(\exists x)((Cx \cdot Dx) \cdot Px)$, because this would say that there is something which is *both* a cat and a dog, and is pedigreed. In summary, in *universal* propositions with compound subjects (such as "cats and dogs"), symbolize with a *disjunctive subject,* and in *existential* propositions, use a *conjunction of existential sentences.*

Subjects containing *modifiers* should be symbolized with the modifiers *conjoined* to the subject proper. For example, in the sentence "Some white cats with blue eyes are part Siamese," the subject is "white cats with blue eyes," and if we let Cx be "x is a cat," Wx be "x is white," Bx be "x has blue eyes," we can symbolize the subject as $Cx \cdot Wx \cdot Bx$. Then the whole sentence would be symbolized as $(\exists x)((Cx \cdot Wx \cdot Bx) \cdot Sx)$. There are various ways of modifying the subject in English. Sometimes the adjectives are simply stated, as in "white cat," sometimes prepositional phrases are used, as in "cats with blue eyes," and sometimes phrases such as "that" or "which" are used. For instance, "Cats *that don't like humans* have probably been mistreated" should be symbolized as $(x)((Cx \cdot \sim Hx) \supset Mx)$, where Hx means "x likes humans" and Mx means "x has probably been mistreated." Of course, the modifying phrase itself may have a compound structure, as in "Dogs *that will bite if provoked* are not good housepets." This sentence could be symbolized as $(x)((Dx \cdot (Px \supset Bx)) \supset \sim Gx)$, which would be read literally as "For any x, if x is a dog that will bite if provoked, then x is not a good housepet." No matter what the structure of the modifying phrase, however, it should be *conjoined* to the subject.

An example that combines all sorts of complexity is "Plum trees and apple trees that have not been either pruned or sprayed will not bear well and will attract bugs." We will use the following abbreviations: $Px \equiv x$ is a plum tree; $Ax \equiv x$ is an apple tree; $Ux \equiv x$ has been pruned; $Sx \equiv x$ has been sprayed; $Wx \equiv x$ will bear well; and $Bx \equiv x$ will attract bugs. We can then symbolize the sentence as $(x)(((Px \lor Ax) \cdot \sim (Ux \lor Sx)) \supset (\sim Wx \cdot Bx))$, which would be read literally as "For any x, if x is either a plum tree or an apple tree, and has not been either pruned or sprayed, then x will not bear well and will attract bugs."

Notice that despite the occurrence of various operators, the rule of thumb for categorical propositions still holds: *the major operator following a universal quantifier is a conditional and the major operator following the existential quantifier is a conjunction.* This is true whether the quantifier is negated or unnegated.

One word of caution: some predicates in English cannot be construed as applying to individuals, and hence do not fit our quantifier pattern. For instance, the sentence "Gorillas are becoming extinct" cannot be symbolized as $(x)(Gx \supset Bx)$ (where $Bx \equiv x$ is becoming extinct), because this would say that for any (individual) x, if x is a gorilla, then x is becoming extinct. This just doesn't make sense, however, since it is only *species* that become extinct, and not individual animals. The proper analysis of this sentence would take us into a more complicated logic, so we will ignore it, having given the warning.

Another word of caution: sometimes a modifying phrase seems to be a conjunction, but really isn't. If we say "There is a large mouse in here" we ought not to translate this as $(\exists x)((Mx \cdot Lx) \cdot Hx)$, just as we would not translate "There is a small elephant outside" as $(\exists x)((Ex \cdot Sx) \cdot Ox)$. To be a small elephant means to be small *for* an elephant rather than small absolutely, and to be a big mouse is to be big *for* a mouse rather than just big. Otherwise, we would probably have to conclude that the animal outdoors was smaller than the one indoors, which would be absurd. What we need to do in cases like this is to use Mx to mean "x is a *large mouse*," and Ex to mean "x is a *small elephant*."

Now that you know the standard interpretations for complex subjects and predicates, symbolizing most complex sentences should not be difficult, especially if you follow the three-step procedure mentioned in the last unit. First, *identify the form* of the sentence; second, *identify the subject class and predicate class*; and third, *symbolize subject and predicate*. Let us apply this method to the sentence "Bears with cubs are dangerous if approached." Here the sentence is clearly universal, and it is affirmative; thus it is an A proposition, so that the overall form will be $(x)(\underline{\quad} \supset \underline{\quad})$. The subject class is bears with cubs, and the predicate class is things that are dangerous if approached. To symbolize subject and predicate we can use the following abbreviations: $Bx \equiv x$ is a bear; $Cx \equiv x$ has cubs; $Dx \equiv x$ is dangerous; $Ax \equiv x$ is approached. To symbolize the subject we can use the function "x is a bear and x has cubs," which would be abbreviated as $(Bx \cdot Cx)$. The predicate function would be "x is dangerous if approached," which we could symbolize as $(Ax \supset Dx)$. We now simply plug these functions into the appropriate slots in the sentence form to get our symbolization: $(x)((Bx \cdot Cx) \supset (Ax \supset Dx))$. This would be read literally as "For any x, if x is a bear and x has cubs, then if x is approached x is dangerous."

A fairly easy existential proposition is "Some teachers without second jobs are happy but not prosperous." Here the subject is teachers without second jobs, and the predicate is things that are happy but not prosperous. We will use the following abbreviations: $Tx \equiv x$ is a teacher; $Sx \equiv x$ has a second job; $Hx \equiv x$ is happy; $Px \equiv x$ is prosperous. To symbolize the subject function we would use $(Tx \cdot \sim Sx)$, and for the predicate function we have $(Hx \cdot \sim Px)$. The entire symbolization would then be $(\exists x)((Tx \cdot \sim Sx) \cdot (Hx \cdot \sim Px))$, which would be read "There is some x such that x is a teacher and x does not have a second job, and x is happy but not prosperous."

We will discuss negated quantifier sentences in the next section, since these generally have a number of alternative symbolizations.

2. Equivalent Symbolizations

Quantifier sentences with complex subjects and predicates can often be symbolized in a number of different, equivalent ways, both because of our quantifier equivalence rules and because of our sentential replacement rules. Any phrase of the form "neither Fx nor Gx," for instance, can be symbolized either as $\sim (Fx \lor Gx)$ or as $(\sim Fx \cdot \sim Gx)$; "Fx only if Gx" may become $(Fx \supset Gx)$ or $(\sim Gx \supset \sim Fx)$; and any negated quantifier statement will have the two forms exhibited in the quantifier negation equivalences. To complicate matters even further, in many cases a sentence may be rephrased in such a way that subject and predicate adjectives are redistributed. In the sentence "Bears with cubs are dangerous," for instance, we may take the subject as bears with cubs and the predicate as dangerous things, or we may take the subject as bears and the predicate as things that are dangerous if they have cubs, rephrasing the sentence as "Bears are dangerous if they have cubs." Sometimes many of these alternatives will occur in the same sentence, leading to a great many different, but equivalent, symbolizations.

In the A proposition "Bears with cubs are dangerous *only* if approached," for instance, we can take the subject as "bears with cubs" and the predicate as "are dangerous only if approached." We might interpret the predicate as saying that *if* they are dangerous they must have been approached ($Dx \supset Ax$), or, equivalently, that if they are *not* approached then they are not dangerous ($\sim Ax \supset \sim Dx$). Thus the full symbolization could be either $(x)((Bx \cdot Cx) \supset (Dx \supset Ax))$ or $(x)((Bx \cdot Cx) \supset (\sim Ax \supset \sim Dx))$. In addition, we might interpret the subject as "bears" and read the predicate as "if x has cubs, then x is dangerous only if approached." This would give us a symbolization that looks like $(x)(Bx \supset (Cx \supset (Dx \supset Ax)))$, or like $(x)(Bx \supset (Cx \supset (\sim Ax \supset \sim Dx)))$. Notice that the latter two are equivalent to the former two by the propositional rule of Exportation, while the first and second, and the third and fourth, are equivalent by the rule of Contraposition. So we have four different symbolizations almost without having batted an eye. When we add negated quantifiers the situation gets even more interesting.

Suppose we have the E proposition "No English sheepdog is vicious unless it is either mistreated or ill." Here the subject is "English sheepdog" and the predicate (including the negative) is "is not vicious unless it is either mistreated or ill." We could symbolize this, then, as $(x)(Ex \supset (\sim (Mx \lor Ix) \supset \sim Vx))$, which would be read "For any x, if x is an English sheepdog, then x is not vicious if x is neither mistreated nor ill." Equivalently, we could interpret this sentence as saying that if an English sheepdog is vicious, then it must be mistreated or ill, that is, $(x)(Ex \supset (Vx \supset (Mx \lor Ix)))$. (This is equivalent to the first form by Contraposition.) Since it is a "no" statement, we can *also* interpret it as a negated existential, in which case we might rephrase the English as "There is no English sheepdog which is vicious that is not mistreated or ill," that is, $\sim (\exists x)(Ex \cdot Vx \cdot \sim (Mx \lor Ix))$. Again taking the sentence as universal, we could rephrase it as "Any vicious English sheepdog is mistreated or ill," where the subject is vicious English sheepdogs. This would be symbolized: $(x)((Ex \cdot Vx) \supset (Mx \lor Ix))$. Or, to turn it around, we might say "No English sheepdog which is not mistreated or ill is vicious," $(x)((Ex \cdot \sim (Mx \lor Ix)) \supset \sim Vx)$. And if this is not enough to make you feel mistreated, ill or even vicious, there is more. We could symbolize each of the forms which has the negated disjunction $\sim (Mx \lor Ix)$ as $(\sim Mx \cdot \sim Ix)$ instead. This would give us *eight* different ways of symbolizing the sentence, all equivalent and all equally correct, and you might well be able to come up with still others.

It is important that you be able to recognize these equivalences, both of the symbolic form and of the English sentences. If you do not know when one *formula* is equivalent to another, you may think that your answer is wrong just because it differs from the one in the answer section. It may be wrong, of course, but it may also be a correct, equivalent version. It is impossible to list *all* the correct answers, so in many cases you will have to figure out for yourself whether your answer is right. One way to do this is to apply the replacement rules, and see whether one form can be transformed into the other.

Most of the equivalences will involve Quantifier Negation, C.Q.N., Exportation, Contraposition, DeMorgan's, and Double Negation. If you are thoroughly familiar with these rules, you should have little trouble moving back and forth between different equivalent formulations.

In more complex sentences it may not be so easy to pick out subject and predicate, and in such cases one thing to do is *rephrase* the English sentence until you get an equivalent version that is easy to symbolize. In the sentence "No bear with cubs will fail to protect them, unless she is ill," for instance, it is fairly clear that the subject can be interpreted as bears with cubs, but it is not clear how to read the predicate. However, we may rephrase the sentence as "Any bear with cubs will protect them unless she is ill," and this should not be difficult to symbolize. We will use Bx and Cx as before, Ix for "x is ill," and Px for "x protects her cubs"; we can then symbolize the sentence as $(x)((Bx \cdot Cx) \supset (\sim Ix \supset Px))$, which would be read "For any x, if x is a bear with cubs, then if x is not ill x will protect her cubs," which is an appropriate rendering of the original sentence. Note that the sentence above, as we have symbolized it, is in the form of an A; an alternative formulation, then, would be as a negated O, of the form $\sim (\exists x)(\underline{\quad} \cdot \sim \underline{\quad})$. Plugging in the subject and predicate functions, we get $\sim (\exists x)((Bx \cdot Cx) \cdot \sim (\sim Ix \supset Px))$. This is turn is equivalent, by C.E., DeM., and D.N., to $\sim (\exists x)((Bx \cdot Cx) \cdot (\sim Ix \cdot \sim Px))$, which would be read "There does not exist an x such that x is a bear with cubs and x is not ill and x does not protect her cubs," which does mean the same as the other versions. Several other possible symbolizations are given below; you should read them out in their English versions and verify that they mean the same as the original. You should also try to derive them from the original by applying the appropriate replacement rules.

$$(x)((Bx \cdot Cx) \supset (\sim Px \supset Ix))$$
$$(x)((Bx \cdot Cx \cdot \sim Ix) \supset Px)$$
$$(x)(Bx \supset (Cx \supset (\sim Ix \supset Px)))$$
$$(x)((Bx \cdot Cx) \supset (Px \vee Ix))$$

You will, of course, have to be very adept at applying the C.Q.N. rules to more complicated cases. Keep in mind that the equivalence $\sim (x)(\phi x \supset \psi x) \equiv (\exists x)(\phi x \cdot \sim \psi x)$ tells you that, for *any* negated universal sentence, no matter what the subject and predicate phrases, it will be equivalent to an existential sentence with the original subject phrase conjoined with the *negation* of the predicate phrase. Exercise 1, at the end of the unit, will give you practice in making these transformations. Go back and review the C.Q.N. rules, given in Unit 12, if you have forgotten them.

Let us take an example of a negated universal sentence and analyze it in detail. In the sentence "Not every diligent student who doesn't party gets both good grades and a good job" the subject class is diligent students who don't party, and the predicate class is things that get both good grades and a good job. If we use Dx, Sx, Px, Gx, and Jx in the obvious way, we can symbolize the subject as $(Dx \cdot Sx \cdot \sim Px)$ and the predicate as $(Gx \cdot Jx)$. Since the form is negated universal, the skeleton of the formula will be $\sim (x)(\underline{\quad} \supset \underline{\quad})$, and if we then plug in subject and predicate phrases, we get as our symbolization $\sim (x)((Dx \cdot Sx \cdot \sim Px) \supset (Gx \cdot Jx))$. This would be read literally as "It is not the case that for every x, if x is a diligent student who doesn't party, then x gets both good grades and a good job."

The equivalent existential formula is $(\exists x)((Dx \cdot Sx \cdot \sim Px) \cdot \sim (Gx \cdot Jx))$, which says that there are some diligent students who don't party who will not get both good grades and a good job. This clearly means the same thing as the original sentence.

One final word about E propositions. Sentences such as "No wild raccoon will bite unless it is rabid or afraid" are tricky because of the "unless." It is best to symbolize these as *universals* rather than negated existentials, and you should first write out the skeleton and then fill in the subject phrase. For the sentence above this will give $(x)((Wx \cdot Rx) \supset \underline{\quad})$. At this point you want to rephrase the English sentence according to the partial symbolization, and you will need to ask yourself what the predicate should be. Here we have "For any x, if x is a wild raccoon, then . . ." Then what? Looking at the original sentence again, we see that the predicate phrase should be "x will not bite ($\sim Bx$) unless x is rabid (Dx) or afraid (Ax)," which can be symbolized as $\sim (Dx \vee Ax) \supset \sim Bx$. The complete symbolization, then, is $(x)((Wx \cdot Rx) \supset (\sim (Dx \vee Ax) \supset \sim Bx))$, which is read "For any x, if x is a wild raccoon, then if x is neither rabid nor afraid, then x will not bite." An equivalent English sentence would be "Wild raccoons will bite only if they are rabid or afraid," and this would be symbolized by the equivalent formula $(x)((Wx \cdot Rx) \supset (Bx \supset (Dx \lor Ax)))$.

A schematic rendering of these tricky sentences would be "No S is a P, unless Q," and the schematic symbolization would be $(x)(Sx \supset (\sim Qx \supset \sim Px))$. The negative existential version is "There are no S's that are P's but not Q's," which can be symbolized as $\sim (\exists x)(Sx \cdot Px \cdot \sim Qx)$, which is equivalent to the first version.

EXERCISES

1.   Use the appropriate form of C.Q.N. once on the following formulas. Then apply some of your propositional equivalence rules to get other forms. This is to give you some familiarity with the various equivalent symbolizations which complex formulas may have.

*a. $\sim (x)(Fx \supset (Gx \lor Hx))$

 b. $\sim (\exists x)((Fx \cdot Gx) \cdot (Hx \supset Ix))$

*c. $\sim (x)((Fx \cdot \sim Gx) \supset (Hx \cdot \sim Ix))$

 d. $\sim (\exists x)(Fx \cdot ((Hx \lor Ix) \supset (Px \lor Qx)))$

*e. $\sim (x)((Fx \cdot (Gx \lor Hx)) \supset ((Px \cdot Qx) \supset \sim Rx))$

2.   Symbolize the following, using the abbreviations indicated. Remember when you are checking your answers that there will be more than one correct form.

$Px = x$ is a person; $Cx = x$ ate chicken; $Sx = x$ got sick; $Ix = x$ ate rice; $Hx = x$ ate ham salad; $Ox = x$ ate potatoes; $Ex = x$ ate bean salad; $Gx = x$ is good; $Bx = x$ is beautiful; $Wx = x$ will get to heaven; $Rx = x$ gets rich; $Ax = x$ is appreciated; $Fx = x$ is famous; $Yx = x$ is happy.

 *a.   Some people who ate chicken got sick.

  b.   Not everyone who ate rice got sick.

 *c.   Everyone who got sick ate either chicken or ham salad.

  d.   No one who ate ham salad ate chicken.

 *e.   Not everyone who ate chicken and ham salad ate both potatoes and bean salad.

  f.   Some beautiful people are not good and will not get to heaven.

 *g.   No beautiful people who are not good will get to heaven.

  h.   Anyone who is either beautiful or good will either get rich or get to heaven.

 *i.   Some good people are appreciated even though they are not either rich or famous.

  j.   No one will be appreciated unless they are either beautiful or rich.

 *k.   No one who is unappreciated will be happy, unless they are rich.

  l.   Only people who are beautiful, rich, or famous will be appreciated.

*m.   Not only the rich or famous will get to heaven.

  n.   Not everyone who is beautiful and rich is both famous and appreciated.

*o.   No one who is not either beautiful or rich will be either happy or appreciated, but they will get to heaven.

3.   Write down the English sentences which correspond to the following formulas. Use the abbreviations from Exercise 2.

*a. $\sim (\exists x)(Px \cdot Hx \cdot Sx)$

  b. $\sim (x)((Px \cdot Bx \cdot Gx) \supset (Rx \lor Ax))$

*c. $\sim (x)((Px \cdot \sim Ax \cdot \sim Rx) \supset \sim Yx)$

  d. $(x)((Px \cdot ((Rx \cdot Fx) \lor (Ax \cdot Yx))) \supset (Gx \lor Bx))$

*e. $\sim (\exists x)((Px \cdot (Rx \lor Fx)) \cdot (Bx \cdot \sim Gx) \cdot Wx$

4.   Symbolize the following, using the abbreviations below.

$Px \equiv x$ is a painting; $Jx \equiv x$ is by Joe; $Bx \equiv x$ is beautiful $Vx \equiv x$ is valuable; $Gx \equiv x$ is framed in gold; $Dx \equiv x$ is studded with diamonds; $Sx \equiv x$ is a sculpture; $Hx \equiv x$ is by Harvey; $Ux \equiv x$ contains unicorns; $Ax \equiv x$ is admired by the critics; $Lx \equiv x$ is loved by the public; $Cx \equiv x$ is comprehensible; $Mx \equiv x$ contains monsters.

*a.   Some paintings by Joe are beautiful but not valuable.

  b.   No paintings by Joe are valuable unless they are framed in gold.

*c.   All of Joe's paintings are valuable if they are framed in gold and studded with diamonds.

  d.   Not all paintings and sculptures by Joe are both beautiful and valuable.

*e.   No painting by Joe or Harvey is either beautiful or valuable unless it contains unicorns.

  f.   Not all of Harvey's paintings that contain unicorns are admired by the critics and loved by the public.

*g.   Some of Joe's paintings that contain unicorns are incomprehensible and are admired by the critics if and only if they are not loved by the public.

  h.   Any painting by Harvey or Joe that does not contain unicorns or monsters is loved by the public only if it is valuable but incomprehensible.

*i.   No painting by Harvey or Joe is comprehensible unless it is either beautiful and contains unicorns or is valuable and framed in gold.

  j.   Paintings and sculptures by Harvey and Joe that are beautiful only if they contain unicorns and valuable only if they are studded with diamonds are not admired by the critics unless they are not loved by the public.

5.   Symbolize the following, using the abbreviations indicated.

$Bx \equiv x$ is a businessman; $Cx \equiv x$ consults a lawyer; $Lx \equiv x$ has broken a law; $Mx \equiv x$ is a millionaire; $Tx \equiv x$ pays taxes; $Px \equiv x$ is a politician; $Ex \equiv x$ is reelected; $Sx \equiv x$ should have his head examined; $Ax \equiv x$ abuses his powers; $Rx \equiv x$ is respected; $Ix \equiv x$ is impeached; $Ux \equiv x$ is sued; $Vx \equiv x$ votes his conscience.

*a.   Not every businessman who consults a lawyer has broken a law.

b. Some businessmen are millionaires but pay no taxes.

*c. Politicians who pay no taxes will not be reelected.

d. Every businessman who pays taxes but doesn't consult a lawyer should have his head examined.

*e. No politician who has broken the law and abuses his powers will be either respected or reelected.

f. Any politician will be impeached if he either has broken the law or abuses his powers.

*g. No politician will be impeached as long as he pays taxes and does not abuse his powers.

h. Some millionaire businessmen consult a lawyer only if they either have broken the law or are sued.

*i. Any millionaire politician will be reelected as long as he pays taxes, hasn't broken the law, and isn't sued.

j. Some politicians are not reelected even though they are respected and are neither sued nor impeached.

6.   Give at least three different equivalent forms of the following sentences. Be able to say in English, more or less literally, what the formulas say. Use the same abbreviations as in Exercise 5.

*a. No politician will be reelected unless he is respected and votes his conscience.

b. A businessman will consult a lawyer only if he is a millionaire or has broken a law.

*c. No politician will be either impeached or sued as long as he pays his taxes and has not broken the law.

d. Not every politician who has broken the law and pays no taxes is sued or impeached.

*e. No politician or businessman consults a lawyer unless he has either broken a law, is a millionaire, or has abused his powers and has been sued.

Quantifier Form and Truth-Functional Compounds of Quantifier Statements

A. INTRODUCTION

So far, you have been operating without any explicit definition of what a quantifier statement is; this has caused no problems up to this point, but once we get to the next unit, on proofs, it will be extremely important that you be very clear on what is and what is not a quantifier statement. The special rules which we will introduce there for quantifier proofs will work *only* on formulas of quantifier form (just as Hypothetical Syllogism works only on statements in the form of conditionals); thus it is essential that you be able to recognize these forms. So what is the problem, you may wonder. Isn't a quantifier statement just one that contains a quantifier? By no means; quantifier statements can occur as *subformulas* of truth-functional compounds, just as conjunctions can occur as subformulas of disjunctions. We need to distinguish between a formula such as $(\exists x)Fx \cdot (\exists x)Gx$, which is a conjunction, and $(\exists x)(Fx \cdot Gx)$, which is a quantifier statement. (If you can't see the difference, let Fx mean "x is a dog" and Gx mean "x is a cat." Then the former statement says something true: that there are dogs and there are cats. The latter statement, however, says something false, or even absurd, that there is something that is both a dog and a cat.)

What, then, is a quantifier statement? This is the first thing you will learn in this unit. We will then go on to talk about the difference in form and in meaning between quantifier formulas and truth-functional compounds with quantifier formulas as parts, and finally, you will learn to symbolize both sorts of formulas.

B. UNIT 14 OBJECTIVES

1. Be able to state the definition of a quantifier formula.
2. Be able to distinguish between quantifier statements and truth-functional compounds of quantifier statements, and be able to identify the overall form of a statement.
3. Be able, given a variety of such statements, to symbolize quantifier statements and their truth-functional compounds.

C. UNIT 14 TOPICS

1. Quantifier Form

As noted in the introduction, not every formula that contains a quantifier is of quantifier form, just as not every formula that contains the disjunction operator is a disjunction. What, then, should be the definition of a quantifier statement? Intuitively, a quantifier statement is one which says either that *something* has a certain property (whether simple or complex) or that *everything* has a certain property; that is, either it says there is *something* such that ϕx, or that for *everything* ϕx. This seems to indicate that a quantifier statement must begin with a quantifier, and this is correct. *The first condition for a formula to be of quantifier form is that it begin with a quantifier.*

But this is not sufficient; not every statement that begins with a quantifier is actually of quantifier form. $(x)Fx \lor (x)Gx$, for instance, though it begins with a quantifier, is not a quantifier statement but a disjunction. What is the difference between this and the genuine quantifier formula $(x)(Fx \lor Gx)$? To answer this, we need to dust off the notion of the *scope* of a quantifier which we introduced earlier. By definition, *the scope of a quantifier is the first complete formula to follow it*, just as the scope of a negation is the first complete formula following the tilde. In $(x)Fx \lor (x)Gx$, for instance, the first formula following the quantifier is just Fx, so this is as far as the scope reaches. In $(x)(Fx \lor Gx)$, however, the first complete formula is $(Fx \lor Gx)$, since $(Fx$, which includes the left parenthesis, is not a formula at all. We can now state the second condition for a formula to be of quantifier form: *the scope of the initial quantifier must extend to the end of the formula.* Notice that we have said the *initial* quantifier. This is important, since in many, if not most, cases, there will be *some* quantifier whose scope extends to the end of the sentence. In $(x)Fx \lor (x)Gx$, for instance, the scope of the *second* quantifier extends to the end of the sentence, but this doesn't count. To be a quantifier statement rather than, for example, a disjunction, the scope of the *first* quantifier must extend to the end of the formula.

We can now state the definition of quantifier form: *a formula is a quantifier statement, or of quantifier form, if and only if (1) it begins with a quantifier, and (2) the scope of the initial quantifier extends to the end of the formula.* We will use $(x)\phi x$ and $(\exists x)\phi x$ to represent quantifier formulas, where it is understood that, if ϕx is compound, it must be enclosed in parentheses so that the scope of the quantifier extends to the end of the formula.

2. Truth-Functional Compounds and Quantifier Form

As noted in the introduction, a statement that is not itself of quantifier form may contain quantifier statements as parts. Thus $(x)Fx \lor (x)Gx$ is a disjunction, with both disjuncts being universal statements. It is very important to recognize the difference, both in form and meaning, between formulas that are quantifier statements and those that are truth-functional compounds of quantifier statements. Some examples may help make this distinction clear. If we use Mx for "x is male" and Fx for "x is female," and restrict our domain of discourse to mammals, then the universal quantifier formula $(x)(Fx \lor Mx)$ says "Everything is either male or female," which is true. But the formula $(x)Fx \lor (x)Mx$, which looks very much like the first one, is a *disjunction* and says "*Either* everything is male *or* everything is female," which is obviously false. The two formulas $(\exists x)(Fx \cdot Gx)$ and $(\exists x)Fx \cdot (\exists x)Gx$ also look superficially alike, but are very different, as a suitable interpretation will show. The first is an existential statement, since it begins with an existential quantifier whose scope extends to the end of the sentence. The second, however, is a conjunction, with the two conjuncts being existential formulas. It is *not* a quantifier statement, since the scope of the initial quantifier does not extend to the end of the sentence. If we use $Fx \equiv x$ is an odd number, and $Gx \equiv x$ is an even number, the difference becomes apparent, since the first formula, $(\exists x)(Fx \cdot Gx)$, thus interpreted, says that there is a number which is both odd and even, which is obviously false and even absurd. The second formula, $(\exists x)Fx \cdot (\exists x)Gx$, however, says only that there are odd numbers and there are also even numbers, which is obviously true.

We find the same difference in the use of the conditional. $(x)(Fx \supset Gx)$, which is a *universal* proposition, is quite different from $(x)Fx \supset (x)Gx$, which is a *conditional*. We can see this if we use $Fx \equiv x$

never mugs anybody, and $Gx \equiv x$ has never been mugged. Then the first sentence says, roughly, that anyone who never mugs another will never get mugged, which is, unfortunately, false (at least in this society). On the other hand, the second form, $(x)Fx \supset (x)Gx$, says that if all x's never mug anyone, then all x's never get mugged; that is, if nobody mugs another, then nobody gets mugged, which would seem to be true. The existential forms are also different. If we use $Fx \equiv x$ is a fish and $Ux \equiv x$ is a unicorn, then $(\exists x)Fx \supset (\exists x)Ux$, a conditional, is false, since the antecedent, which says there are fish, is true, while the consequent, which says there are unicorns, is false. On the other hand, $(\exists x)(Fx \supset Ux)$ is true, since this very weak existential proposition only asserts that there is something such that *if* it is a fish *then* it is a unicorn, and all we need to make it true is something which is not a fish (since this would make Fx false, and so $Fx \supset Ux$ true). Try working out your own examples for the biconditional.

It should be noted that conjunctions combined with universal statements, and disjunctions combined with existentials, yield equivalent forms. That is, $(x)Fx \cdot (x)Gx$ is equivalent to $(x)(Fx \cdot Gx)$, and $(\exists x)Fx \vee (\exists x)Gx$ is equivalent to $(\exists x)(Fx \vee Gx)$. But they are still different formulas: the first is a conjunction, the second a universal statement, and the third is a disjunction, while the fourth is an existential formula.

So far we have said nothing about negations, but these will be extremely important when we come to proofs. The first condition for a formula to be of quantifier form is that it *begin* with a quantifier, and this *rules out* formulas such as $\sim (x)(Fx \supset Gx)$. *No negated quantifier statement is considered a quantifier formula*; it is a *negation*, one of the five truth-functional compounds. You will be able to use the four quantifier rules *only* on quantifier statements, and this means *not* on negations, so it is very important to be acutely aware of the difference. (On negated formulas, you will use Q.N. or C.Q.N. first.) Again, a formula that begins with a negation cannot begin with a quantifier, so it cannot be a quantifier formula.

By this time it should be fairly easy for you to identify the form of a symbolic expression, at least if you stop to think about it a little. We have now seven different kinds of compound formulas: universals, existentials, conjunctions, disjunctions, negations, conditionals, and biconditionals. Remember that for a formula to be a quantifier statement, that is, a universal or existential statement, it must (1) begin with a quantifier, and (2) have the scope of the *initial* quantifier extending to the end of the sentence. This is fairly easy to check out. $\sim (\exists x)Fx$, for instance, is not a quantifier statement since it does not begin with a quantifier, and $(x)Fx \supset (Fa \cdot Fb)$ is not, because the scope of the quantifier does not go to the end of the sentence. But what if it isn't a quantifier formula? How can you tell what it is? This is just a matter of learning to spot the major operator of the formula, and you have had plenty of practice with this in propositional logic. The procedure is the same here. The placement of parentheses is what determines the major operator—the overall form of the sentence—and by this time you probably recognize it almost at a glance. You should be able to tell right off, for instance, that $(x)Fx \lor ((\exists x)Gx \cdot (\exists x)Hx)$ is a *disjunction*, that $(\exists x)Fx \cdot \sim (\exists x)Gx$ is a *conjunction*, and that $\sim (x)Fx \supset \sim (x)Gx$ is a *conditional*.

At this point, it might be appropriate to run through all the kinds of formulas we have encountered so far, and to round out the collection with a few we have not yet mentioned. In Unit 10, the first unit in quantifier logic, we talked about *singular sentences*, those that contain names. Simple singular sentences are our simplest units, and will be symbolized in the form *Fa*. In Unit 11 we discussed *simple quantifier formulas*, those that have a quantifier prefixed to a simple function or a negated simple function, such as $(\exists x)Fx$ and $(\exists x) \sim Gx$; we also considered the negations of such formulas. In Unit 12 we discussed *categorical propositions* and their negations, and in Unit 13 *more complex forms of categorical propositions*. All these formulas, other than the simple singular ones, were either quantifier formulas or their negations. We have seen in this unit that there may also be *truth-functional compounds of quantifier formulas*; and there are still other possibilities. Formulas such as $(x)(Fx \cdot Gx)$, for instance, are perfectly good quantifier formulas; they are just not categorical. Furthermore, we may combine any of the sorts of formulas we have with our truth-functional operators. The following are all perfectly good formulas; the form is indicated to the right.

$Fa \supset (x)(Fx \equiv Gx)$	Conditional
$(x)(Fx \equiv \sim Gx)$	Universal
$\sim (\exists x)(Fx \supset Hx)$	Negation
$(x)Fx \equiv \sim Hx$	Biconditional
$(x)Fx \cdot ((x)Gx \lor (x)Hx)$	Conjunction
$\sim (x)Fx \supset \sim (x)Gx$	Conditional
$(\exists x)(Fx \cdot Gx) \cdot Hx$	Conjunction
$(x)(Fx \supset Gx) \lor Ha$	Disjunction
$\sim ((Fa \cdot Ha) \supset (\exists x)(Fx \cdot Hx))$	Negation
$(\exists x)((Fx \supset Hx) \lor (Gx \equiv \sim Hx))$	Existential

Exercise 2 at the end of the unit, which you should now complete, will give you further practice in recognizing the forms of such propositions.

3. Symbolizing Truth-Functional Compounds

You should by now have little trouble in identifying the form of a given symbolic expression and in distinguishing between quantifier statements, disjunctions, conjunctions, and so on. You now need to practice going from the *English* sentence to the symbolism, where you are not told in advance whether the sentence is a quantifier statement or not. What you need to do is to analyze the structure of the English sentence to determine whether it is a quantifier statement or a truth-functional compound. Here, where you will have a mixture of the different kinds of sentences, symbolization techniques will be a combination of what you have learned in sentential logic and what you have learned so far in predicate logic. *The first thing to do is to determine the overall structure of the sentence*—universal, existential, disjunction, conjunction, negation, etc. Once you have done this, you can then go on to analyze the parts: if it is a disjunction, for example, symbolize the disjuncts; if it is a universal statement, pick out and symbolize subject and predicate. A few examples should be sufficient to give you the idea. Obvious abbreviations will be used.

The sentence "Not every politician is a crook and not every clergyman is honest" has the overall form of a conjunction, since the "and" joins two complete sentences. The first conjunct is the negation of a universal statement, and so is the second. Thus the overall form will be $\sim (x)(\underline{\quad} \supset \underline{\quad}) \cdot \sim (x)(\underline{\quad} \supset \underline{\quad})$. The next thing to do is to identify and symbolize the subjects and predicates of the two conjuncts, and at this point that should not be hard for you. Once we do this, the whole sentence is symbolized as $\sim (x)(Px \supset Cx) \cdot \sim (x)(Lx \supset Hx)$. The sentence "If nobody comes to the party then there is someone who will not be happy" is a conditional, with an E statement as the antecedent and an O statement as the consequent. If we use $Px \equiv x$ is a person, $Ax \equiv x$ comes to the party, and $Hx \equiv x$ is happy, then the whole symbolization would be $(x)(Px \supset \sim Ax) \supset (\exists x)(Px \cdot \sim Hx)$. We can also have truth-functional compounds of quantifier statements and singular propositions, or truth-functional compounds of singular propositions alone. The sentence "If everyone passed the exam, then Mary got an A and got into law school" is a conditional, with a universal for antecedent and conjunction of singular propositions as consequent. One way of symbolizing this would be $(x)(Px \supset Ex) \supset (Am \cdot Lm)$. In the following exercises you will have all sorts of combinations, and the best thing to do is simply analyze the structure bit by bit until you get the end result. There will be categorical propositions, singular statements, and simple quantifier statements, as well as truth-functional compounds, so be very careful in analyzing structure.

EXERCISES

1. Give interpretations of *your own* to show that the following pairs of formulas do *not* mean the same. Make one sentence true and the other false.

a. $(x)Fx \lor (x)Gx$ and $(x)(Fx \lor Gx)$

b. $(\exists x)(Fx \cdot Gx)$ and $(\exists x)Fx \cdot (\exists x)Gx$

c. $(x)Fx \supset (x)Gx$ and $(x)(Fx \supset Gx)$

d. $\sim (\exists x)(Fx \cdot Gx)$ and $(\exists x) \sim (Fx \cdot Gx)$

2. Which of the following are quantifier statements? For those that are not, indicate what their form is.

*a. $\sim (x)(Fx \supset Gx)$

b. $(\exists x)(Fx \cdot Gx) \cdot Hx$

*c. $(x)((Fx \lor Gx) \supset \sim (Hx \lor Ix))$

d. $(\exists x)((Fx \cdot Gx) \cdot \sim (\sim Hx \lor Ix))$

*e. $(\exists x)(Fx \cdot Gx) \supset (\exists x)(Fx \cdot Hx)$

f. $(x)(Fx \supset Gx) \cdot (Hx \lor Ix)$

*g. $(x)(Fx \equiv (Hx \cdot Ix))$

h. $\sim (x)Fx \cdot \sim (x)Gx$

*i. $(\exists x) \sim (Fx \cdot Gx)$

j. $(x)Fx \supset (Gx \cdot Hx)$

*k. $(x)(Fx \supset Gx) \lor (x)Hx$

l. $\sim (x)Fx \cdot (x)Gx$

*m. $\sim ((\exists x)Fx \lor (\exists x)Gx)$

n. $(x)((Fx \cdot Gx) \supset (Hx \lor Ix)) \lor (Px \lor Gx)$

*o. $(\exists x)((Fx \lor Gx) \equiv (Gx \lor Ix))$

p. $(\exists x)(Fx \cdot Gx) \cdot (\exists x)Hx$

*q. $\sim (x)((Fx \lor Gx) \cdot \sim Hx) \supset Ix$

r. $(\exists x)((Fx \cdot Gx) \supset (Hx \cdot Ix))$

*s. $\sim (\exists x)Fx \lor (x)(Gx \supset Hx)$

t. $(x) \sim (Fx \lor ((Hx \cdot Jx) \supset Ix))$

3. Symbolize the following, using the abbreviations indicated. These sentences may be of any form, including singular, simple quantifications, categorical propositions, or truth-functional compounds. Analyze the overall structure before you do anything else.

$Px \equiv x$ is a politician; $Rx \equiv x$ is respected; $Ex \equiv x$ is reelected; $Ax \equiv x$ abuses his powers; $Ix \equiv x$ is impeached; $Mx \equiv x$ is a millionaire; $Vx \equiv x$ votes his conscience; $Hx \equiv x$ is honest; $Bx \equiv x$ takes bribes; $Lx \equiv x$ is lying; $Sx \equiv x$ is a person; $j =$ John; $r =$ Richard.

*a.   There are some politicians who do not abuse their powers, and some who do.

b.   Not every politician who is a millionaire abuses his powers.

*c.   If every politician votes his conscience, then none will be impeached.

d.   Some politicians who do not vote their conscience are respected and reelected.

*e.   Either no politician takes bribes or none is respected.

f.   Politicians are respected if and only if they do not take bribes.

*g. If John is not honest then there are no honest politicians.

h. John does not take bribes but he does abuse his power, and will be impeached.

*i. If Richard does not take bribes or abuse his power, then he will not only not be impeached but will be reelected.

j. If John and Richard both take bribes, then not all politicians are honest.

*k. John will take a bribe only if all politicians take bribes.

l. Either someone is lying, or John and Richard are taking bribes.

*m. There are some politicians who are either taking bribes or lying, and they will be impeached.

4. Write down the English sentences which correspond to the following formulas. Use the abbreviations from Exercise 3.

*a. $(\exists x)(Hx \cdot Px) \supset \sim Bj$

b. $\sim (x)(Px \supset Hx) \cdot \sim (x)(Px \supset \sim Hx)$

*c. $\sim (\exists x)(Px \cdot Rx) \supset (\exists x)(Px \cdot Bx)$

d. $\sim (Bj \vee Br) \supset ((Mj \cdot Mr) \vee (Hj \cdot Hr))$

*e. $(\sim Rj \cdot \sim Ej) \supset (Aj \vee \sim Hj \vee \sim Mj)$

f. $(x)((Px \cdot Hx) \supset Rx) \vee (\exists x)(Px \cdot Hx \cdot \sim Ex)$

*g. $(x)((Px \cdot \sim Hx \cdot Bx) \supset (Mx \cdot \sim Rx \cdot \sim Vx))$

h. $\sim (\exists x)(Px \cdot Vx) \supset \sim (\exists x)(Px \cdot (Rx \vee Ex))$

*i. $\sim (\exists x)(Px \cdot Lx \cdot Rx) \cdot \sim (x)((Px \cdot Lx \cdot Bx) \supset \sim Ex)$

j. $((Bj \cdot Ar) \vee (Lr \cdot \sim Vj)) \supset (\sim (\exists x)(Px \cdot Hx) \vee (\exists x)(Sx \cdot Lx))$

UNIT 15
Proofs in Predicate Logic

A. INTRODUCTION

In the introduction to quantifier logic, it was noted that there are some arguments that cannot be proved using just the methods of sentential logic, but which require an analysis of the internal structures of the sentences involved. You have now learned how to do this analysis, how to break down sentences into their component parts, and at this point you are ready to learn how to construct proofs.

There is actually very little new that you will have to learn here; most of the business of proof construction in quantifier logic uses things you already know, such as the rules for sentential logic, the quantifier negation equivalences, and the definition of a quantifier formula. The only thing you still need to do is to learn the four rules for using quantifier formulas in inferences. Once you learn these, the process of constructing proofs should not be difficult.

B. UNIT 15 OBJECTIVES

1. Be able to state the four quantifier rules, with all necessary restrictions.
2. Be able to explain why the restrictions are necessary.

259

3. Be able to construct proofs of arguments containing just quantifier statements or their negations.

4. Be able to construct proofs of arguments containing truth-functional compounds of quantifier statements.

5. Be able to construct proofs of theorems in predicate logic.

C. UNIT 15 TOPICS

1. Preliminary Statement of the Four Quantifier Rules

As noted in Unit 10, the argument "All cats are mammals, and all mammals are vertebrates, so all cats are vertebrates" cannot be proved given just the methods of sentential logic, since it would have to be symbolized as p, q /∴ r, which is not a valid argument form. You now know, however, that the argument can be symbolized in quantifier logic as $(x)(Cx \supset Mx)$, $(x)(Mx \supset Vx)$ /∴ $(x)(Cx \supset Vx)$. This looks a good bit more promising, since the rule of Hypothetical Syllogism seems to connect the premises and conclusion. However, this rule is *not* applicable to the argument as it stands, since premises and conclusion are not in the form of conditionals, but rather are universal statements. What, then, needs to be done?

Notice that the *propositional functions* of the premises and conclusion *are* in the form of conditionals, so if we could get them, or formulas like them, standing alone, we would be able to use the rule of H.S. Thus, what we need is a rule that will allow us to *drop* the quantifiers and infer instances, and a rule that will allow us to *add on* the quantifier again once we have the instance we want. We would then be able to go from $(x)(Cx \supset Mx)$ and $(x)(Mx \supset Vx)$ to some particular instance, perhaps $(Ca \supset Ma)$ and $(Ma \supset Va)$. We could then apply the rule of H.S. to get $(Ca \supset Va)$, and then apply the rule that allows us to go back to the universal statement. In this way we could infer the desired conclusion, $(x)(Cx \supset Vx)$.

The rule that will allow us to drop universal quantifiers and infer instances of the universal formula will be called, naturally enough, *Universal Instantiation (U.I.)*, and will have the form $(x)\phi x$ /∴ ϕa, where a can be any name. Since a universal statement $(x)\phi x$ says that for *any x*, for *anything*, ϕ is true of that thing, we ought to be able to infer that ϕ is true of a, of b, of c, and so on. That is, we should be able to infer any *instance* ϕa, ϕb, ϕc, . . . , and this is exactly what the rule of U.I. allows us to do.

We will also have a rule for adding on the universal quantifier, for going from an instance ϕa to the universal formula $(x)\phi x$. This will be called the rule of *Universal Generalization (U.G.)*, and will be stated in roughly the form ϕa /∴ $(x)\phi x$. (The rules will be stated more precisely, and explained more fully, in Sections 3 and 4.) Unlike the rule of U.I., however, *this rule can be used only in very special circumstances*, only with special kinds of instances. *We cannot always go from an instance to the universal statement*; it would be absurd, for example, to infer that because the Earth contains intelligent life, *Le*, therefore everything contains intelligent life, $(x)Lx$. What we need to do is to state the rule and then *restrict its application* to those cases in which it is clearly justified. This will take some rather lengthy explanation, which will be given in Section 4, but for now, we can say that in cases like the above, where the premises are all universal, it is correct to infer the universal statement again at the end.

There are parallel rules for the existential quantifier. The rule of *Existential Instantiation (E.I.)* will allow us to go from an existential formula to some instance, and will be stated in much the same form as the rule of U.I.: $(\exists x)\phi x /\therefore \phi a$. We need to be careful with this rule, however; like the rule of U.G. (not U.I.), *it cannot be used in all instances, but will have to be restricted*. We cannot infer from an existential formula any instance we wish, since the existential statement $(\exists x)\phi x$ says only that *at least one* individual has the property ϕ, which certainly does not guarantee that any individual we choose will have the property ϕ. We will go into the details of how to pick a correct instance in Section 4, and will only note for now that *if* we choose properly, we are permitted to go from an existential statement to an instance.

The rule of *Existential Generalization (E.G.)*, which will permit us to go from a particular instance ϕa to the existential formula $(\exists x)\phi x$, will be stated as follows: $\phi a /\therefore (\exists x)\phi x$. Like the rule of U.I. (not U.G.), it is relatively unrestricted. If we have already demonstrated that a particular individual has the property ϕ, then we are certainly justified, without further ado, in concluding that *something* has the property ϕ, that is, $(\exists x)\phi x$.

A summary of the four quantifier rules is given below. Note that this is only a *preliminary* statement of the rules, without the restrictions. *The rule of U.G., in particular, will look a little different in its restricted form.* The rules will be given in full in Sections 3 and 4.

U.I.	U.G.	E.I.	E.G.
$(x)\phi x$	ϕa	$(\exists x)\phi x$	ϕa
$/\therefore \phi a$	$/\therefore (x)\phi x$	$/\therefore \phi a$	$/\therefore (\exists x)\phi x$

At this point, we can use our rules to construct a proof of the following argument: "All U.S. presidents have been ambitious, and some U.S. presidents have been Quakers, so some Quakers have been ambitious."

a.
1. $(x)(Px \supset Ax)$ Pr.
2. $(\exists x)(Px \cdot Qx)$ Pr. $/\therefore (\exists x)(Qx \cdot Ax)$
3. $Pa \cdot Qa$ E.I. 2

4. $Pa \supset Aa$	U.I. 1
5. Pa	Simp. 3
6. Aa	M.P. 4,5
7. Qa	Simp. 3
8. $Qa \cdot Aa$	Conj. 7,6
9. $(\exists x)(Qx \cdot Ax)$	E.G. 8

Notice that the first move is to *drop the quantifiers* by applying the instantiation rules. We then *derive an instance of the conclusion* we want, and then *apply the appropriate generalization rule* to derive the quantified formula. To clarify this process we will discuss the concept of an *instance* of a general formula.

2. Instances of General Formulas

As you have seen, the four quantifier rules go either from general formulas (those that are of universal or existential form) to instances (in the instantiation rules), or from instances to general formulas (in the generalization rules). You already know that a general formula is one that begins with a quantifier whose scope extends clear to the end of the formula. We need now to state precisely the meaning of "instance." *An instance of a general formula is the result of deleting the initial quantifier and replacing every "matching" variable in the propositional function uniformly with some individual constant.* (The "matching" variables are simply those that are bound by the quantifier, that is, the ones that have the same letter as the quantifier.) The relationship between a general formula and its instances is illustrated below; the singular statements underneath the general formula are all instances of that formula.

a. $(x)((Fx \lor Gx) \supset Hx)$ b. $(\exists x)((Fx \cdot \sim Gx) \lor (Hx \cdot \sim Ix))$

$(Fa \lor Ga) \supset Ha$ $(Fa \cdot \sim Ga) \lor (Ha \cdot \sim Ia)$

$(Fb \lor Gb) \supset Hb$ $(Fd \cdot \sim Gd) \lor (Hd \cdot \sim Id)$

$(Fc \lor Gc) \supset Hc$ $(Fe \cdot \sim Ge) \lor (He \cdot \sim Ie)$

$(Fg \lor Gg) \supset Hg$ $(Fh \cdot \sim Gh) \lor (Hh \cdot \sim Ih)$

It is clearly a simple matter to recognize instances of general formulas such as those above, which contain only a single variable. It should be noted, however, that there may also be general formulas that contain individual constants and/or other quantified variables. $(x)((Fx \cdot Gb) \supset Hx)$ would be an example of a formula containing a constant. Instances of this formula would include $((Fa \cdot Gb) \supset Ha)$, $((Fb \cdot Gb) \supset Hb)$, and $((Fc \cdot Gb) \supset Hc)$. The principle is the same: drop the quantifier and replace each variable bound by the deleted quantifier with a constant. *Do not* replace the *existing* constant with another constant; replace *only* the variables bound by the initial quantifier. Formulas containing other quantified variables will be discussed in Units 17 through 20. It is extremely important to remember that (1) *Every* matching variable must be replaced, and (2) The *same* constant must be used for each matching variable.

To designate *instances* of the general formulas $(x)\phi x$ and $(\exists x)\phi x$, we will use expressions such as ϕa. We may define ϕa as follows: *where ϕx is any propositional function on x, simple or complex, ϕa is the result of replacing every 'x' in ϕx with an 'a'. That is, ϕa is a formula just like ϕx, except that every occurrence of 'x' in ϕx has been replaced by an 'a'.* Since an instance of a general formula is simply the result of deleting the initial quantifier and replacing all the x's with a's, it is clear that ϕa *is an instance of $(x)\phi x$ or $(\exists x)\phi x$.*

In applying the four quantifier rules, we will use the special notation a/x, to indicate that a is the instance letter that is being substituted for the variable x in the propositional function ϕx. We will include this notation as a part of the justification in each instantiation or generalization step, so that it is always clear what constant is being substituted for what variable.

3. The Rules of U.I. and E.G.

As noted earlier, a universal formula states that *every* individual has a certain property; it should certainly be correct, then, to infer from a universal statement any particular instance. If everything is affected by gravity, $(x)Gx$, then John is affected by gravity, Gj, and if all U.S. presidents have been men, $(x)(Px \supset Mx)$, then if Coolidge was a U.S. president, then Coolidge was a man, $Pc \supset Mc$. We may state the rule of U.I. formally as follows:

U.I.

$$\frac{(x)\phi x}{\therefore\ \phi a}$$

From a universal formula we may infer any instance.

The rule of E.G. is similarly uncomplicated, and its justification is obvious. If we have an instance of a formula, ϕa, this means that the individual named by a has the property ϕ. Since we *have* an individual with the property ϕ, we are surely justified in inferring that *there is something* with the property ϕ, that is, $(\exists x)\phi x$. We may state the rule formally as follows:

E.G.

$$\frac{\phi a}{\therefore\ (\exists x)\phi x}$$

Given any instance, we may infer the corresponding existential formula.

Let us now construct a very simple proof, using these two rules. Our premises will be (1) Everything is useful, and (2) Anything which is useful is good; and our conclusion will be that something is good.

b. 1. $(x)Ux$ Pr.
    2. $(x)(Ux \supset Gx)$ Pr. /∴ $(\exists x)Gx$
    3. Ua U.I. 1, a/x
    4. $Ua \supset Ga$ U.I. 2, a/x
    5. Ga M.P. 3,4
    6. $(\exists x)Gx$ E.G. 5, a/x

4. The Rules of E.I. and U.G.; Flagging Restrictions

As noted earlier, the rules of E.I. and U.G. must be rather heavily restricted, and it will be easier to learn these restrictions if you can see *why* they are needed. The point is really very simple: without the restrictions the rules are not valid. Used indiscriminately, they would lead to logical errors, invalid inferences.

If we had an unrestricted rule of E.I., for instance, it would be an easy matter, given the true premises that there are odd numbers and there are even numbers, to prove the false conclusion that there are numbers that are both odd and even! The "proof" would be as follows:

c. 1. $(\exists x)(Ox \cdot Nx)$ Pr.
    2. $(\exists x)(Ex \cdot Nx)$ Pr. /∴ $(\exists x)(Ox \cdot Ex \cdot Nx)$
    3. $Oa \cdot Na$ E.I. 1, a/x
    4. $Ea \cdot Na$ E.I. 2, a/x
    5. Oa Simp. 3
    6. $Oa \cdot Ea \cdot Na$ Conj. 5,4
    7. $(\exists x)(Ox \cdot Ex \cdot Nx)$ E.G. 6, a/x

Where have we gone wrong here? It is fairly obvious that it is in the fourth step, where we have used a to stand for the even number, even though we had used a to stand for the odd number in the previous step. *What we need is a restriction that will prevent us from using the same instance letter for two different existential propositions.* Just because we know that there are ɸ's and there are ψ's, we are not permitted to say that there is some *one* thing which is both a ɸ and a ψ.

Without restrictions on U.G. we could also make the absurd inference cited earlier, from "The Earth contains intelligent life" to "Everything contains intelligent life." We would need only two steps:

d. 1. Le Pr. /∴ $(x)Lx$
    2. $(x)Lx$ U.G. 1, e/x

We could also make erroneous inferences from existential propositions to universal propositions, such as "There are cats with stripes /∴ Everything is a cat with stripes." For this we would need only three steps:

e. 1. $(\exists x)(Cx \cdot Sx)$ Pr. /∴ $(x)(Cx \cdot Sx)$
    2. $Ca \cdot Sa$ E.I. 1, a/x
    3. $(x)(Cx \cdot Sx)$ U.G. 2, a/x

Clearly, this will not do; *we obviously need restrictions on U.G. as well, to make sure we do not infer universal statements either from specific instances or from instances derived from existential formulas.* What can we do to prevent such glaring logical errors?

There are a number of different formulations of the quantifier rules which would help us get around these difficulties. One thing we might do would be simply to lay down a restriction for each problem that might come up, but this would be a rather cumbersome, *ad hoc* approach, and it might turn out that there were problems we hadn't anticipated, requiring further restrictions. What we will do instead is to use a method which is simple to use, quite elegant, and which covers all the problems at once, instead of piece by piece.

We will prevent the kind of erroneous inferences mentioned above, as well as any others that might crop up, by a series of *flagging* restrictions on the instance letters used in the rules of E.I. and U.G. *We will require that the instance letter in any use of E.I. or U.G. be flagged*, and we will then impose three further restrictions on the use of flagged letters. There is nothing mysterious about the process of flagging; to flag a letter, we simply *declare* that it is being flagged, and thereafter make sure that we observe all the conditions that are imposed on flagged letters. To flag a letter is just something like noting that it requires special attention.

In using the rule of E.I., we will flag the instance letter by writing "flag a" (or "flag b" or "flag c") as a part of the justification for that step, and the full statement of the rule of E.I. will be just like our earlier, preliminary version, except that we note that the instance letter must be flagged. The rule of E.I. will then be the following:

E.I.

$$\frac{(\exists x)\phi x}{\therefore \phi a}$$ provided we flag a Given an existential formula, we may infer an instance, provided we flag the instance letter.

Now, what sorts of conditions must be placed on the flagged letters in order to prevent the logical errors mentioned above? Remember that the invalid inference in proof c, to the conclusion that there are numbers that are both odd and even, was possible because we used the *same* instance letter for both existential premises. Thus, what we need is a restriction that will guarantee that *we never use the same instance letter for more than one existential statement. We can do this by requiring that every flagged letter be new to the proof.* It should be noted that one consequence of this restriction is that *the same letter may not be flagged more than once in a proof.* Thus, if you check over your proof and find, for instance, that you have flagged a twice, you will know immediately that you have made an error. The formal restriction is stated below.

R₁ A letter being flagged must be new to the proof, that is, it may not
appear, either in a formula or as a letter being flagged, previous to the
step in which it gets flagged.

Notice how this restriction blocks the inference from "There are odd
numbers and there are even numbers" to "There are numbers that are
both odd and even."

1. $(\exists x)(Ox \cdot Nx)$ Pr.
2. $(\exists x)(Ex \cdot Nx)$ Pr. /∴ $(\exists x)(Ox \cdot Ex \cdot Nx)$
3. $Oa \cdot Na$ E.I. 1, a/x (flag a)
4. $Ea \cdot Na$ E.I. 2, a/x (flag a)   **Error**

We cannot infer $Ea \cdot Na$ at step 4, because the instance letter is not new
to the proof; it has already made an appearance at step 3.

It is important to remember in doing these proofs that *an essential
part of the justification for the E.I. step is a declaration that the instance
letter is being flagged. This declaration must be included with each E.I.
step, or the step is incorrect.*

We can now carry out a proof using the fully restricted rule of E.I.,
as well as the unrestricted rules of U.I. and E.G.

f. 1. $(\exists x)(Fx \cdot \sim Gx)$ Pr.
    2. $(x)(Hx \supset Gx)$ Pr. /∴ $(\exists x)(Fx \cdot \sim Hx)$
    3. $Fa \cdot \sim Ga$ E.I. 1, a/x (flag a)
    4. $Ha \supset Ga$ U.I. 2, a/x
    5. Fa Simp. 3
    6. $\sim Ga$ Simp. 3
    7. $\sim Ha$ M.T. 6,4
    8. $Fa \cdot \sim Ha$ Conj. 5,7
    9. $(\exists x)(Fx \cdot \sim Hx)$ E.G. 8, a/x

Notice that in using the rule of E.I. at step 3, we have not only cited the
rule but also included in our justification the substitution notation a/x,
as well as the declaration that we are flagging a.

Before we go on to discuss U.G., we need to emphasize a few points
of strategy in doing quantifier proofs. First, because of the way the rules
are stated, *you should almost always do your E.I. instantiations first.* If
you apply U.I. first, the instance letter will already have been used, and
you will have to choose another letter for the E.I. step. If we had inferred
$Ha \supset Ga$ by U.I. at step 3 above, for instance, we would have had to infer
$Fb \cdot \sim Gb$, or $Fc \cdot \sim Gc$, or some other instance, and *we would then not
have been able to combine these formulas with $Ha \supset Ga$ to derive other
steps. You may never use M.T. on premises such as $Ha \supset Ga$ and $\sim Gb$,*
since the second premise is *not* the negation of the consequent of

Ha ⊃ Ga. Gb is *not* the same statement as *Ga. In general, you will not be able to combine steps with your sentential rules if these steps contain different instance letters.* To avoid getting different instance letters you should, in general, use E.I. first.

Another point which should now be emphasized is that, *in using the unrestricted rules of U.I. and E.G., the instance letter is not flagged and need not be new to the proof;* it may have already appeared. Thus, it is perfectly correct to infer *Ha ⊃ Ga* at step 4 in problem f above, even though the letter *a* was already used in step 3. Keeping these things in mind, we may now move on to our last quantifier rule, U.G.

The rule of U.G. must, of course, also be restricted; as mentioned earlier, we cannot infer a universal proposition from just any old instance. We cannot infer that everything is made of paper, for instance, just because this page is made of paper. Rather, if we infer a formula $(x)\phi x$, we must be sure that the function ϕx holds for *every* individual, and this can happen only with very special instances.

The kinds of instances from which it is proper to infer universal propositions are those that would hold as well for any other individual. In practice, this usually means those that were derived entirely from universal propositions. What we may *not* do is to use U.G. on an instance that is derived from an existential proposition or on a contingent singular statement, such as "The Earth contains intelligent life." (As we will see in the final section, we may use U.G. on tautologous singular statements such as $(Fa \supset Fa)$.)

To guarantee that we have a genuinely universal instance, from which it is legitimate to infer a universal proposition, *we will require that the instance for U.G. be derived within a special kind of subproof,* which we will call a *flagged subproof.* It is called a "flagged subproof" because *it will always begin with a declaration that the instance letter over which we will be generalizing is being flagged.* In other words, the first step of the U.G. subproof will be the expression "flag *a*" (or "flag *b*" or "flag *c*"). This step will be called the "flagging step," and the justification will be "Flagging Step, Universal Generalization," which we will abbreviate as F.S. (U.G.). If the instance letter is *a*, we will call the subproof an "*a*-flagged" subproof, if it is *b*, a "*b*-flagged" subproof, and so on. This flagged subproof will "isolate" the instance from unwanted outside influences such as existential propositions and contingent singular statements, thus ensuring that when we derive our instance it will be one that is dependent *only* on statements that are true of everything, so that we may correctly infer the universal proposition. The flagged subproof, in other words, will serve as a kind of "quarantine," protecting the instance from "contamination" by existential or contingent statements, and keeping it "pure" for the eventual universal generalization.

We may now state the rule of U.G. as follows:

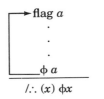

If ϕa is the last step in an a-flagged subproof, then we may infer the universal proposition, $(x)\phi x$.

/∴ (x) ϕx

We will illustrate this rule with a proof for the very simple argument cited in Section 1: "All cats are mammals, and all mammals are vertebrates, so all cats are vertebrates."

g.　1. $(x)(Cx \supset Mx)$　　　Pr.
　　2. $(x)(Mx \supset Vx)$　　Pr. /∴ $(x)(Cx \supset Vx)$
　　3. flag a　　　　　　　F.S. (U.G.)
　　4. 　$Ca \supset Ma$　　　U.I. 1, a/x
　　5. 　$Ma \supset Va$　　　U.I. 2, a/x
　　6. 　$Ca \supset Va$　　　H.S. 4,5
　　7. $(x)(Cx \supset Vx)$　　U.G. 6, a/x

Here we were able to derive the instance $(Ca \supset Va)$ within the a-flagged subproof, therefore we are permitted to infer the universal conclusion $(x)(Cx \supset Vx)$.

The flagged subproofs will be used in the same way as subproofs for C.P. and I.P.; they will be set in, as we have done above, and the flagging step, which operates much like an assumption, will be discharged at the step in which U.G. is applied. The only differences are that for U.G. the first step in the subproof is a flagging step, rather than a formula, and in our justification for U.G. we will cite only the last step in the subproof, rather than all the steps in the subproof. We will use the same kind of scope marker as we used for C.P. and I.P., and the *a-flagged scope marker*, as we will call it, will extend from the flagging declaration through the instance over which we will generalize.

It often happens that the instance for a universal proposition is a conditional, as in the problem above. This means that the best strategy will often be to use C.P. to derive the instance (though above we could just use H.S.), and this will mean setting up *two* subproofs, the first for U.G. and the second for C.P. This is what happens in the proof below, where we need to use C.P. to get the instance and U.G. to derive the universal conclusion.

h.　1. $(x)((Fx \lor Gx) \supset Hx)$　　Pr.
　　2. $(x)(Hx \supset (Ix \cdot Sx))$　　Pr. /∴ $(x)(Fx \supset Sx)$
　　3. flag a　　　　　　　　　F.S. (U.G.)
　　4. 　Fa　　　　　　　　　Assp. (C.P.)
　　5. 　$(Fa \lor Ga) \supset Ha$　　U.I. 1, a/x

6.	$Ha \supset (Ia \cdot Sa)$	U.I. 2, a/x
7.	$Fa \lor Ga$	Add. 4
8.	Ha	M.P. 5,7
9.	$Ia \cdot Sa$	M.P. 6,8
10.	Sa	Simp. 9
11.	$Fa \supset Sa$	C.P. 4–10
12.	$(x)(Fx \supset Sx)$	U.G. 11, a/x

The flagging restrictions we have imposed so far keep us from making erroneous inferences from existential to universal propositions. If you now try to do proof e, for example, you will find, if you observe Restriction 1, that it is not possible.

It was pointed out earlier that it would be incorrect to infer a universal statement from a specific instance; the argument "The Earth contains intelligent life, so everything contains intelligent life" is obviously invalid. It is also incorrect to infer some specific instance from an existential proposition. From the true premise "Some planets have rings," for instance, we may not infer the false specific instance "The Earth is a planet with rings." In order to block these two kinds of erroneous inferences we need to add another flagging restriction, which reads as follows:

R₂ A flagged letter may not appear either in the premises or in the conclusion of a proof.

The simple proof from the Earth containing intelligent life to everything containing intelligent life, Le /∴ $(x)Lx$, is then blocked, because in order to use U.G. to derive $(x)Lx$, we would have to set up an e-flagged subproof. Since e appears in the premise, however, this is not legitimate, so it is not possible to use U.G. on the premise Le, or on any other premise containing a constant. If we tried to derive "The Earth is a planet with rings" from the true premise that some planets have rings, we would be blocked because we would have the constant e appearing in the *conclusion* of the proof, which is also prohibited by R₂.

One final restriction will finish our quantifier rules. We include this only for the sake of completeness, since it will really play no role until we get to relational proofs in Unit 18.

R₃ A flagged letter may not appear outside the subproof in which it gets flagged.

This means, among other things, that if we do an E.I. step *within* a U.G. subproof (we would have to use a different letter, of course), the instance letter for that step of E.I. may not appear anywhere in the proof except within that U.G. subproof. We will see in Unit 18 how this restriction blocks invalid inferences that would otherwise be possible in relational logic.

At this point, before we go on to discuss proof strategies, we must make a number of important observations about the use of the quantifier rules. First, note that we have used *names* rather than variables in doing all our instantiations. *You should never use individual variables as instance letters.* This will be particularly important in relational logic. It is simply a feature of this proof system that *we have two different kinds of individual letters*, which play different roles and which should not be confused.

Second, and extremely important, *you may use the instantiation rules only on formulas which are of quantifier form—those which begin with a quantifier whose scope extends to the end of the sentence.* This means that we *cannot* use U.I. on a formula such as $(x)Fx \supset (x)Gx$, since this is in the form of a conditional, not of a quantifier formula. Nor could we use U.I. on $(x)Fx \supset Gx$, or on $(x)Fx \supset Fa$, since in neither of these cases does the scope of the quantifier extend to the end of the sentence. Since to be of quantifier form a statement must *begin* with a quantifier, *we also cannot use the instantiation rules on formulas such as* $\sim(x)(Fx \supset Gx)$, *as these begin with a negation sign. For all negated quantifier statements, you will first have to use Q.N. or C.Q.N.* to get the equivalent quantifier formula, and only then can you use U.I. or E.I. In the case above you could use C.Q.N. to get $(\exists x)(Fx \cdot \sim Gx)$, and then use *E.I.* to get $(Fa \cdot \sim Ga)$.

The same thing holds in reverse for the generalization rules. In using U.G. or E.G. the *conclusion* must always be a formula which begins with a quantifier, whose scope extends to the end of the formula. This means that $(x)Fx \supset (x)Gx$ *cannot* be the conclusion of U.G. (though it can be the conclusion of C.P.), and $(\exists x)Fx \cdot Gx$ cannot be the conclusion of E.G. (though it can be the result of Conj.). Nor can we infer formulas such as $\sim(\exists x)(Fx \cdot Gx)$ by U.G. or E.G. We would first have to derive its equivalent, $(x)(Fx \supset \sim Gx)$, by U.G., and then use C.Q.N.

Another thing that must be mentioned is that we need, at this point, to revise our definition of "proof" once again, since we have added another kind of step, the flagging step. The revised definition would read simply "A proof is a sequence of steps such that each one is either a premise, an assumption, a flagging step, or follows from previous steps according to the given rules of inference, and such that the last step in the sequence is the desired conclusion."

5. Constructing Proofs for "Pure" Quantifier Arguments

In this section we will be concerned only with "pure" quantifier arguments, those whose premises and conclusions are all either quantifier statements or their negations. In the next section we will talk about arguments containing truth-functional compounds of quantifier statements, the sort of formula you encountered in the last unit. In the "pure" quantifier proofs, since you are starting with quantified statements, the first thing to do will be to *drop the quantifiers using the instantiation rules*. You will then be left with instances of the premises. What you then need to do is *derive an instance of the conclusion*. Once you have this instance you need just one more step; *use one of the generalization rules to get the quantified conclusion*. An example of such a proof is worked out below. (You may need more steps if negations are involved; we will discuss these cases below.)

j.
	1. $(x)(Fx \supset (Gx \lor Hx))$	Pr.
	2. $(\exists x)(Fx \cdot \sim Hx)$	Pr. /∴ $(\exists x)(Fx \cdot Gx)$
	3. $Fa \cdot \sim Ha$	E.I. 2, a/x (flag a)
	4. $Fa \supset (Ga \lor Ha)$	U.I. 1, a/x
	5. Fa	Simp. 3
	6. $Ga \lor Ha$	M.P. 4,5
	7. $\sim Ha$	Simp. 3
	8. Ga	D.S. 6,7
	9. $Fa \cdot Ga$	Conj. 5,8
	10. $(\exists x)(Fx \cdot Gx)$	E.G. 9, a/x

At steps 3 and 4 we dropped the quantifiers, in steps 5 through 9 we derived an instance of the conclusion, and at step 10 we added the quantifier back on. Remember always to flag the instance letter when using E.I.

A more complex example, using U.G., is worked out below. One thing to keep in mind, with arguments that have universal conclusions, is that the propositional function of the conclusion will often be a conditional. This means that you may have to use C.P. to derive an instance of the conclusion, which means that there will often be two nested subproofs for such arguments. In the following problem the outside subproof will be for U.G.; the first step in this subproof will be the flagging step, and the last step will be the instance of the conclusion. The inner subproof will be for C.P., and so the first step of the subproof will be the antecedent of the conditional and the last step the consequent.

k.
	1. $(x)((Fx \lor Gx) \supset \sim (Hx \lor Ix))$	Pr.
	2. $(x)(Zx \supset (Hx \cdot Wx))$	Pr. /∴ $(x)(Gx \supset \sim Zx)$
	3. ⎡flag a	F.S. (U.G.)
	4.	$(Fa \lor Ga) \supset \sim (Ha \lor Ia)$	U.I. 1, a/x
	5.	$Za \supset (Ha \cdot Wa)$	U.I. 2, a/x
	6.	⎡Ga	Assp. (C.P.)
	7.	$Fa \lor Ga$	Add. 6
	8.	$\sim (Ha \lor Ia)$	M.P. 4,7
	9.	$\sim Ha \cdot \sim Ia$	DeM. 8
	10.	$\sim Ha$	Simp. 9
	11.	$\sim Ha \lor \sim Wa$	Add. 10
	12.	$\sim (Ha \cdot Wa)$	DeM. 11

13. | |____ $\sim Za$ M.T. 5,12
14. | $\overline{Ga \supset \sim Za}$ C.P. 6–13
15. $(x)(Gx \supset \sim Zx)$ U.G. 14, a/x

Notice that we terminate the C.P. subproof after step 13, and the U.G. subproof after step 14. You must always apply C.P. and U.G. at *different* steps, just as with C.P. and I.P.; you may never terminate two subproofs in the same step.

You may be wondering where you *begin* a U.G. subproof. Probably the best strategy is to begin it as soon as you can. One thing you may *not* do is to start it *after* you have made an instantiation using the letter to be generalized over, since then the letter, which must be flagged, will already have appeared in the proof.

Remember that the rules of U.I. and E.G. are unrestricted; from a universal formula we may infer *any* instance, and from any instance we may infer an existential proposition. This means that, unlike E.I. and U.G., we *may* use U.I. and E.G. on constants. The following is a rather simple example; notice that since the instance letter here is r, the substitution notation is r/x, rather than a/x.

1. 1. $Mr \cdot Cr$ Pr.
    2. $(x)(Cx \supset Ex)$ Pr. /∴ $(\exists x)(Mx \cdot Ex)$
    3. $Cr \supset Er$ U.I. 2, r/x
    4. Mr Simp. 1
    5. Cr Simp. 1
    6. Er M.P. 3,5
    7. $Mr \cdot Er$ Conj. 4,6
    8. $(\exists x)(Mx \cdot Ex)$ E.G. 7, r/x

Of course, not all proofs are this simple. In proofs that have *negated* quantifier statements as premises or conclusions, you will need to add a few more steps. As emphasized earlier, you *cannot* use the four quantifier rules on negated statements. What you must do is to use Q.N. or C.Q.N. to go from the negated statement to its unnegated, equivalent quantifier formula. To brush up your memory, these rules are stated below. If you have not already learned them, that is, *memorized* them, do so now, since you will be using them frequently from now on. Q.N. rules are for *any* negated quantifier statements; the ϕx stands for any propositional function, whether simple or complex. The C.Q.N. rules are for negated *categorical* propositions (or their more complex instances). Here the ϕx and the ψx stand for simple or complex subject and predicate phrases. You can cite Q.N. or C.Q.N. for any of the forms listed.

Q.N. Rules

1. $\sim (x)\phi x :: (\exists x) \sim \phi x$
2. $\sim (\exists x)\phi x :: (x) \sim \phi x$
3. $\sim (x) \sim \phi x :: (\exists x)\phi x$
4. $\sim (\exists x) \sim \phi x :: (x)\phi x$

C.Q.N. Rules

1. $\sim (x)(\phi x \supset \psi x) :: (\exists x)(\phi x \cdot \sim \psi x)$
2. $\sim (\exists x)(\phi x \cdot \psi x) :: (x)(\phi x \supset \sim \psi x)$
3. $\sim (x)(\phi x \supset \sim \psi x) :: (\exists x)(\phi x \cdot \psi x)$
4. $\sim (\exists x)(\phi x \cdot \sim \psi x) :: (x)(\phi x \supset \psi x)$

An example of an argument involving negated quantifiers is worked out below. Notice that the C.Q.N. steps must be done *before using instantiation*, at the very beginning of the proof, and *after using generalization*, at the very end of the proof.

m.

1.	$\sim (x)(Fx \supset (Gx \lor Hx))$	Pr.
2.	$(x)(Ax \supset Gx)$	Pr. /∴ $\sim (x)(Fx \supset Ax)$
3.	$(\exists x)(Fx \cdot \sim (Gx \lor Hx))$	C.Q.N. 1
4.	$Fa \cdot \sim (Ga \lor Ha)$	E.I. 3, a/x (flag a)
5.	$Aa \supset Ga$	U.I. 2, a/x
6.	Fa	Simp. 4
7.	$\sim (Ga \lor Ha)$	Simp. 4
8.	$\sim Ga \cdot \sim Ha$	DeM. 7
9.	$\sim Ga$	Simp. 8
10.	$\sim Aa$	M.T. 5,9
11.	$Fa \cdot \sim Aa$	Conj. 6,10
12.	$(\exists x)(Fx \cdot \sim Ax)$	E.G. 11, a/x
13.	$\sim (x)(Fx \supset Ax)$	C.Q.N. 12

The strategy for proving arguments whose conclusions are negations is a little different from those in which the conclusion is a quantifier statement. In the latter case, as noted earlier, you are to find an instance of the conclusion, and then use one of the generalization rules. If you have a negated conclusion, however, this won't work; it doesn't even make sense to talk about an *instance* of a *negation*. What you need to do in planning your strategy is to figure out first what statement the conclusion is *equivalent* to. In problem m the conclusion is a negated universal that is equivalent to, and can be derived from, the existential formula $(\exists x)(Fx \cdot \sim Ax)$. Thus, what we need is an instance of this existential formula, that is, $(Fa \cdot \sim Aa)$. Then we can use E.G., and then C.Q.N. to get the conclusion. Again, *with negated conclusions, get an instance of the equivalent, unnegated quantifier formula, use U.G. or E.G., and then use Q.N. or C.Q.N. to derive your conclusion.*

In general your proofs will have the following structure, after the premises are stated:

(Possibly) Q.N. or C.Q.N.

Instantiation rules

Propositional rules to get instance

Generalization rule

(Possibly) Q.N. or C.Q.N.

One thing you may find a bit tricky is the use of C.Q.N. on complex propositions. The first form of C.Q.N., for instance, is stated as: $\sim (x)(\phi x \supset \psi x) :: (\exists x)(\phi x \cdot \sim \psi x)$. Remember that ϕx and ψx can stand for any subjects and predicates, simple or complex, and that the rule says that if not all of the S is P, then some S is not P. Applied to a particular case, the rule might look like this: $\sim (x)((Fx \cdot (Gx \vee Hx)) \supset (Jx \equiv Sx))$ $/\therefore (\exists x)((Fx \cdot (Gx \vee Hx)) \cdot \sim (Jx \equiv Sx))$. A complex negated existential proposition would be the following: $\sim (\exists x)((Fx \cdot Gx) \cdot ((Hx \vee Jx) \supset \sim Sx))$, from which we could derive, by C.Q.N., $(x)((Fx \cdot Gx) \supset \sim ((Hx \vee Jx) \supset \sim Sx))$. If in doubt, you might try underlining the subjects and predicates to make sure you have the transformations right.

We will work out one more, rather complex example, which involves C.Q.N. as well as the U.G. rule, for which we will need a flagged subproof.

n.   1. $(x)((Fx \cdot Gx) \supset (Hx \vee \sim (Ix \vee Jx)))$ Pr.
        2. $\sim (\exists x)(Fx \cdot \sim Gx)$ Pr.
        3. $\sim (\exists x)(Hx \cdot \sim (Ix \cdot \sim Zx))$ Pr. $/\therefore \sim (\exists x)(Fx \cdot \sim (Hx \equiv Ix))$

If you are staring at this in blank despair, it is time for a few reminders on strategy. Remember that your first few steps in problems with negated quantifier statements as premises should be applications of Q.N. or C.Q.N. This would give us steps 4 and 5:

    4. $(x)(Fx \supset Gx)$ C.Q.N. 2
    5. $(x)(Hx \supset (Ix \cdot \sim Zx))$ C.Q.N. 3

Since the conclusion is also a negated quantifier statement, it will have to be derived by Q.N. or C.Q.N., so we will first have to get its equivalent form, which is $(x)(Fx \supset (Hx \equiv Ix))$. Since this is a universal formula, it will be derived by the use of U.G. This means that we must begin a flagged subproof before doing any instantiations, and must then derive the instance $Fa \supset (Ha \equiv Ia)$ within the flagged subproof. Thus we know what the next four steps must be: a flagging declaration and three instantiations.

    6.   flag a F.S. (U.G.)
    7. $(Fa \cdot Ga) \supset (Ha \vee \sim (Ia \vee Ja))$ U.I. 1, a/x
    8. $Fa \supset Ga$ U.I. 4, a/x
    9. $Ha \supset (Ia \cdot \sim Za)$ U.I. 5, a/x

Now that we have our instances, we must derive $Fa \supset (Ha \equiv Ia)$, and since this is a conditional, the best strategy is to use C.P., which means assuming Fa. Given Fa we will want to derive $(Ha \equiv Ia)$, and since this is a *bi*conditional, we can try to get the two conditionals $(Ha \supset Ia)$ and $(Ia \supset Ha)$ separately, and then conjoin them. This means setting up two more C.P. subproofs. Thus, the next few steps will be the following:

    10. Fa Assp. (C.P.)
    11. Ha Assp. (C.P.)
    12. $Ia \cdot \sim Za$ M.P. 9,11
    13. Ia Simp. 12
    14. $Ha \supset Ia$ C.P. 11–13

We now need to assume Ia and try to derive Ha, to get $(Ia \supset Ha)$ by C.P. Once we assume Ia, we will use step 7 to get Ha, which means we will need $(Fa \cdot Ga)$ and $\sim \sim (Ia \lor Ja)$. This is simply a matter of using sentential rules, and once the subproof is completed, the rest of the proof goes quickly.

15.	Ia	Assp. (C.P.)
16.	Ga	M.P. 8,10
17.	$Fa \cdot Ga$	Conj. 10,16
18.	$Ha \lor \sim (Ia \lor Ja)$	M.P. 7,17
19.	$Ia \lor Ja$	Add. 15
20.	$\sim \sim (Ia \lor Ja)$	D.N. 19
21.	Ha	D.S. 18,20
22.	$Ia \supset Ha$	C.P. 15–21
23.	$(Ha \supset Ia) \cdot (Ia \supset Ha)$	Conj. 14,22
24.	$Ha \equiv Ia$	B.E. 23
25.	$Fa \supset (Ha \equiv Ia)$	C.P. 10–24
26.	$(x)(Fx \supset (Hx \equiv Ix))$	U.G. 25, a/x
27.	$\sim (\exists x)(Fx \cdot \sim (Hx \equiv Ix))$	C.Q.N. 26

It is very important to remember, when doing complex proofs such as the one above, which have many subproofs one inside the other, that *once you have discharged an assumption or a flagging step, you may not use that assumption or any step within that subproof again.* Thus, having assumed Ia at step 15, we may *not* get Ha immediately from step 11, since Ha occurs in a subproof which was terminated at step 13.

6. Constructing Proofs for Arguments Containing Truth-Functional Compounds

In Unit 14, you practiced symbolizing sentences that turned out to be truth-functional compounds of quantifier expressions, such as $(x)Fx \supset (x)Gx$. Proofs involving these sorts of formulas require a little more ingenuity than those involving just quantifier expressions or their negations. Here there is no uniform strategy such as dropping quantifiers, getting an instance, and adding the quantifiers back on at the end. The reason for this is that with a premise such as $(x)Fx \supset (x)Gx$, which is *not* a universal statement, we cannot use Universal Instantiation. We may use Modus Ponens, Modus Tollens, or other rules involving conditionals, but not U.I. We might even use Q.N., since the equivalence rules are applicable to any part of a formula, but we may not use U.I. because this is applicable *only* to formulas that are of universal form. Thus, strategy turns out to be somewhat more complex, and requires more thought than with the previous sorts of proofs. Perhaps the best way to see this is by taking an example. Suppose we have as premises (1) $(x)(Fx \supset Gx)$, and (2) $(x)Gx \supset (x)Hx$; and from these we are to derive $(x)Fx \supset (x)Hx$. We could get $(Fa \supset Ga)$ by U.I. from 1, but we *cannot* get $(Ga \supset Ha)$ by U.I. from premise 2, since premise 2 is not a universal formula but a conditional, and we could not get $(x)Fx \supset (x)Hx$ by U.G. from $Fa \supset Ha$, since the conclusion is not universal. What, then, should be our strategy? The best way to work it out is to revert again to the method of "working backwards." Look at the form of the conclusion, figure out what would be needed as a next-to-last step, then what would be needed to get the next-to-next-to-last step, and so on. In this case the conclusion is a *conditional*, so the wisest approach would be to try to use Conditional Proof. This would mean assuming $(x)Fx$ and trying to derive $(x)Hx$, so we might set up a skeleton of the proof as follows:

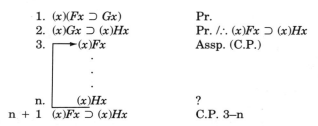

1.	$(x)(Fx \supset Gx)$	Pr.
2.	$(x)Gx \supset (x)Hx$	Pr. /∴ $(x)Fx \supset (x)Hx$
3.	$(x)Fx$	Assp. (C.P.)
	.	
	.	
	.	
n.	$(x)Hx$	?
n + 1	$(x)Fx \supset (x)Hx$	C.P. 3–n

Now the question is, having assumed $(x)Fx$ how do we get $(x)Hx$? Well, it could obviously come from step 2 by M.P. *if* we could get $(x)Gx$. Now we must figure out how to get $(x)Gx$. We do have $(x)Fx$, and we have $(x)(Fx \supset Gx)$ but we *cannot* just use M.P. here, since $(x)Gx$ does not occur as the consequent of a conditional. Rather, what we must do is to derive $(x)Gx$ by U.G., which means deriving Ga within an a-flagged subproof. The completed proof would go as follows:

o.	1.	$(x)(Fx \supset Gx)$	Pr.
	2.	$(x)Gx \supset (x)Hx$	Pr. /∴ $(x)Fx \supset (x)Hx$
	3.	$(x)Fx$	Assp. (C.P.)
	4.	flag a	F.S. (U.G.)
	5.	Fa	U.I. 3, a/x
	6.	$Fa \supset Ga$	U.I. 1, a/x
	7.	Ga	M.P. 5,6
	8.	$(x)Gx$	U.G. 7, a/x
	9.	$(x)Hx$	M.P. 2,8
	10.	$(x)Fx \supset (x)Hx$	C.P. 3–9

Notice that the structure of this proof, in which we are to prove a conditional, is different from the structure of proofs in which we are proving a universal statement, such as problems g, h, and k, in this unit. Since the end result in problem o is a *conditional*, our outside subproof is for C.P. We then want to derive a universal statement *within* the C.P. subproof, so the subproof for U.G. comes *inside* the C.P. subproof. By contrast, in proof k, for example, the end result was to be a *universal* statement, so our first subproof, the *outer* structure, was an a-flagged subproof for

U.G. Within that subproof we made our C.P. assumption, since we wanted to derive the conditional, $Ga \supset \sim Za$, *before* we did the U.G. The outlines of the two proofs are as follows:

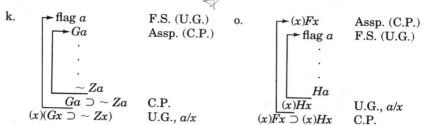

In doing these proofs, in which you have truth-functional compounds of quantifier statements, it will be particularly important to be acutely aware of the structure of your conclusion, and to plan strategy accordingly. Remember that you *must not use U.I.* on a formula such as $(x)Fx \supset (x)Gx$, since it is *not* a universal statement but rather is a conditional. Nor, if you use U.G. on $(Fa \supset Ga)$, can you derive $(x)Fx \supset (x)Gx$, since it is *not* a universal statement. You would have to derive this latter formula by C.P., or perhaps H.S. or C.E., since it is a conditional. We will do one last example to illustrate the procedure for problems containing truth-functional compounds.

p.

1.	$(x)Fx \lor (x) \sim Gx$	Pr.
2.	$(x)(Fx \supset Jx)$	Pr.
3.	$\sim (\exists x)(Hx \cdot Jx)$	Pr. /∴ $\sim (x) \sim Hx \supset \sim (\exists x)Gx$
4.	$\sim (x) \sim Hx$	Assp. (C.P.)
5.	$(\exists x)Hx$	Q.N. 4
6.	Ha	E.I. 5, a/x (flag a)
7.	$(x)(Hx \supset \sim Jx)$	C.Q.N. 3
8.	$Ha \supset \sim Ja$	U.I. 7, a/x
9.	$\sim Ja$	M.P. 6,8
10.	$Fa \supset Ja$	U.I. 2, a/x
11.	$\sim Fa$	M.T. 9,10
12.	$(\exists x) \sim Fx$	E.G. 11, a/x
13.	$\sim (x)Fx$	Q.N. 12
14.	$(x) \sim Gx$	D.S. 1,13
15.	$\sim (\exists x)Gx$	Q.N. 14
16.	$\sim (x) \sim Hx \supset \sim (\exists x)Gx$	C.P. 4–15

7. Constructing Proofs of Quantifier Theorems

There is nothing new at all in this section; we are simply combining what you learned in Unit 9 about proving theorems with what you have now learned about quantifier proofs. Just as in sentential logic, there are certain formulas in predicate logic which are always true, and always able to be proved, without premises. These are the theorems

of predicate logic. A very simple example of such a theorem would be
$(\exists x)(Fx \cdot Gx) \supset ((\exists x)Fx \cdot (\exists x)Gx)$. The proof follows.

q. 1. ┌─► $(\exists x)(Fx \cdot Gx)$ Assp. (C.P.)
    2. │ $Fa \cdot Ga$ E.I. 1, a/x (flag a)
    3. │ Fa Simp. 2
    4. │ Ga Simp. 2
    5. │ $(\exists x)Fx$ E.G. 3, a/x
    6. │ $(\exists x)Gx$ E.G. 4, a/x
    7. └── $(\exists x)Fx \cdot (\exists x)Gx$ Conj. 5,6
    8. $(\exists x)(Fx \cdot Gx) \supset ((\exists x)Fx \cdot (\exists)Gx)$ C.P. 1–7

Where there are universal formulas to be proved, of course, we will
have to use our flagged subproofs and perhaps other assumptions as well.
A somewhat more complex example, in which this occurs, is below. Notice
that, since it is a conditional, our *first* assumption will be for C.P.; only
then do we set up our flagged subproof for U.G. The theorem to be proved
is $((x)Fx \lor (x)Gx) \supset (x)(Fx \lor Gx)$.

From our assumption at step 1 we want to infer the consequent of
the conditional to be proved, $(x)(Fx \lor Gx)$. Since the latter is a universal
statement, it will be derived by U.G., and thus we need a flagged subproof.
The *instance* of this universal formula will be $(Fa \lor Ga)$, so this is what
we need to aim at. Notice, however, that we are *not* permitted to infer
$(Fa \lor Ga)$ from step 1 by U.I., since step 1 is not a universal statement.
Rather, it is a disjunction, so appropriate rules might be D.S. or C.E. How,
then, do we get $(Fa \lor Ga)$? At this point, if you are stumped, you should
remember your old friend I.P.; whenever you get stuck in a proof, a promising approach is to assume the *opposite* of what you want, try to get a
contradiction, and so derive what you need by I.P. If we use this approach
here, we would assume $\sim (Fa \lor Ga)$ at step 3, and then try to derive a
contradiction. As it turns out, this is not difficult.

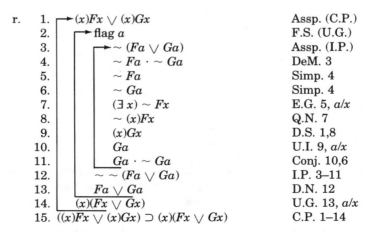

r. 1. ┌─► $(x)Fx \lor (x)Gx$ Assp. (C.P.)
    2. │ ┌─► flag a F.S. (U.G.)
    3. │ │ ┌─► $\sim (Fa \lor Ga)$ Assp. (I.P.)
    4. │ │ │ $\sim Fa \cdot \sim Ga$ DeM. 3
    5. │ │ │ $\sim Fa$ Simp. 4
    6. │ │ │ $\sim Ga$ Simp. 4
    7. │ │ │ $(\exists x) \sim Fx$ E.G. 5, a/x
    8. │ │ │ $\sim (x)Fx$ Q.N. 7
    9. │ │ │ $(x)Gx$ D.S. 1,8
   10. │ │ │ Ga U.I. 9, a/x
   11. │ │ └── $Ga \cdot \sim Ga$ Conj. 10,6
   12. │ │ $\sim \sim (Fa \lor Ga)$ I.P. 3–11
   13. │ └── $Fa \lor Ga$ D.N. 12
   14. └── $(x)(Fx \lor Gx)$ U.G. 13, a/x
   15. $((x)Fx \lor (x)Gx) \supset (x)(Fx \lor Gx)$ C.P. 1–14

More examples of theorems are given in Exercise 5 at the end of the unit. As noted, there is nothing new to these proofs; they are simply proofs which require no premises.

This completes our exposition of the proof method for quantifier logic. The beauty of these rules is that they will be completely adequate for the proofs we will be constructing in relational logic. Many rule systems use one set of rules for one-variable logic, which we have been discussing here, and another set for relational logic, which we will begin in Unit 17. We will need a few more rules for making inferences using identity statements, but the *quantifier* rules are now complete. Be sure to do the exercises at the end of the unit. If you can do most of those without too much trouble, you have a good understanding of the proof method for predicate logic.

STATEMENT OF THE QUANTIFIER RULES, WITH ALL NECESSARY RESTRICTIONS

A. Preliminary definitions

  1. ϕx is a propositional function on x, simple or complex. If complex, it is assumed that it is enclosed in parentheses, so that the scope of any prefixed quantifier extends to the end of the formula.

  2. ϕa is a formula just like ϕx, except that every occurrence of x in ϕx has been replaced by an a.

  3. An instance of a general formula is the result of deleting the initial quantifier and replacing each variable bound by that quantifier uniformly with some name.

  4. An a-flagged subproof is a subproof that begins with the words "flag a," and ends with some instance containing a.

B. The four quantifier rules

Universal Instantiation (U.I.)

$(x)\phi x$

$/\therefore \phi a$

Existential Instantiation (E.I.)

$(\exists x)\phi x$ provided

$/\therefore \phi a$ we flag a

Universal Generalization (U.G.)

$/\therefore (x)\phi x$

Existential Generalization (E.G.)

ϕa

$/\therefore (\exists x)\phi x$

C. Flagging restrictions

R_1 A letter being flagged must be new to the proof, that is, it may not appear, either in a formula or as a letter being flagged, previous to the step in which it gets flagged.

R_2 A flagged letter may not appear either in the premises or in the conclusion of a proof.

R_3 A flagged letter may not appear outside the subproof in which it gets flagged.

EXERCISES

*1. Which of the following are correct applications of the rule cited? (Ignore flagging restrictions.)

a.	$(\exists x)Fx \cdot (\exists x)Gx \;/\therefore\; Fa \cdot Ga$	E.I.
b.	$(x)(Fx \supset (Gx \lor Hx)) \;/\therefore\; Fa \supset (Ga \lor Ha)$	U.I.
c.	$(\exists x)(Fx \supset Gx) \;/\therefore\; Fa \supset Ga$	E.I.
d.	$(x)(Fx \cdot Gx) \supset (Hx \cdot Ix) \;/\therefore\; (Fa \cdot Ga) \supset (Ha \cdot Ia)$	U.I.
e.	$Fb \lor Gb \;/\therefore\; (\exists x)(Fx \lor Gx)$	E.G.
f.	$(x)((Fx \cdot Gb) \supset (Hx \cdot Ib)) \;/\therefore\; (Fa \cdot Gb) \supset (Ha \cdot Ib)$	U.I.
g.	$\sim (Fb \supset Gb) \;/\therefore\; \sim (x)(Fx \supset Gx)$	U.G.
h.	$Fb \lor Gc \;/\therefore\; (\exists x)(Fx \lor Gc)$	E.G.
i.	$\sim (x)Fx \;/\therefore\; \sim Fa$	U.I.
j.	$(\exists x) \sim (Fx \supset Gx) \;/\therefore\; \sim (Fb \supset Gb)$	E.I.
k.	$Fa \supset (Ga \supset Ha) \;/\therefore\; (x)Fx \supset (x)(Gx \supset Hx)$	U.G.
l.	$(\exists x)(Fb \lor (Gx \supset Hc)) \;/\therefore\; Fb \lor (Ga \supset Hc)$	E.I.
m.	$(\exists x)Fx \cdot (Gx \lor Hx) \;/\therefore\; Fa \cdot (Ga \lor Ha)$	E.I.
n.	$(x)((Fa \cdot Gx) \supset (Hx \lor Ic)) \;/\therefore\; (Fa \cdot Ga) \supset (Ha \lor Ic)$	U.I.
o.	$(x)((Fa \cdot Gx) \supset (Hx \lor Ic)) \;/\therefore\; (Fa \cdot Ga) \supset (Hc \lor Ic)$	U.I.
p.	$(x)((Fa \cdot Gx) \supset (Hx \lor Ic)) \;/\therefore\; (Fa \cdot Gb) \supset (Hb \lor Ic)$	U.I.
q.	$(x)((Fa \cdot Gx) \supset (Hx \lor Ic)) \;/\therefore\; (Fc \cdot Gc) \supset (Hc \lor Ic)$	U.I.
r.	$\sim Fa \supset \sim Ga \;/\therefore\; \sim (x)Fx \supset \sim (x)Gx$	U.G.
s.	$\sim Fa \supset \sim Ga \;/\therefore\; (x)(\sim Fx \supset \sim Gx)$	U.G.
t.	$\sim (Fa \cdot Ga) \;/\therefore\; \sim (\exists x)(Fx \cdot Gx)$	E.G.

2. Construct proofs for the following arguments.

a. $(x)(Cx \supset Dx),\ (x)(Ex \supset \sim Dx) \;/\therefore\; (x)(Ex \supset \sim Cx)$

*b. $(x)(Fx \supset \sim Gx),\ (\exists x)(Hx \cdot Gx) \;/\therefore\; \sim (x)(Hx \supset Fx)$

c. $(x)(Kx \supset Lx),\ (x)((Kx \cdot Lx) \supset Mx) \;/\therefore\; (x)(Kx \supset Mx)$

d. $(x)(Sx \supset (Tx \supset Ux)),\ (x)(Ux \supset (Vx \cdot Wx)) \;/\therefore\; (x)((Sx \cdot Tx) \supset Vx)$

*e. $(x)(Tx \supset (Fx \cdot Dx)),\ \sim (x)(Tx \supset \sim Bx) \;/\therefore\; \sim (x)(Dx \supset \sim Bx)$

f. $(x)((Ax \lor Bx) \supset (Cx \cdot Dx)),\ (x)((Cx \lor Dx) \supset (Ax \cdot Bx)) \;/\therefore\; (x)(Ax \equiv Cx)$

g. $\sim (\exists x)(Fx \cdot Gx),\ (x)(Zx \supset (Gx \lor Hx)),\ (\exists x)(Fx \cdot Zx) \;/\therefore\; \sim (x)(Fx \supset \sim Hx)$

*h. $(x)((Bx \lor Wx) \supset ((Ax \lor Fx) \supset Sx)) \;/\therefore\; (x)((Bx \cdot Fx) \supset Sx)$

i. $(x)(Cx \supset (Fx \lor Nx)),\ (x)(Fx \supset Bx),\ \sim (x)(Cx \supset Bx) \;/\therefore\; \sim (x)(Cx \supset \sim Nx)$

*j. $(\exists x)(Cx \cdot (\sim Sx \supset (Vx \vee Wx)))$, $(x)(Vx \supset \sim Cx)$, $\sim (\exists x)(Wx \cdot Cx)$
/∴ $(\exists x)(Sx \cdot \sim Wx)$

*k. $\sim (\exists x)(Fx \cdot \sim (Gx \cdot Hx))$, $(x)((Gx \vee Sx) \supset Zx)$, $\sim (\exists x)(Zx \cdot Ax)$
/∴ $(x)(Fx \supset \sim Ax)$

l. $(x)(Fx \supset (Hx \vee Ix))$, $\sim (x)(Fx \supset Hx)$, $(x)(Ix \supset (\sim Zx \equiv Hx))$
/∴ $\sim (x)(Fx \supset \sim Zx)$

m. $(x)(Fx \supset (Bx \equiv \sim Tx))$, $\sim (x)(Fx \supset (Bx \vee Cx))$, $\sim (\exists x)(Tx \cdot \sim (Dx \supset Cx))$
/∴ $(\exists x)(Fx \cdot \sim (Cx \vee Dx))$

n. $\sim (\exists x)((Ax \cdot Bx) \cdot \sim Cx)$, $\sim (\exists x)(Ax \cdot \sim Bx)$, $(x)(Cx \supset \sim (Sx \vee Tx))$
/∴ $\sim (\exists x)(Ax \cdot Tx)$

*o. $(x)((Ax \vee Bx) \supset (Gx \cdot \sim Hx))$, $(x)(Gx \supset Hx)$, $(x)((Dx \cdot Ex) \supset Bx)$,
$\sim (\exists x)(Px \cdot \sim Ex)$ /∴ $\sim (\exists x)(Px \cdot (Ax \vee Dx))$

p. $\sim (\exists x)(Ax \cdot Bx)$, $\sim (\exists x)(Ax \cdot \sim Cx)$, $\sim (x)(Ax \supset (Fx \vee Gx))$, $(x)(Hx \supset Gx)$
/∴ $\sim (x)(Cx \supset (Bx \vee Hx))$

q. $(x)((Ax \vee Bx) \supset \sim (Gx \cdot \sim Hx))$, $\sim (\exists x)(Ax \cdot Dx)$, $\sim (\exists x)(Hx \cdot \sim Dx)$
$(x)(Fx \supset (Gx \equiv Wx))$ /∴ $\sim (\exists x)((Ax \cdot Fx) \cdot Wx)$

r. $(x)(Ax \equiv \sim Cx)$, $\sim (\exists x)(Sx \cdot Cx \cdot Ex \cdot \sim (Ax \vee Bx))$
/∴ $\sim (\exists x)(Ex \cdot \sim Bx \cdot Sx \cdot \sim Ax)$

s. $(x)((Ax \equiv Bx) \supset (Cx \supset (Zx \vee Wx)))$, $\sim (\exists x)(Cx \cdot (Ex \supset Wx))$,
$\sim (\exists x)(Ex \cdot Cx \cdot Zx)$, $(\exists x)(Cx \cdot (\sim Bx \equiv \sim (Dx \supset Zx)))$,
$\sim (\exists x)(Cx \cdot \sim (Wx \vee Dx))$ /∴ $\sim (x)(Ax \supset Zx)$

t. $\sim (\exists x)(Px \cdot Qx \cdot (Rx \equiv Tx))$, $(x)(Px \equiv (\sim Sx \vee (Ax \supset Bx)))$,
$\sim (\exists x)(Ax \cdot \sim Sx)$, $\sim (\exists x)(Qx \cdot \sim (Px \vee Tx))$, $(x)((Px \equiv Rx) \supset Ax)$
/∴ $(x)(Qx \supset (Tx \vee Bx))$

*3. Symbolize the following and then construct proofs for them.

a. Doctors and lawyers are well-paid professionals. No well-paid doctor eats at McDonald's, and no professional shops at Woolworth's. Therefore, no doctor eats at McDonald's or shops at Woolworth's. (Use Dx, Lx, Wx, Px, Mx, and Sx for "x shops at Woolworth's.")

b. Anyone who repairs his own car is highly skilled and saves a lot of money on repairs. Some people who repair their own cars have menial jobs. Therefore, some people with menial jobs are highly skilled. ($Px \equiv x$ is a person; Rx, Hx, $Sx \equiv x$ saves money; $Mx \equiv x$ has a menial job)

c. Some policemen are forced to moonlight (take a second job). No individual who works at two jobs can be fully alert on the job. A policeman who is not fully alert on the job will make errors of judgment. So, some policemen will make errors of judgment. (Px; $Jx \equiv x$ takes two jobs; $Ax \equiv x$ is fully alert on the job; $Ex \equiv x$ makes errors)

d. Some juveniles who commit minor offenses are thrown into prison, and any juvenile thrown into prison is exposed to all sorts of hardened criminals. A juvenile who is exposed to all sorts of hardened criminals will become bitter and learn more techniques for committing crimes. Any individual who learns more techniques for committing crimes is a menace to society, if he is bitter. Therefore, some juveniles who commit minor offenses will be menaces to society. (Use Jx, Cx, Px, Ex, Bx, Tx, and Mx)

4.   Construct proofs for the following problems, which contain truth-functional compounds.

  a. $(x)(Fx \supset Gx)$, $\sim ((\exists x)Gx \lor (\exists x)Hx)$ /∴ $\sim (\exists x)Fx$

  b. $(x)(Fx \supset \sim Gx) \supset (x)(Fx \supset \sim Hx)$ /∴ $(\exists x)(Fx \cdot Hx) \supset (\exists x)(Fx \cdot Gx)$

  c. $(x)((Fx \lor Gx) \supset Hx)$, $(x)Hx \supset (x)Sx$ /∴ $(x)Fx \supset (x)Sx$

  d. $(x)Fx \lor (x)Gx$, $(x)(Fx \supset Hx)$ /∴ $(\exists x) \sim Gx \supset (x)Hx$

*e. $(\exists x)Fx \supset (x)(Gx \lor Hx)$, $(x)(Fx \supset \sim Gx)$ /∴ $\sim (\exists x)Hx \supset (x) \sim Fx$

*f. $(\exists x)Fx \supset (x)(Hx \supset \sim Jx)$, $\sim (\exists x)(Sx \cdot \sim Jx)$ /∴ $(x)Fx \supset \sim (\exists x)(Hx \cdot Sx)$

  g. $(x)(Fx \supset Gx) \supset (x)(Fx \supset Hx)$, $(\exists x)(Fx \cdot \sim Hx)$, $(\exists x) \sim Gx \supset \sim (x)Jx$
    /∴ $(\exists x) \sim Jx$

*h. $\sim ((x)Fx \lor (\exists x)Gx)$, $(x)(Hx \supset Fx)$, $((\exists x) \sim Hx \cdot (x)Sx) \supset (\exists x)Gx$
    /∴ $(\exists x) \sim Sx$

  i. $((\exists x)Fx \lor (\exists x)Gx) \supset (\exists x) \sim Hx$, $(x)(Hx \lor Px)$, $\sim (\exists x)(Px \cdot Qx)$
    /∴ $(x)Qx \supset (x) \sim Fx$

*j. $\sim ((\exists x) \sim Fx \cdot (\exists x)(Gx \cdot \sim Hx))$, $(x)Hx \supset \sim (\exists x)(Zx \cdot Wx)$, $\sim (x)(Wx \supset Fx)$
    /∴ $\sim (\exists x) \sim Zx \supset \sim (x)Gx$

5.   Construct proofs for the following theorems of predicate logic.

*a. $(x)(Fx \supset Gx) \supset (\sim (x)Gx \supset \sim (x)Fx)$

  b. $\sim (\exists x)(Fx \cdot Gx) \supset ((x)Fx \supset \sim (\exists x)Gx)$

*c. $(x)(Fx \cdot Gx) \equiv ((x)Fx \cdot (x)Gx)$

*d. $\sim ((\exists x)Fx \lor (\exists x) \sim Gx) \supset (x)(Fx \supset Hx)$

  e. $((\exists x)Fx \lor (\exists x)Gx) \equiv (\exists x)(Fx \lor Gx)$

  f. $(x)(Fx \supset \sim (Gx \lor Hx)) \supset \sim (\exists x)(Fx \cdot Hx)$

UNIT 16
Invalidity in Quantifier Logic

A. INTRODUCTION

As we saw at the end of Unit 9, the proof method can only be used to show that arguments are valid, not that they are invalid. If we do find a proof (and have done it correctly), then we can be sure that the premises imply the conclusion, but if we do *not* find a proof we cannot infer anything. It *may* be that there is no proof (that is, the argument is invalid), but it may also be that we have not tried hard enough, or are just tired, bored, or lacking in ingenuity. Failure to come up with a proof cannot be used as evidence that the argument is invalid. (Indeed, there are cases in the history of mathematics where proofs were not found until hundreds of years after a problem was posed!) Remember also that we cannot infer that an argument is invalid even if we are able to prove the *negation* of the conclusion, since we may be able to prove *both* the conclusion and its negation if the premises are inconsistent. What can we do, then, to show that an argument is invalid in quantifier logic?

What we need to do is revert again to the semantic method and reintroduce the notion of a *counterexample*—an instance with true premises and a false conclusion. Only if we can find such an instance are we justified in concluding that the argument is invalid. The methods of constructing counterexamples for quantifier arguments will be the topic of this unit.

There will be two such methods, the "natural interpretation" method, in which you look for *meanings* of the predicate letters which will yield true premises and a false conclusion, and the "model universe" method, in which you find an artificial world that can be interpreted so as to make premises true and conclusion false.

B. UNIT 16 OBJECTIVES

1. Be able to demonstrate the invalidity of quantifier arguments, using the "natural interpretation" method.
2. Learn the definition of truth for the universal and existential quantifiers.
3. Be able to demonstrate the invalidity of quantifier arguments, using the "model universe" method.

C. UNIT 16 TOPICS

1. The Natural Interpretation Method

A counterexample in quantifier logic is the same as in sentential logic: an instance of an argument form that has all the premises true but the conclusion false. In demonstrating that an argument is invalid, then, we need to *determine its form* and then *try to find an instance of the form with true premises and a false conclusion.*

To represent the *form* of an argument in predicate logic we will simply use its symbolization.[1] The aim of the natural interpretation method is to find an interpretation, an assignment of meanings to the predicate letters in the symbolization, such that all the premises turn out to be true with a false conclusion. The procedure will consist of two steps: (1) specifying the domain of discourse, the set of objects over which the bound variables range, the things to which the formulas could be referring, and (2) reinterpreting the predicate letters by assigning to them propositional functions that make the premises true and the conclusion false.

We can illustrate this procedure with the following argument: "All communists are in favor of socialized medicine and all socialists are in favor of socialized medicine, so all socialists are communists." This could be symbolized as $(x)(Cx \supset Fx)$, $(x)(Sx \supset Fx)$ /∴ $(x)(Sx \supset Cx)$. We can show that this argument is invalid by letting the domain be human beings, and

---

[1] It would be possible to introduce predicate variables to represent the form, but the arrangement of quantifiers and operators would be exactly the same, so we will bypass this extra complication.

assigning meanings as follows: $Cx \equiv x$ is a normal man, $Sx \equiv x$ is a normal woman, $Fx \equiv x$ has a brain. The two premises, thus interpreted, say that all normal men have brains and all normal women have brains, and both are true. The conclusion, however, says that all normal women are normal men, and this is false. Thus, we have an instance of the argument form, an interpretation of the formulas, in which premises are true and conclusion is false, which shows that the form is invalid.

It is essential to begin this process by stating the domain of discourse, the set of objects in which the predicates will be interpreted. This is because a sentence with interpreted predicates may be true in one domain but false in another. If Wx means "x walks before the age of six months" and Px means "x is precocious," for instance, then the formula $(x)(Wx \supset Px)$, which would mean "Any creature that walks before the age of six months is precocious," would be *true* in the domain of humans but *false* in the domain of cats. You may take as your domain any set of objects, including the entire universe (the universal domain), but it is usually easier to find counterexamples by restricting the domain to a specific set, such as humans, animals, or numbers.

Once you have selected the domain, you then pick out propositional functions to assign to the predicate letters, and you must make sure that these functions *make sense when applied to the individuals in the domain*. If your domain is numbers, for instance, it will not do to pick as propositional functions "x is red" or "x is cold," since the predicates "red" and "cold" are simply not applicable to numbers. Also, keep in mind that you must be *consistent* in your assignments of functions to variables; if you assign "x is a dog" to Ax at one point in an argument form, you must also assign "x is a dog" to every other occurrence of Ax in that form. One thing that may make your task easier in certain cases is that you *may* use the *same* propositional function for different predicate letters.

Another thing, which seems obvious but needs to be said, is that you must make sure that the interpretation you come up with really is a substitution instance of the argument form. Don't interpret the formula $(x)((Ax \cdot Bx) \supset Cx)$, for instance, as "Anything that *either* flies *or* swims is an animal." Be especially careful with E propositions, of the form $(x)(Fx \supset \sim Gx)$; read these as "No F is a G," rather than "All F are not G," since the latter is ambiguous.

Finally, you must make sure that your interpretation is one in which *there is no question* that the premises are all true and the conclusion false; an interpretation in which this is in doubt cannot be considered a counterexample. It would not do, for instance, to use the following as an interpretation for the argument form we considered earlier: $(x)(Ax \supset Bx)$, $(x)(Cx \supset Bx) / \therefore (x)(Cx \supset Ax)$. "All Baptists believe evolution is false, and all Mormons believe evolution is false, so all Mormons are Baptists." The conclusion is certainly false but the premises are false as well, so it does not serve as a counterexample to the argument form. Again, *you must be sure in your interpretation that premises are all true and conclusion is false.*

In order to make sure that you are not overlooking anything, you should use the following systematic procedure in doing your problems: (1) State the domain of discourse. (2) Interpret the predicate letters by assigning some English predicate to each; that is, assign to each expression such as Ax or Bx some propositional function such as "x is a dog." (3) Write out the fully interpreted English sentences, with the meanings you have assigned to the predicate letters plugged into the appropriate slots. At this point, if you have come up with a counterexample, it should be evident that premises are true and conclusion false. Just so there is no question, however, you should complete a fourth and final task: (4) Justify your truth value assignments. In writing down your answers, always spell everything out in detail; do not just give the meanings of the predicate letters and expect the reader to figure out what the sentences mean, and whether they are true or false! Writing everything out will help you make sure that you do in fact have a counterexample; if you are too sketchy, you may overlook some crucial factor that makes your answer wrong.

We will illustrate this four-step procedure with the following problem, and you can use this as a model for your answers. To be very explicit, we will number the premises and conclusion as (1), (2), and (3). The form to be shown invalid is: (1) $(x)(Ax \supset Bx)$, (2) $(\exists x)(Ax \cdot Cx)$ /∴ (3) $(x)(Cx \supset Bx)$.

MODEL ANSWER

1. The domain is the entire universe.
2. $Ax \equiv x$ is a cat; $Bx \equiv x$ is a mammal; $Cx \equiv x$ has four legs.
3. (1) All cats are mammals. (2) Some cats have four legs. (3) All four-legged things are mammals.
4. (1) is a biological truth. (2) is true because my cat has four legs. (3) is false because there are some four-legged things that are not mammals, for instance, alligators.

A slightly more complex argument would be the following, which we will do rather informally: $(x)((Ax \lor Bx) \supset \sim Cx)$, $(x)((Dx \cdot Ex) \supset Cx)$ /∴ $(x)(Dx \supset \sim Ax)$. We can show that this is invalid by using the following interpretation, taking the set of human beings as our domain: $Ax \equiv x$ has been a U.S. president; $Bx \equiv x$ has been a U.S. vice-president; $Cx \equiv x$ is female; $Dx \equiv x$ is a parent; $Ex \equiv x$ has given birth. The first premise then reads "No president or vice-president of the United States has been female." This is, unfortunately, true. The second premise reads "Any parent who has given birth is female." This is also true (although biological science keeps making progress). The conclusion, however, says "No parent has been president of the United States," and this, of course, is false. Thus we have a counterexample, and so the form is invalid.

The natural interpretation method can also be used on arguments containing individual constants; we simply assign to the constant some name so that, in conjunction with the interpretation of the predicates, the premises turn out to be true and the conclusion false. In Unit 15 we noted that the following argument could not be proved because of restriction 2, that a flagged variable may not appear in the conclusion of a proof: $(\exists x)Fx$ /∴ Fa. Using the natural interpretation method, we can now see very easily that the argument is invalid. If we let Fx mean "x is a former U.S. president," and let a stand for Florence Nightingale, then the premise is true, since it says that former U.S. presidents exist (Nixon is one), but the conclusion is false, since it says that Florence Nightingale is a former U.S. president.

A more challenging example would be the following: $(x)(Mx \supset \sim Wx)$, $(\exists x)(Ux \cdot Mx)$, Wa /∴ $(\exists x)(Ux \cdot \sim Mx)$. Here if we use Mx for "x is male," Wx for "x is female," Ux for "x was a U.S. president," and a for Betty Ford, then we have a counterexample. The premises are all true, since they say, respectively, "No males are females," "Some U.S. presidents were male," and "Betty Ford is female." The conclusion, however, says that some U.S. presidents were *not* male, and this is false. Thus the argument is invalid.

The natural interpretation method can be quite entertaining: you can think up all sorts of crazy examples to show invalidity. Give your imagination free rein. There are some drawbacks to this method, however, that make it unsuitable for all cases. In the first place, you do have to use your imagination; there is no precise method, no algorithm, for coming up with a counterexample. If your imagination runs dry, you are just out of luck. Furthermore, for some argument forms the premises and conclusion are so complex that it would tax even the liveliest imagination to come up with an interpretation that works. Look at some of the proofs in the last unit, for instance. Finally, of course, you do have to be sure that the premises are true and the conclusion false in the interpreted formulas. Since this method does not always work well, it is sometimes necessary to resort to the "model universe" method, for which there is an algorithm (in principle), and which thus requires almost no ingenuity or creativity. We will discuss the model universe method in Section 3, but first it will be necessary to explain the truth conditions for quantifier statements. This we will do in the next section.

2. Truth Conditions for Quantifier Statements

In order to understand the model universe method, you must know something about the semantics, or truth conditions, for quantifiers; otherwise the procedure is not going to make much sense. Keeping in mind that what we need to do is to find instances with *true* premises and a *false* conclusion, it would behoove us now to find out what it means for a quantifier statement to be true or false. (This was taken for granted in the preceding section.)

The definition of truth for quantified statements is just what you would expect given the meanings of "all" and "some," except that truth or falsity is always relative to a particular domain. (The sentence "Everything has a heart," for instance, is true in the domain of mammals, but false in the domain of all living organisms.) Since a universal statement $(x)\phi x$ says that *each* individual (in the domain) has the property ϕ, it will naturally be true only if each individual does in fact have the property. This means that *each of the instances ϕa, ϕb, ϕc, . . . is true*, where a, b, c are names of the individuals in the domain. $(x)\phi x$ *will be false just in case at least one instance is false.*[2]

Similarly, since an existential statement $(\exists x)\phi x$ says that *at least one* individual has the property, ϕ, *it will be true just in case at least one of the instances ϕa, ϕb, ϕc, . . . is true*. Conversely, $(\exists x)\phi x$ *will be false if and only if every one of its instances if false.*

There is a very strong analogy between the universal quantifier and conjunction. $(x)\phi x$ is true if and only if ϕa, ϕb, ϕc, . . . are all true, but this amounts to saying that the *conjunction* $(\phi a \cdot \phi b \cdot \phi c$ . . .) is true. Also, there is an analogy between the existential quantifier and disjunction, since $(\exists x)\phi x$ is true just in case at least one of ϕa or ϕb or ϕc . . . is true, which means that the disjunction $(\phi a \lor \phi b \lor \phi c$ . . .) is true. The universal and existential statements are not the *same* as conjunctions and disjunctions, because the number of their elements is indefinite, but we do have a kind of quasi-equivalence which we might state as follows:

$$(x)\phi x \approx (\phi a \cdot \phi b \cdot \phi c \; . \; . \; .)$$
$$(\exists x)\phi x \approx (\phi a \lor \phi b \lor \phi c \; . \; . \; .)$$

This analogy between universal statements and conjunctions, and existential statements and disjunctions, will be essential when we come to the model universe method in the next section.

3. The Model Universe Method

For a quantifier argument form to be *valid*, it must be truth-functionally valid in *any* domain whatsoever, since logic is supposed to

---

[2] This is the substitution interpretation of the quantifiers, and is adequate for our purposes. There is another kind of semantics for quantifiers, however, which is more complex and which may be needed for very large domains. This is a topic which should be discussed in a more advanced course, so we will merely note its existence here, and settle for our simpler semantics.

be applicable to any subject matter. This means that if an argument form is *invalid* in *some specific* domain, it is invalid, period, just as a sentential argument form is invalid if it has a single counterexample.

The *model universe* method makes use of the fact that it takes only a single domain to show that an argument form is invalid, and it uses highly restricted, artificial domains, in which it is relatively easy to come up with a counterexample. In using this method we *pick out the domain*, which will sometimes consist of only a single individual (though more often two or three); *rewrite the quantifier statements in that domain*, making use of the definition of truth given above; and then *try to find, within that rewritten argument form, an instance with true premises and a false conclusion*. This may sound a little mysterious in the abstract, so let us proceed to specifics.

As usual, the first thing we must do in finding an interpretation for an argument form is to state the domain of discourse. Now, the strange thing about the model universe method is that *it does not matter what the individuals are in the domain*; the only thing that counts is *how many* individuals are in the domain.[3] This saves us a lot of work, since we do not have to rack our brains trying to figure out whether we should use Reagan, Nixon, the dog down the street, the number 5, etc. All we have to do is pick the right *number* of individuals for our domain, and we may then arbitrarily designate them as *a*, *b*, *c*, and so on. It is customary to indicate the domain—the set of objects—by enclosing the names in braces, so a one-member domain would be designated by {*a*}, a two-member domain by {*a*, *b*}, a three-member domain by {*a*, *b*, *c*}, etc. (The question of how we know how many individuals to include in the domain will be discussed later.)

Once we have our domain, we *apply the definition of truth for quantifier statements within that domain*, making use of the analogy between universal statements and conjunctions and between existential statements and disjunctions. Since a universal statement is true if and only if each of its instances is true, *we rewrite the universal formulas as conjunctions of all the instances in the domain*, and since an existential statement is true if and only if at least one of its instances is true, *we rewrite the existential formulas as disjunctions of all the instances in the domain*. Since the rewritten formulas are formulas of *sentential logic*, we may simply apply the truth table method to see whether there is an instance with true premises and a false conclusion.

We will go through a number of examples to illustrate the process of rewriting quantified formulas within particular domains. Suppose our domain is the two-member set {*a*, *b*}, and we have the universal formula (*x*)*Fx*. *To say that everything is an F, within the domain* {*a*, b}, *is simply*

---

[3] This is a fact that has been formally proved about our quantifier logic.

to say that a is an F and that b is an F, that is, *Fa and Fb*. Thus the rewritten formula would be $(Fa \cdot Fb)$. If the universal formula were $(x)((Fx \lor Gx) \supset Hx)$, the rewritten version would be the conjunction $((Fa \lor Ga) \supset Ha) \cdot ((Fb \lor Gb) \supset Hb)$. If the domain were $\{a, b, c, d\}$, then the first formula would be rewritten as $(Fa \cdot Fb \cdot Fc \cdot Fd)$.

Again, *universal formulas will be rewritten as conjunctions of all the instances in the domain*. What if there is just *one* instance—if our domain is just $\{a\}$? Then to say *everything* has the property F, $(x)Fx$, is simply to say Fa, since a is all there is in the domain. $(x)((Fx \lor Gx) \supset Hx)$ would be rewritten as $((Fa \lor Ga) \supset Ha)$ in a domain with just one individual.

Existential formulas will be rewritten as disjunctions of all the instances in the domain. Thus if our domain is $\{a, b\}$, the formula $(\exists x)Fx$ will be rewritten as $(Fa \lor Fb)$. The formula $(\exists x)(Fx \cdot Gx)$ will come out as $(Fa \cdot Ga) \lor (Fb \cdot Gb)$, and the more complex formula $(\exists x)((Fx \cdot \sim Gx) \cdot (Hx \supset \sim Jx))$ will be rendered as $((Fa \cdot \sim Ga) \cdot (Ha \supset \sim Ja)) \lor ((Fb \cdot \sim Gb) \cdot (Hb \supset \sim Jb))$. If our domain is $\{a, b, c, d\}$, the first formula, $(\exists x)Fx$, will be rewritten as $(Fa \lor Fb \lor Fc \lor Fd)$, and the second, $(\exists x)(Fx \cdot Gx)$, will be rewritten as $(Fa \cdot Ga) \lor (Fb \cdot Gb) \lor (Fc \cdot Gc) \lor (Fd \cdot Gd)$. What happens if there is only one member of the domain? If we say *something* has the property F, and a is the only thing in the domain, then it must be a that has the property F, so we rewrite $(\exists x)Fx$ in the domain $\{a\}$ simply as Fa. The formula $(\exists x)(Fx \cdot Gx)$ would be rewritten in $\{a\}$ as $(Fa \cdot Ga)$. Note that within a domain of just one individual, the existential and universal statements mean exactly the same; they both say that the single individual has the property.

Once we have a domain, and have reformulated the quantifier statements within that domain, we can proceed to try to find a counterexample, an instance with all true premises but a false conclusion. This is essentially just an exercise in the short truth table method in sentential logic, since what we end up with, once we do the rewriting, is a sentential argument form. Let us take an example to illustrate. Suppose I argue that some quarks are massless and some neutrinos are massless, so some quarks are neutrinos. This could be symbolized as $(\exists x)(Qx \cdot Mx)$, $(\exists x)(Nx \cdot Mx) / \therefore (\exists x)(Qx \cdot Nx)$. Now suppose you think there is something quirky about this argument and set about to show that it is invalid. You pick (for reasons which will be discussed later) a domain with two individuals, $\{a, b\}$, and within this domain the reformulation of the argument form would be as follows: $(Qa \cdot Ma) \lor (Qb \cdot Mb)$, $(Na \cdot Ma) \lor (Nb \cdot Mb) / \therefore (Qa \cdot Na) \lor (Qb \cdot Nb)$. Now if we make Qa and Nb false, the conclusion is false, since both disjuncts turn out to be false. But we can still make the premises true by making Na and Ma, and Qb and Mb true. Thus we have a counterexample in this domain, so the argument is invalid.

In some cases, the argument will be invalid even in a domain with just one individual. If someone argues, for instance, that all communists opposed the war in Vietnam and all communists wants to abolish private property, so anyone who opposed the war in Vietnam wants to abolish private property, you could show the argument invalid as follows. It could be formalized as $(x)(Cx \supset Bx)$, $(x)(Cx \supset Ax)$ /∴ $(x)(Bx \supset Ax)$, and rewriting the argument in a domain with just one individual, $\{a\}$, would give us $(Ca \supset Ba)$, $(Ca \supset Aa)$ /∴ $(Ba \supset Aa)$. This formulation can be invalidated by making Ba true and Aa and Ca false. This gives us true premises and a false conclusion, so the argument form is invalid. If an argument form is not invalid in a domain with just one individual, you must, of course, go on to test larger domains.

When you are using the short truth table method on a particular interpretation, looking for a value assignment which gives true premises with a false conclusion, remember that you cannot stop and conclude that there is no counterexample in that domain just because the first assignment fails. You must go through *all* possible ways of making premises true and conclusion false before you conclude that none of them works. If you *do* find an invalidating instance right off, fine; then you can stop, knowing that the argument is invalid. However, especially with these rather complex arguments, there will be many possible ways of getting counterexamples, so if at first you don't succeed, try, try again. It would probably be a good idea at this point to go back and review the section in sentential logic on the short truth table method.

In writing out your answers to model universe invalidity problems, you should follow a four-step procedure analogous to that of the natural interpretation method. Here the four steps will be:

1. Pick a domain.
2. Rewrite the quantifier statements, changing universal formulas to conjunctions of all instances in the domain and existential formulas to disjunctions of all instances in the domain.
3. Interpret the predicate letters by assigning truth values to each of the simple sentences, such as Fa, Fb, and so on.
4. Indicate how the truth table computations on these rewritten formulas yield true premises with a false conclusion.

An example of a model answer to a model universe invalidity problem would be the following, where the argument form is: (1) $(\exists x)(Fx \cdot Gx)$ (2) $(\exists x)(Gx \cdot Hx)$ /∴ (3) $(\exists x)(Fx \cdot Hx)$.

1. Domain = $\{a, b\}$.
2. (1) $(Fa \cdot Ga) \lor (Fb \cdot Gb)$ (2) $(Ga \cdot Ha) \lor (Gb \cdot Hb)$
   /∴ (3) $(Fa \cdot Ha) \lor (Fb \cdot Hb)$
3. $Fa = T$, $Ga = T$, $Ha = F$, $Fb = F$, $Gb = T$, $Hb = T$

4. (1) is true because $(Fa \cdot Ga)$ is true. (2) is true because $(Gb \cdot Hb)$ is true. (3) is false because both $(Fa \cdot Ha)$ and $(Fb \cdot Hb)$ are false, since Ha and Fb are false.

You might also combine 2, 3, and 4 by simply exhibiting the truth table computation as follows:

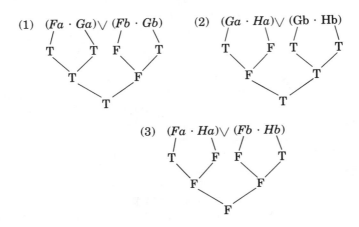

(1) $(Fa \cdot Ga) \lor (Fb \cdot Gb)$

(2) $(Ga \cdot Ha) \lor (Gb \cdot Hb)$

(3) $(Fa \cdot Ha) \lor (Fb \cdot Hb)$

We must now say a few words about proving invalidity for argument forms that contain truth-functional compounds, such as $(x)Fx \supset (x)Gx$ $/\therefore (x)(Fx \supset Gx)$. In cases such as this, simply rewrite each individual *subformula* which is of quantifier form according to your chosen domain, and then apply the short truth table method. In a domain of two individuals, for instance, the reformulation of $(x)Fx$ would be simply $Fa \cdot Fb$, and the reformulation of $(x)Gx$ would be $Ga \cdot Gb$. We know that rewriting $(x)(Fx \supset Gx)$ would yield $(Fa \supset Ga) \cdot (Fb \supset Gb)$. If we put this all together, we have the following rewrite of the argument form above for the domain $\{a, b\}$:

$$(Fa \cdot Fb) \supset (Ga \cdot Gb) \quad /\therefore (Fa \supset Ga) \cdot (Fb \supset Gb)$$

It is then not difficult to find a counterexample. If we make Fa true and Ga false, we have a false conclusion. We can then make the premise true by making Fb false, since this means that the antecedent of the conditional, $(Fa \cdot Fb)$, is false, so that the conditional itself is true. Thus we have a true premise with a false conclusion in the domain $\{a, b\}$, and so the argument form is invalid. Notice that in a domain with just one individual, $\{a\}$, the form would have no counterexample, since the rewritten argument would be simply $(Fa \supset Ga) /\therefore (Fa \supset Ga)$.

The proper format for the answer to the problem above would be the following:

1. Domain = {*a, b*}

2, 3, 4. (*Fa · Fb*) ⊃ (*Ga · Gb*)   /∴ (*Fa ⊃ Ga*) · (*Fb ⊃ Gb*)

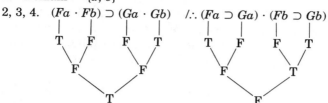

There will be several invalidity problems containing truth-functional compounds in the exercises. The method is the same as for those containing only "pure" quantifier statements; just rewrite each quantifier *subformula* according to your chosen domain, and then look for a counterexample.

The model universe method also works for arguments containing constants. Just pick a domain containing the individual denoted by the constant (let '*a*' denote *a*); you will generally need a domain with at least one other individual. We will illustrate this method with the same arguments we used for the natural interpretation method; note that we rewrite *only* the formulas containing quantifiers.

a. (∃ *x*)*Fx* /∴ *Fa*

 1. The domain is {*a, b*}
 2. *Fa* ∨ *Fb* /∴ *Fa*
 3. *Fa* = F, *Fb* = T
 4. By truth table computations the premise is true and the conclusion is false.

b. (*x*)(*Mx* ⊃ ~ *Wx*), (∃ *x*)(*Ux · Mx*), *Wa* /∴ (∃ *x*)(*Ux · ~ Mx*)

 1. The domain is {*a, b*}

2, 3, 4. (*Ma* ⊃ ~ *Wa*) · (*Mb* ⊃ ~ *Wb*), (*Ua · Ma*) ∨ (*Ub · Mb*), *Wa*

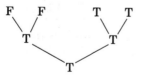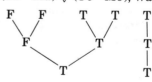

/∴ (*Ua · ~ Ma*) ∨ (*Ub · ~ Mb*)

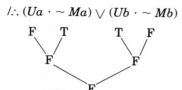

You may be wondering at this point how we know which domain to pick; why one with two individuals in the last example, rather than one, or three, or ten? There are some rules of thumb which you will want to keep in mind. In the first place, if the propositional functions in the argument, with the quantifiers removed, are truth-functionally invalid, then you will need a domain of only one individual. The following would be an example, although it is a fairly complex argument form:

$$(x)((Fx \cdot Gx) \supset \sim (Hx \lor Ix)), (x)(Gx \supset ((Px \lor Qx) \supset Hx)) \: / \therefore \: (x)(Fx \supset \sim Qx).$$

(All you need to do is make Fx and Qx true, and Gx false.) If the form is not obviously invalid in this way, then move on to a domain with two individuals, then three, then four, etc. You *will* come to an end, as we will explain in the next paragraph.

There is an upper limit on the size of the domain you need to test. After a certain point, if you have found no counterexample, then you can conclude that there *is* no counterexample and that the argument form is therefore valid. This upper limit is determined by the number of *predicate letters* in the argument form. *Where n is the number of different predicate letters in an argument form, the largest domain you need to test is one with 2^n individuals.* If there is no counterexample in this domain, then there is no counterexample in any larger domain, so if you have tested all domains up to that point you may conclude that the argument form is valid.

Another relevant fact is that if there *is* a counterexample in a domain with m individuals, then there is also a counterexample in all larger domains. It follows by contraposition (and C.Q.N.) that if there is no counterexample in a domain with m individuals, then there is also no counterexample in any *smaller* domain. Thus if there is no counterexample in a domain with 2^n individuals (where n is again the number of different predicate letters in the argument form), then there can be no counterexample in any smaller domain *or* in any larger domain; that is, there is no counterexample in *any* domain, and so the argument form is valid. This means that theoretically you need only test one domain, the one with 2^n individuals. In practice, however, this would be rather unwieldy, and it is generally quicker and easier to start with small domains and work your way up.

However, even if it does not have much practical value, it is of great theoretical interest that we need test only the one domain. Given that we have only a finite number of predicate letters, the number 2^n will always be finite, and this means that *we have a decision procedure for one-variable predicate logic.* The number of individuals we need to test will be finite, and the reformulation of the quantifier formulas will thus yield a finite sentential argument form which can be tested according to the truth table method.

Interestingly enough, when we come to relational logic—logic which requires the use of more than one variable—we will no longer have a decision procedure. Some arguments will be invalid only in a domain with an *infinite* number of individuals, so obviously we cannot test for invalidity by trying 1, then 2, then 3, and so on. We would never get to the invalidating domain! And, of course, there is no mechanical method for coming up with the proof either, so it may well be that with some arguments you have no way of knowing whether they are valid or invalid (although they must be one or the other). You may not be able to find a proof, but from that you cannot conclude that there *isn't* any proof, since you simply might not have hit on the right combination. And you may not be able to come up with a counterexample either, but you can't conclude from that that there *is* no counterexample. You will be left up in the air! This is precisely what it means for a system *not* to have a decision procedure; there is no foolproof mechanical method which will tell for every argument whether it is valid or invalid. In fact, if there were decision procedures for every part of logic and mathematics, it would be a pretty dull business. What is exciting about these disciplines, at least at the upper levels, is that they require insight, ingenuity, creativity, and even genius. We will begin to get into these more advanced levels in the next few sections, where we introduce relational predicate logic.

DEFINITIONS

1. The universal formula **(x)φx is true in a domain** if and only if each of the instances φa, φb, φc for individuals in that domain is true. (x)φx is **false in a domain** if and only if at least one of the instances is false.
2. The existential formula **(∃ x) φx is true in a domain** if and only if at least one of the instances φa, φb, φc for individuals in that domain is true. (∃ x)φx is **false in a domain** if and only if every one of the instances is false.
3. An **argument in quantifier logic is valid** if and only if there is no counterexample in any domain. It is invalid if and only if there is some domain in which there is a counterexample.

EXERCISES

1. Rewrite the following in a domain of three individuals.

*a.   (x)(Fx ∨ Gx)

 b.   (∃ x)(Fx · (Gx ∨ Hx))

*c.   (x)(Fx ∨ (Gx · Hx))

 d.   (x)Fx ⊃ (∃ x)Gx

*e. $(x)(Fx \lor Gx) \supset (\exists x)(Hx \cdot Ix)$

 f. $(\exists x)(Fx \cdot Gx) \supset ((\exists x)Fx \cdot (\exists x)Gx)$

2.   Use the natural interpretation method to show that the following arguments are invalid.

*a. $(x)(Ax \supset Bx), (\exists x)Ax /\therefore (x)Bx$

 b. $(x)(Ax \supset Bx), (x)(Bx \supset Cx) /\therefore (x)(Cx \supset Ax)$

*c. $(\exists x)(Fx \cdot Gx), (x)(Gx \supset Hx) /\therefore (x)(Fx \supset Hx)$

 d. $(\exists x)(Fx \cdot Gx), (\exists x)(Gx \cdot Hx) /\therefore (\exists x)(Fx \cdot Hx)$

*e. $(x)(Fx \supset \sim Gx), (\exists x)(Hx \cdot Gx) /\therefore (x)(Fx \supset \sim Hx)$

 f. $(x)(Fx \supset Gx), (\exists x)(Hx \cdot Gx), (\exists x)(Hx \cdot \sim Gx) /\therefore (x)(Fx \supset \sim Hx)$

*g. $(x))((Fx \cdot Gx) \supset Hx), (x)(Hx \supset Jx) /\therefore (x)(Fx \supset Jx)$

 h. $(x))((Fx \lor Gx) \supset Ax), (\exists x)(Fx \cdot Bx) /\therefore (x)(Bx \supset Ax)$

*i. $(\exists x)Fx \cdot (\exists x)Gx \cdot (\exists x)Hx /\therefore (\exists x)(Fx \cdot Gx \cdot Hx)$

 j. $(x)(Fx \lor Gx \lor Hx) /\therefore (x)Fx \lor (x)Gx \lor (x)Hx$

*k. $(x)Fx \supset (x)(Gx \supset Hx) /\therefore (x)((Fx \cdot Gx) \supset Hx)$

 l. $Fa, \sim Gb /\therefore (\exists x)(Fx \cdot \sim Gx)$

 m. $(x)(Fx \supset (Gx \lor Hx)), Fa, \sim Hb /\therefore (\exists x)(Fx \cdot Gx)$

 n. $(x)(Gx \supset Hx), (\exists x)(Gx \cdot Fx), Fa /\therefore Fa \cdot Ha$

*3.   Use the model universe method to show that the above are invalid.

*4.   Use the model universe method to show invalidity for the arguments that turned out invalid in Exercise 2 in Unit 1.

5.   For each of the arguments below, decide on the basis of your now well-developed logical intuitions whether it is valid or invalid. If you think it is valid, try to construct a proof; if you think it is invalid, try to construct a counterexample, using either (or both) the natural interpretation method or the model universe method.

*a.   Any senator who votes with the banking lobby has his own interests at heart. Every senator from Texas has his own interests at heart, so every senator from Texas will vote with the banking lobby. (Use Sx, Bx, Ix, and Tx)

 b.   Anyone who repairs his own car is highly skilled. Some professors repair their own cars. So some professors are highly skilled. (Use Rx, Sx, Px)

*c.   Anyone who repairs his own car is highly skilled, and some who are highly skilled are highly paid. No elementary teacher is highly paid. Therefore no elementary teacher repairs his own car. (Use Rx, Sx, Px, Tx)

 d.   Anyone who has time and patience can repair his own car. Some people do not repair their own cars. So some people do not have patience. ($Px \equiv x$ is a person, Tx, $Ax \equiv x$ has patience, Rx)

PART THREE: RELATIONAL PREDICATE LOGIC

UNIT 17

Symbolization in Relational Predicate Logic

A. INTRODUCTION

As we saw in Unit 10, at the very beginning of quantifier logic, there are certain arguments which are clearly valid but which cannot be shown to be valid using just the resources of sentential logic. Thus, it was necessary to introduce a more complex system of logic in which we analyzed the simple sentences into their components. Now that you know how to do this kind of analysis, how to symbolize one-variable quantifier statements and determine whether the arguments containing them are valid or invalid, you are going to be rewarded for your efforts by learning about arguments for which these more elaborate methods are *not* sufficient! Just as there are arguments whose validity cannot be demonstrated using only sentential logic, so are there arguments whose validity cannot be shown using only the resources of one-variable predicate logic. For such arguments we will have to move to a still deeper level of analysis.

An example of such an argument would be the following: "Some people don't like any dogs, and puppies are dogs, so some people don't like puppies." This is obviously a valid argument (however deplorable), but if we try to demonstrate its validity in monadic predicate logic by constructing a proof for it, we get nowhere. It would have to be symbolized something like the following: $(\exists\ x)(Px\ \cdot\ \sim\ Lx)$, $(x)(Ux \supset Dx)$ $/\therefore(\exists\ x)(Px\ \cdot\ \sim\ Kx)$,where $Px \equiv x$ is a person, $Lx \equiv x$ likes some dogs, $Ux \equiv x$ is a puppy, $Dx \equiv x$ is a dog, and $Kx \equiv x$ likes puppies. Notice that with this symbolization, which is the best we can do in one-variable predicate logic, there is no way to derive the conclusion from the premises. The argument is valid, however, and the question is, how do we show that it is valid? Well, we do have the universal connection between puppies and dogs in premise 2, and the first premise says that some people *like no dogs*. The conclusion says that some people *like no puppies*, and what we need to do is to symbolize the "liking" predicate in such a way that we can then make the connection between not liking dogs and not liking puppies, given the link between puppies and dogs.

We will effect this connection by introducing *relational predicates*, that is, predicates that state a relation between two or more individuals instead of just stating a property of one individual. The "liking" predicate, for instance, is clearly *not* just a one-place predicate; we don't say just that one thing *likes*, but rather that one thing likes *another*. Once we have such relational predicates available, it will be a simple matter to construct a proof for the argument above.

In this unit we first introduce relational predicates, and you will learn to symbolize singular sentences containing these predicates. We then discuss quantification over these simple sentences, which will be a little more involved than in one-variable predicate logic because we will have more than one quantifier to reckon with. We then discuss quantifier negation for many-variable predicates, and in Sections 4 and 5 we turn to "categorical" relational statements. What you will need to learn is indicated below.

B. UNIT 17 OBJECTIVES

1. Be able to recognize polyadic, in other words, many place or relational predicates, and be able to symbolize polyadic singular sentences.
2. Learn to symbolize simple quantifier sentences containing relational predicates, including those with more than one quantifier.
3. Be able to apply the quantifier negation rules to quantified relational formulas.
4. Learn to symbolize English sentences consisting of simple quantifier statements or their negations.
5. Learn to symbolize "categorical" relational statements.

C. UNIT 17 TOPICS

1. Relational Predicates and Singular Sentences

As noted above, there are certain predicates that state a relation between two or more individuals, rather than just asserting a property of a single individual. Examples of such predicates would be "is taller than," "loves," "is a brother of," "lives next door to," "is a better bridge player than," and so on. Singular sentences using these predicates, with the names italicized, would be "*John* is taller than *Mary*," "*Superman* loves *Lois*," "*Billy* is a brother of *Jimmy*," "*Nixon* lives next door to *Carter*," and "*Goren* is a better bridge player than *Bobby Fischer*." These are all *two-place* predicates because it takes two names, in addition to the predicate, to complete the sentences. Examples of sentences containing three-place predicates would be "*Chicago* is between *Detroit* and *Minneapolis*," "*John* told *Mary* about *Bob*," and "*Kissinger* introduced the *Shah* to *Nixon*." An example of a four-place relational sentence would be "*New York* is farther from *San Francisco* than *Minneapolis* is from *Dallas*." It is also possible to have five-place, six-place, and in general *n*-place predicates, for any *n*, but the ones for which we will have the most use are the two and three-place predicates.

As in one-variable predicate logic we will represent predicates by propositional functions, only here, since there will be more than one name in the singular sentence, there will be more than one variable in the propositional function. *Polyadic*, or *many-place* functions are those that contain two or more *different* variables. More specifically, a *two-place* function is one that contains two different variables, a *three-place* function is one that contains three different variables, and so on.

Some examples of many-place propositional functions are below.

x is taller than y.

x loves y.

x is a brother of y.

x is between y and z.

x told y about z.

x introduced y to z.

x is further from y than z is from w.

An *n*-place propositional function will become a singular sentence when *all* the variables are replaced with names. It is important to remember that there must be *n* names put in for the *n* variables in order to complete the singular sentence. If there are only two names, for instance, put in for a three-place function, the result is a one-place function, not a singular sentence.

We will abbreviate propositional functions in much the same way as we did in one-variable logic, except that here we will have more than one variable following the capital predicate letter. We will normally put the variables in alphabetical order following the predicate letter; thus we can abbreviate "x loves y" as Lxy, "x is taller than y" as Txy, "x is between y and z" as $Bxyz$, and so on. (More will be said shortly about the importance of the *order* in which the variables appear in the function.) As in one-variable predicate logic, you will generally be given the symbolizations for the functions; only in Exercise 1 will you have to identify and abbreviate the functions yourself.

Symbolizing purely singular sentences (those without any quantifiers) in relational predicate logic is much the same as in monadic predicate logic: first you must isolate and abbreviate the propositional function, and then put in for the variables the appropriate individual constants. The function for "Russia sold Alaska to the United States," for example, would be "x sold y to z," and could be symbolized as $Sxyz$. Since we normally use the first letter of a name as its abbreviation, the sentence then could be symbolized as $Srau$. The sentence "Beth despises Andrew" contains only a two-placed function, and could be symbolized as Dba, where Dxy abbreviates "x despises y." There are several examples of relational singular sentences below; the propositional functions are written below the sentences, and the abbreviations for each are to the right.

1.	John loves Mary. x loves y	Ljm Lxy
2.	Andrew works for the Pentagon. x works for y	Wap Wxy
3.	Texas is south of Minnesota x is south of y	Stm Sxy
4.	Rockefeller owns the Chase Manhattan. x owns y	Orc Oxy
5.	Richard Burton gave Elizabeth Taylor the Hope Diamond. x gave y to z	$Grhe$ $Gxyz$
6.	Elizabeth Taylor gave George the Bulgarian Emerald in exchange for the Hope Diamond. x gave y to z in exchange for w	$Gebgh$ $Gxyzw$

As noted earlier, the *order* of the variables and individual constants is extremely important in symbolizations in relational logic. When you are given abbreviations for the propositional functions, such as Sxy for "x is south of y" (example 3 above), *the variables will almost always be placed in alphabetical order in the abbreviated function, and the order of the variables in the unabbreviated function will tell you how the individuals are related.* In example 3, for instance, the *first-named* individual in the symbolization is supposed to be the one that is to the south of the second-named individual. Thus, we symbolize that Texas is south of Minnesota by *Stm* rather than *Smt. Smt* would say, erroneously, that Minnesota is south of *Texas.* Example 5 is particularly instructive on this issue; here the abbreviation indicates that the first constant stands for the thing *doing the giving*, the second constant stands for the thing *being given*, and the third stands for the *recipient* of the gift. Thus the symbolization must be *Grhe*, rather than *Greh*, which you might be tempted to put down if you didn't look carefully at the abbreviation.

One place in which you need to be especially careful about order is in sentences in the *passive* voice, such as "John *is loved* by Mary." We could here take the function as "*x* loves *y*," which we could abbreviate as *Lxy*. We would then symbolize the sentence as *Lmj*, rather than *Ljm*, since *Lmj* would say that Mary loves John, which means the same as saying that John is loved by Mary. *Ljm* says that John loves Mary, which is, of course, quite a different proposition.

There are a few things you need to watch in addition to being careful about the order of the letters. You may, for instance, be given *compound* sentences, such as "John loves Mary and Beth." It is important to see that this is a *conjunction* of two two-place relational sentences, rather than a three-place relational sentence. What it really says is that John loves Mary *and* John loves Beth, so it would be symbolized as (*Ljm · Ljb*). You would *not* symbolize it as *Ljm · b*; this is not a proper formula, and makes no sense in our logical system. Operators are always placed between two *formulas*, and *never* between two names. Nor would it be correct simply to *run on* the three names, as in *Ljmb*. (Somehow students find this very tempting; be on guard against it.) This is wrong because "loves" is a *two-place* predicate, not a three-place predicate, and so to have a proper sentence there must be only two individual constants following the predicate letter.

It may sometimes happen in relational predicate logic that you need the *same* individual constant in more than one place following the predicate letter; that is, the same constant may go in for two (or more) different variables. If *Lxy* means "*x* loves *y*" again, we can symbolize "John loves *himself*" as *Ljj*, where '*j*' stands for both the thing doing the loving and the thing being loved. Similarly, if *Txyz* means "*x* teaches *y* to *z*," we could symbolize "Mary taught *herself* computer science" as *Tmcm*. If *Gxyz* means "*x* gave *y* to *z*" we could symbolize "John gave *himself* to God" as *Gjjg*. Such sentences, which often use the words "himself" or "herself," are called *reflexive*; there will be some examples of such reflexive sentences among your exercises at the end of the unit.

At this point, you should be able to complete Exercises 1 and 2 at the end of the unit; be sure to check your answers to see whether you are understanding the material.

2. Multiple Quantifiers

Given propositional functions, we can get sentences either by replacing the free variables with constants or by prefixing an appropriate quantifier. In relational logic, since we have more than one variable, we have several possible combinations. We may replace both (or all) variables with constants, yielding a pure singular proposition. We may replace one of the variables by a constant, and quantify over the other (for a two-place predicate), or we may quantify over *all* the variables. The possibilities, while not endless, are certainly numerous, even for a two-place function. For three or four-place functions, of course, there are even more combinations.

Let us take one example, and go through a number of the various kinds of sentences we can get by combining constants, existential quantifiers, and universal quantifiers. We will let Lxy mean "x loves y," and we will use 'j' and 'm' again for John and Mary. *We will here restrict our domain to human beings*, so that it will not be necessary to use the predicate "x is a person." Given the interpretations above, of course, the formulas Ljm and Lmj symbolize "John loves Mary" and "Mary loves John," respectively. Suppose we prefix a quantifier in place of one of the constants, for instance, $(\exists x)Lxj$. What would this mean? Well, it begins with an existential quantifier, and this means that the overall form of the sentence is existential.[1] Read literally, it says "There is some x such that x loves John," or, in other words, "Somebody loves John" (since our universe is humans). If we use the universal quantifier instead, to get $(x)Lxj$, this would obviously say "For every x, x loves John," or, in other words, "Everyone loves John." We can also quantify over the second individual letter, and put in the constant 'j' for the first letter. In this case we would get the two formulas $(\exists x)Ljx$ and $(x)Ljx$. In these formulas, since 'j' is the first letter, John is the first-named individual, and so it is John who is doing the loving, of someone and everyone, respectively. Thus the two sentences say "John loves someone" and "John loves everyone." If we were to read the first formula, $(\exists x)Ljx$, literally, we would have "There is an x such that John loves x, and the second formula, $(x)Ljx$, would read "For every x, John loves x," and these are clearly equivalent to our more colloquial versions above. Thus we have four possibilities for one quantifier (assuming we use the same constant in each case):

---

[1] Of course, it is also a singular sentence, since it contains a constant, but when we talk of the *form* of a sentence we mean the arrangement of its quantifiers and sentential operators.

$(\exists x)Lxj$	Someone loves John.
$(x)Lxj$	Everyone loves John.
$(\exists x)Ljx$	John loves someone.
$(x)Ljx$	John loves everyone.

In general, when you are given a quantified sentence, especially with relational predicates, you should get in the habit of reading it *literally*, beginning with the first symbol, as we have done above. You will always begin at the far left, and for sentences of quantifier form, *it is the leftmost quantifier which will determine the form of the sentence.* This will be especially important where you have more than one quantifier, as in the examples in the next paragraphs.

Suppose we have *two* quantifiers in front of our function, instead of a constant at one place. We can have either both universal, both existential, or a combination of universal and existential. In addition, the order of the two quantifiers may be reversed. This gives us *eight* different possibilities, although, as we shall see, two of the pairs are equivalent. These eight possible combinations are listed below.

$(x)(y)Lxy$	$(x)(\exists y)Lxy$
$(y)(x)Lxy$	$(\exists y)(x)Lxy$
$(\exists x)(\exists y)Lxy$	$(\exists x)(y)Lxy$
$(\exists y)(\exists x)Lxy$	$(y)(\exists x)Lxy$

Let us take the unmixed universal cases first: $(x)(y)Lxy$ and $(y)(x)Lxy$. The first would be read literally as "For every x and for every y, x loves y," or in other words, "Every x loves every y," that is, "Everybody loves everybody." Suppose we switch the quantifiers to get the second formula. Here we have "For every y and for every x, x loves y," that is, "Every y is loved by every x," that is, once again, "Everybody loves everybody." Thus the two formulas are equivalent. This will generally be the case: if you have contiguous universal quantifiers, it does not make any difference in what order they appear. $(x)(y)(z)Bxyz$ means the same as $(z)(y)(x)Bxyz$ or $(y)(z)(x)Bxyz$, and so on. (This is related to the fact that we have commutation for conjunction.)

We have the same equivalence between two uses of the existential quantifier. Let us work out the meanings of $(\exists x)(\exists y)Lxy$ and $(\exists y)(\exists x)Lxy$. The first would be read literally as, "There is some x such that there is some y such that x loves y," or in other words, "Some x loves some y." The second formula would be read "There is some y such that there is some x such that x loves y," or in other words, "Some y is loved by some x." In both cases we could simply say "Somebody loves somebody." In general, contiguous existential quantifiers may also be exchanged without changing the meaning of the sentence. (This is related to the fact that we have commutation for disjunction.) $(\exists x)(\exists y)$ $(\exists z)(\exists w)Gxyzw$ will mean exactly the same as $(\exists w)(\exists y)(\exists x)(\exists z)Gxyzw$, or any other permutation of the existential quantifiers.

The situation changes, however, when we have a *mixture* of universal and existential quantifiers; here changing the order at the front will make an enormous difference in meaning, so you must learn very carefully how to figure out what these sentences say. Suppose we have the sentence $(x)(\exists y)Lxy$. What does this mean? Since it begins with a *universal* quantifier, it is a universal sentence, and we should begin by saying "For every x." Now, what are we saying about all these x's? Well, the rest of the sentence is $(\exists y)Lxy$ (a propositional function on x). Read literally, *this* portion says "There is some y such that x loves y," that is, "x loves somebody." Putting it together, we get "For every x, there is some y such that x loves y," or "For every x, x loves somebody"; in other words, "Everybody loves somebody." (No one is entirely heartless.)

Now let us switch the quantifiers, to get $(\exists y)(x)Lxy$. What would this mean? Notice first that it is an existential proposition: we would begin by saying "There is some y such that." Now, what about this y? What is the rest of the formula saying about y? If we read it literally we get "For every x, x loves y," or "Everybody loves y." Lucky y! We might call him or her the "Universal lovee," the one whom everyone loves. If we read the entire sentence literally, starting again from the far left, we get "There is some y such that every x loves y," or "There is some person whom everyone loves." Notice that this says something *very* different from the previous sentence.

We still have two possibilities: in the above sentences we had universal x and existential y; we can also have existential x and universal y, for the following two formulas: $(\exists x)(y)Lxy$ and $(y)(\exists x)Lxy$. The first of these is an existential formula, and would be read "There is some x such that for all y, x loves y." In other words, there is some one human being who loves everyone. This fictional creature we might call "the universal lover," who needs to be distinguished from the "universal lovee" above, the one whom everyone loved. This is one who *loves* everyone, rather than one who *is loved* by everyone. A more colloquial version of this formula, $(\exists x)(y)Lxy$, would be simply "There is someone who loves everyone."

The other formula, $(y)(\exists x)Lxy$, the last in our series, is universal, and so would begin with "For all y." Now, what is being claimed about all the y's? The rest of the formula says "There is some x such that x loves y"; in other words, "Someone loves y." So the sentence as a whole says, "For every y, some x loves y," or, more colloquially, "For every person, there is someone who loves that person." Notice that this formula, $(y)(\exists x)Lxy$, is very different from another we did above, $(x)(\exists y)Lxy$. The former, as we have just seen, says that everyone *has* someone who loves him or her—that everyone *has* a lover—while the latter says that everyone *loves* someone—that everyone *is* a lover.

As an aside, notice that the sentence "Somebody loves everybody" is ambiguous in English. It may mean that there is some one person who loves everyone (our third version) or that for every person, there is someone (not necessarily the same one) who loves that person (our fourth version). Such ambiguities are common in English; quantifier logic should help you sort them out.

The four formulas with mixed quantifiers, and their interpretations, are repeated below.

$(x)(\exists y)Lxy$	Everyone has the property of loving (at least) someone.
$(\exists y)(x)Lxy$	There is someone (some one person) who is loved by everyone.
$(\exists x)(y)Lxy$	There is someone who has the property of loving everyone.
$(y)(\exists y)Lxy$	Everyone is loved by someone.

Thus, changing the order of existential and universal quantifiers makes an enormous difference to the meaning (and truth-value) of a sentence. To make this crystal clear, let us take one more example. We will use Fxy for "x is the (biological) father of y," restricting our domain to humans.

$(y)(\exists x)Fxy$ means "For any person y, there is some x such that x is the biological father of y;" in other words. "Everyone has a biological father." This is true. $(\exists x)(y)Fxy$, on the other hand, means "There is some one x such that x is the biological father of *all* (human) y's," and this, of course, is false.

An example of a three-place propositional function would be "x introduced y to z" which we could abbreviate as $Ixyz$. If we use j and m again, and a for Andrew, some singular sentences, with their symbolizations, would be the following: "John introduced Mary to Andrew" ($Ijma$); "Mary introduced John to Andrew ($Imja$); and "Andrew introduced Mary to John" ($Iamj$). (This should again indicate the importance of the order of the individual letters). Suppose we add quantifiers, as in $(\exists x)Imjx$. This would mean "There is some x such that Mary introduced John to x;" in other words, "Mary introduced John to someone." $(\exists y)Iyma$ would mean "There is some y such that y introduced Mary to Andrew," that is, "Someone introduced Mary to Andrew."

Some examples with two quantifiers, and one constant, would be the following:

$(\exists x)(\exists y)Ixyj$	Someone introduced somebody to John.
$(\exists x)(\exists y)Ixmy$	Someone introduced Mary to somebody.

$(\exists y)(z)Imyz$	Mary introduced someone (some one person) to everyone.
$(z)(\exists y)Imyz$	Mary introduced someone to everyone. (That is, for everyone, there was someone Mary introduced to them).

There are many other possibilities as well, and we will not cover them all. You should work out more examples on your own, remembering that the form of the sentence is determined by the leftmost quantifier, and that you should read the formula literally before you try to give a more colloquial version.

For our two-place function Lxy, above, we gave eight possible combinations of quantifiers, which yielded six different meanings. But haven't we left something out? What about the eight possible combinations with the function Lyx, with the variables reversed? Interestingly enough, these versions add nothing new. To see this, let us compare the two formulas $(\exists x)(y)Lxy$ and $(\exists y)(x)Lyx$. The first was one we covered earlier, which we said meant "There is someone who loves everybody." What does the second formula say? If we read it literally we get "There is some y such that for every x, y loves x." Put colloquially, what this says is that someone loves everyone, exactly what the first formula said! If you look closely at the two formulas you may be able to see why they say the same thing. In both cases we have an *existential quantifier first*, followed by a universal quantifier; and in both cases the variable immediately after the predicate letter, which is the one that designates the lover, is the one correlated with the existential quantifier, while the second variable after the predicate letter is in both cases correlated with the universal quantifier. So both formulas say *there is someone* who *loves everyone*. It is clear from the following diagram that the pattern is the same in both formulas.

The two formulas are *identical, except that their bound variables are systematically interchanged*. In cases like this, where the formulas are the same except for a change of bound variable, the formulas are equivalent. They have exactly the same form. Thus, if you happen to use different variables than the ones in the answers, your answer is not necessarily wrong. If it has the *same form* as the answer (or an equivalent form) and only differs (systematically) in the bound variables, then it is correct.

So far we have been talking about going from the *formula* to the English sentence. What about the converse, going from the sentence to the formula symbolizing the English sentence? First figure out the *form* of the sentence, whether purely singular, existential, or universal, and then put together the parts. If it is purely singular, with no quantifiers involved at all, then your task will be easy; just plug in the appropriate constants to the propositional function. If it is an existential formula, you will, of course, begin with an existential quantifier. You should then try to figure out what is being said about the thing referred to by the quantifier. The sentence "Some things make John happy," for instance, is existential, so you would begin with $(\exists x)$. The question is, what is being said about that x? The answer in this case is that x makes John happy, which we might symbolize as Hxj. The sentence would then be completely symbolized as $(\exists x)Hxj$, and would be read as "There is an x such that x makes John happy." You should be able to see that this is a correct interpretation of the English sentence. If it is a universal sentence, you do the same. In the sentence "Everything exerts a gravitational pull on the Earth," for instance, the overall form is clearly universal, so we may begin with (x). Then the question is, what is being asserted about all those x's, and the answer is that they exert a gravitational pull on the Earth, which we could symbolize as Gxe. We could symbolize the whole sentence, then, as $(x)Gxe$, which would be read "For every x, x exerts a gravitational pull on the Earth," clearly a correct reading of the original English sentence.

In cases where there is more than one quantifier you may have a little more difficulty, but here again you should first isolate the overall form of the sentence, whether universal or existential, and write down the appropriate quantifier. Then figure out what is being said about the objects referred to by the quantifier, which in these cases will require more quantifiers. *Symbolize* what is being said about the objects, using whatever other quantifiers you need, put it all together, and you should have your symbolization.

In the sentence "Everything is older than something," for example, the quantifier that determines the form of the sentence is clearly universal, so we may begin by writing down (x). Then we ask again what is being said about all those x's. The answer is that each x *is older than something*. Then the question is how we symbolize "x is older than something." Well, this expression is existential; it says *there is something*, say y, such that x is older than y. This could be symbolized as $(\exists y)Oxy$. Putting this all together we get $(x)(\exists y)Oxy$. Reading it back literally, to be sure we have it right, we get "For every x, there is some y such that x is older than y," which does capture the sense of the original English sentence.

Remember that it is the leftmost quantifier which determines the overall form of the sentence, and that, with mixed quantifiers, if you change the order of the quantifiers, you change the meaning of the sentence. It is thus essential that you begin with the proper quantifier. You can determine the overall form of the sentence in much the same way as in one-variable logic: look for key words such as "all," "every," "some," and "somebody." Of course, with mixed quantifiers, you will have *both* phrases occurring in the sentence, and you will have to use your sense of English to figure out which is the determining phrase. With mixed quantifiers it is very often the *first* appearing quantifier.

Where there are more than two variables the situation gets even more interesting. We will take one final example in which there are three quantifiers involved: "Someone told nasty stories about someone to everyone." Here the overall form is clearly existential, so we begin with $(\exists x)$. Now what is said about x? Something rather unsavory, that x told nasty stores about someone to everyone. Now, how do we symbolize this repellent property of x? Well, it too is existential; it says that *there was someone* such that x told nasty stories about that person to everyone. This, then, would be symbolized by another existential, so we can continue with $(\exists y)$. Now what is true about poor y? That x told nasty stories about y to *everyone*, and this, of course, is universal. We could say "$(z)x$ told nasty stories about y to z." If we use $Nxyz$ for "x told nasty stories about y to z," then this last function could be symbolized as $(z)Nxyz$. Putting the whole thing together we have $(\exists x)(\exists y)(z)Nxyz$. Read literally, it would say "There is some x such that there is some y such that for every z, x told nasty stories about y to z." Or, a little more colloquially. "There is some x and some y such that x told nasty stories about y to everyone.

At this point you should be able to complete Exercise 3 at the end of the unit, which will give you more practice in going from *formulas* to English sentences, and Exercise 4, which will give you practice in symbolizing English sentences. There are no negations involved in these exercises; we will introduce negated quantifiers for relational logic in the next section.

3. Quantifier Negation

In applying the quantifier negation rules to multiply quantified formulas, the procedure is the same as before. You have the rules $\sim (x)\phi x :: (\exists x) \sim \phi x$ and $\sim (\exists x)\phi x :: (x) \sim \phi x$, which tell you that negated existential formulas are equivalent to universals and negated universals to existentials. Thus you could use Q.N. on the formula $\sim (x)(\exists y)Fxy$ to derive $(\exists x) \sim (\exists y)Fxy$. *You must apply the Q.N. rules to only one negated quantifier at a time.* A few other examples would be the following (keeping in mind that since the rules are replacement rules they may be used from right to left as well as from left to right):

1.	$\sim (\exists x)(\exists y) \sim Fxy$	/∴	$(x) \sim (\exists y) \sim Fxy$
2.	$\sim (x)(y) \sim Fxy$	/∴	$(\exists x) \sim (y) \sim Fxy$
3.	$(\exists x) \sim (y) \sim Fxy$	/∴	$\sim (x)(y) \sim Fxy$
4.	$(x) \sim (\exists y)Fxy$	/∴	$\sim (\exists x)(\exists y) \sim Fxy$

Keep in mind also that the replacement rules may be used on *subformulas* as well as the main formula, so that the following could be inferred from the *conclusions* of the above four inferences:

1. $(x)(y) \sim \sim Fxy$
2. $(\exists x)(\exists y) \sim \sim \sim Fxy$
3. $\sim (x) \sim (\exists y)Fxy$
4. $\sim (\exists x) \sim (y)Fxy$

Of course, we also have two other forms of the rule, which can be stated as follows: $\sim (x) \sim \phi x :: (\exists x)\phi x$ and $\sim (\exists x) \sim \phi x :: (x)\phi x$. Using these other forms of the rule, we could make the following inferences:

1. $\sim (\exists x) \sim (y)Fxy$ $/\therefore$ $(x)(y)Fxy$
2. $\sim (x) \sim \sim (\exists y)Fxy$ $/\therefore$ $(\exists x) \sim (\exists y)Fxy$
3. $\sim (x) \sim (\exists y) \sim Fxy$ $/\therefore$ $\sim (x)(y) \, Fxy \text{ or } (\exists x)(\exists y) \sim Fxy$
4. $\sim (\exists x) \sim (y) \sim Fxy$ $/\therefore$ $\sim (\exists x)(\exists y)Fxy \text{ or } (x)(y) \sim Fxy$
5. $(x)(y)Fxy$ $/\therefore$ $(x) \sim (\exists y) \sim Fxy$

If you use several applications of the Q.N. rules, you can "run a negation through" a multiply quantified statement from the beginning of a formula to the final function (and vice versa). Two instances of this process are given below. (Read these from top to bottom, rather than from left to right.)

a. 1. $\sim (x)(y)(z)Fxyz$ b. 1. $\sim (\exists x)(y)(\exists z)Fxyz$
    2. $(\exists x) \sim (y)(z)Fxyz$ 2. $(x) \sim (y)(\exists z)Fxyz$
    3. $(\exists x)(\exists y) \sim (z)Fxyz$ 3. $(x)(\exists y) \sim (\exists z)Fxyz$
    4. $(\exists x)(\exists y)(\exists z) \sim Fxyz$ 4. $(x)(\exists y)(z) \sim Fxyz$

Notice in comparing steps 1 and 4 that, in both a and b, *all the quantifiers change*, and the negation moves from the very front of the formula to just in front of the final function. Exercise 5, at the end of the unit, will give you practice in making these Q.N. transformations.

In the paragraphs above we were simply applying the Q.N. rules to uninterpreted formulas; we must now discuss English sentences containing negated quantifiers and how they should be symbolized. It will be obvious enough when a sentence contains a negated quantifier; again, you will have phrases such as "not all," "not every," or "not everyone," which will indicate a *negated universal*, and phrases such as "no," "none," "no one," "nothing," and so on, which will indicate a *negated existential*. Symbolizations for negated quantifiers, and equivalences between negated quantifiers, will be the same as before; what is different is that you may have more than one quantifier.

Let us begin, however, with sentences involving just a single quantifier, such as "Mary does not like everyone." (Again we will restrict our domain to people.) This is a *negated universal*, so you could begin your symbolization with $\sim (x)$, "It is not the case that for every x," and the proper function would be "Mary likes x." This is easily symbolized as Lmx, and putting the two parts together we have $\sim (x)Lmx$. Notice that this is equivalent, as all negated quantifier expressions will be, to another quantifier statement, namely, $(\exists x) \sim Lmx$. Let us read this literally and see whether it is equivalent to the original sentence. This second quantifier statement says "There is some x such that Mary does not like x," or in other words, "There is someone Mary doesn't like," and this is indeed equivalent to "Mary does not like everyone."

Another example involving only a single quantifier would be "No one likes Richard." Here the phrase "no one" signals a *negated existential*, so we will begin our symbolization with $\sim (\exists x)$. This is read as "There is no x such that" and what should follow it then is "x likes Richard." This latter phrase can be symbolized as Lxr, and putting the two parts together we have $\sim (\exists x)Lxr$, "There is no x such that x likes Richard," and this means simply that no one likes Richard. Note that this would be equivalent by Q.N. to $(x) \sim Lxr$, which would be read "For every x, x does not like (dislikes) Richard," and this again captures the sense of the English sentence.

With more than one quantifier, of course, the situation gets a little more complex, but here again you should first isolate the *basic form* of the sentence, write down the initial negated quantifier, and then figure out what should follow it. Suppose we have the sentence "No one likes everyone." Here the form is indicated by the phrase "no one," so we may begin with $\sim (\exists x)$, "There is no x such that." Now we ask ourselves what there aren't any of, and in this case the answer is people who like everyone; that is, there is no x such that x *likes everyone*. The italicized portion is our propositional function, which is not difficult to symbolize. It is universal, and we can simply say $(y)Lxy$, "For every y, x likes y." Putting the two parts together again, we have $\sim (\exists x)(y)Lxy$, which would be read "There is no x such that for every y, x likes y."

The formula above, of course, has equivalents. If we apply Q.N. once, we get $(x) \sim (y)Lxy$. This would be read as "For every x, x does not like everyone," which does mean the same as "No one likes everyone." If we apply Q.N. again we get $(x)(\exists y) \sim Lxy$, which would be read "For every x there is some y such that x does not like y," or "For every x, there is someone whom x does not like," and again, this is equivalent to saying "No one likes everyone."

Let us take one more example of a two-place function before we go on to three-place functions. "Not everyone gave a present to somebody" is a negated universal; we are saying "It is not the case that for every x, x gave a present to someone." Put this way, it should be easy to see the symbolization: $\sim (x)(\exists y)Gxy$. Using Q.N. once, we get $(\exists x) \sim (\exists y)Gxy$, which would be read "There is some x such that there is no y such that x gave a present to y," or "There are some people who gave a present to no one." Using Q.N. again, we get $(\exists x)(y) \sim Gxy$, which says "There is some x such that for every y, x did not give a present to y," or "There is someone who did not give a present to anyone," which does mean the same as our original sentence.

In the following examples we will use $Ixyz$ for "x introduced y to z," and instead of explaining everything in detail, we will just list several examples with their equivalents and their symbolizations. Be sure to study these examples carefully, and to read out literally the symbolized formulas in each case.

1. No one introduced anybody to Richard. $\sim (\exists x)(\exists y)Ixyr$
   For everyone, they introduced no one to $(x) \sim (\exists y)Ixyr$
   Richard.

2. Not everybody introduced someone to John. $\sim (x)(\exists y)Ixyj$
   Some people introduced no one to John. $(\exists x) \sim (\exists y)Ixyj$ or
                                               $(\exists x)(y) \sim Ixyj$

3. John did not introduce anyone to anybody. $\sim (\exists x)(\exists y)Ijxy$
   For everyone, John did not introduce them $(x) \sim (\exists y)Ijxy$ or
   to anyone. $(x)(y) \sim Ijxy$

4. No one introduced anybody to anyone. $\sim (\exists x)(\exists y)(\exists z)Ixyz$
   For everyone, they were not introduced to $(x) \sim (\exists y)(\exists z)Iyxz$
   anyone.
   For every x and y, no one introduced x to y. $(x)(y) \sim (\exists z)Izxy$

Notice that the order of the function letters is different in each case in 4. See if you can figure out why the formulas are still all equivalent. There are many examples in Exercise 6 of this sort of problem. You should now try to do these exercises.

4. "Categorical" Relational Statements; Complex Subjects and Predicates

There are many relational sentences which are merely complex versions of categorical propositions, in which the subject or predicate, or both, contain additional quantifiers. We can symbolize these sentences by using our old three-step procedure of (1) identifying the form of the sentence, (2) identifying subject and predicate of the English sentence, and (3) symbolizing the subject and predicate. Since you are already familiar with the first two steps, we discuss in this section the procedures for symbolizing complex subjects and predicates.

The sentence "Some dogs wear collars" is clearly an I proposition, and could be paraphrased as "There is some x such that x is a dog and x wears a collar." The subject phrase "x is a dog" presents no difficulty; we can symbolize it simply as Dx. The predicate phrase "x wears a collar" requires a little more analysis. Of course, in one-variable logic we could simply have used, say, Cx for "x wears a collar," and then symbolized the whole as $(\exists x)(Dx \cdot Cx)$. But here we are trying to spell out all the complexities, and we should notice that within the predicate there is an object mentioned, the collar, which stands in a particular relation to the dog— namely, the dog wears the collar. The most perspicuous analysis, then, is to use a separate predicate phrase for being a collar, and then use a relational predicate to indicate the wearing relationship between the dog and the collar. In general, in symbolizing relational sentences, you should have a different one-variable function for each class of objects mentioned, and a relational predicate for each relationship. Let us use Cx for "x is a collar" and Wxy for "x wears y." We can then symbolize the predicate, "x wears a collar," as $(\exists y)(Cy \cdot Wxy)$, and the entire sentence as $(\exists x)(Dx \cdot (\exists y)(Cy \cdot Wxy))$. This would be read as "There is an x such that x is a dog, and there is a y such that y is a collar, and x wears y." In other words, the dog x wears the collar y, which is the intent of the English sentence.

Notice that we have stated the abbreviation for the function Cx using an x, but that we have used a y when we came to the symbolization. The reason for this is that the vocabulary, the list of symbolizations for the propositional functions, is always given uniformly in terms of x for a one-place function, x, y for a two-place function, and so on. However, *in symbolizing*, we *cannot* use x for the different classes. *We must use different variables for each kind of thing mentioned*, or we would generate hopeless confusion. If we tried to use x for both the dog and the collar in the sentence above, for instance, we would have $(\exists x)(Dx \cdot (\exists x)(Cx \cdot Wxx))$, which, if it makes any sense at all, says that there is a dog which is also a collar, and which wears itself. Don't make this kind of silly mistake; always use a *different* variable for each different kind of object mentioned in the sentence, even though your vocabulary, or abbreviations for the functions, will use the same variable.

Let us take another, slightly more complex example: "Any dog that wears a collar has an owner." This could be paraphrased as "For every x, if x is a dog that wears a collar, then x has an owner." Here the subject phrase is "x is a dog that wears a collar" and the predicate phrase is "x has an owner." The subject phrase, since it contains a modifier, will be symbolized by using a conjunction, just as in one-variable logic. That is, we can break it down into "x is a dog and x wears a collar." We can symbolize "x is a dog" as Dx, and "x wears a collar" just the same as above: $(\exists y)(Cy \cdot Wxy)$. The entire subject phrase, then, would be symbolized as $Dx \cdot (\exists y)(Cy \cdot Wxy)$. The predicate phrase says that x *has an owner*. What does this mean, in logical terms? The word "an" indicates an existential construction, and we could paraphrase this predicate as "*there is an owner of x,*" or "there is some z such that z owns x." For this, all we need is the relational predicate Oxy for "x owns y," and we can then symbolize the predicate phrase, that x has an owner, simply as $(\exists z)Ozx$, that is, there is a z such that z owns x. (Here we do not need a third *class* term—a third one-place predicate—because the sentence doesn't specify what kind of owner the dog has. Perhaps it is a human being, but it might be a corporation!) Finally, the sentence is basically an *A* proposition, so the overall form will be $(x)(\underline{\quad} \supset \underline{\quad})$. We can put the whole thing together, plugging in subject and predicate, in the following way: $(x)((Dx \cdot (\exists y)(Cy \cdot Wxy)) \supset (\exists z)Ozx)$. This would be read literally as "For every x, if x is a dog and there is a y such that y is a collar and x wears y, then there is a z such that z owns x."

You should get in the habit of first paraphrasing the sentence to be symbolized, indicating the quantifier by the phrase "for every x" or "for some x," and then stating the subject and predicate functions in terms of the variable x. You will then be able to see more clearly what needs to be done in order to symbolize the subject and predicate phrases. We will do a few of these paraphrases here, just to give you the idea: (The symbolizations will be done later.)

1. Some dogs with owners who don't feed them catch mice.
   There is some x such that x is a dog and x has an owner who doesn't feed x, and x catches mice.

2. Some teachers with bright students give no failing grades.
   There is an x such that x is a teacher and x has bright students, and x gives no failing grades.

3. Any bright student with good grades will get a scholarship.
   For any x, if x is a student and x is bright and x has good grades, then x will get a scholarship.

4. No student who does not get good grades will either get a scholarship or graduate cum laude.
   There is no x such that x is a student and x does not get good grades and x either gets a scholarship or graduates cum laude.

There are many different kinds of phrases that will be symbolized by using relational functions (plus quantifiers). Transitive verbs, of course, will be symbolized in this way. Examples of such verbs, represented here by the appropriate propositional functions, would be the following: "x loves y," "x kicked y," "x delights y," "x knows y," "x believes y," "x thinks about y," and so on. It would obviously be impossible to give a complete list; we would have to reproduce a good part of the dictionary! But watch for verbs like this, and symbolize them accordingly.

Another kind of phrase that is symbolized with a relational function is the *comparative*. Examples of comparatives are "x is taller than y," "x is faster than y," "x has darker eyes than y," "x lives better than y," "x has a lower income than y," and so on. Comparatives, of course, are often indicated by the suffix "er"; this may tip you off that you have a relational predicate.

Other kinds of phrases that indicate the use of relational predicates are the *genitive* and *possessive* cases. Examples of the former would be "*x* is *y*'s wife," "*x* is *y*'s brother," "*x* is *y*'s friend," etc. Such phrases can be rendered by relational functions such as "*x* is the wife of *y*," "*x* is the brother of *y*," "*x* is a friend of *y*," and so on. There is no suggestion here of ownership (marriage laws, fortuntely, have been substantially revised in the past century); the phrase simply states a relationship between two individuals. With *possessives*, however, we do have a kind of ownership. Examples would be "*x* is *y*'s car," "*x* is *y*'s house," "*x* is *y*'s pet," and so on. These phrases could be symbolized using *Oxy* for "*x* owns *y*," and a one-variable predicate to indicate the kind of thing that is owned. We might symbolize the above, for instance, as $(Cx \cdot Oyx)$, $(Hx \cdot Oyx)$, and $(Px \cdot Oyx)$, respectively.

Another very common kind of phrase which is symbolized by a relational function is the prepositional phrase. Examples would be "*x* is in *y*," "*x* is with *y*," "*x* comes after *y*," "*x* bought *y* from *z*," and so on. Phrases such as "with a collar," which we encountered earlier, can, as we saw then, be symbolized by using the phrase "*x* has *y*." The function "*x* is a man with an expensive car" could thus be symbolized as $Mx \cdot (\exists y)(Cy \cdot Ey \cdot Hxy)$.

Notice that, in symbolizing subject and predicate phrases that require relational functions, we have always included a *quantifier* using *y*. It would be *incorrect* to symbolize "*x* wears a collar," just as $Cy \cdot Wxy$. The reason for this is that in the end, when our sentence is completely symbolized, we must not have any free variables left over, and if we fail to quantify over *y*, in the end *y* will be free. (*x*, of course, will be quantified over by our initial quantifier.) Why are we not allowed to have free variables in our symbolization? Simply because we are supposed to be symbolizing *sentences*, expressions which will be true or false. If we have a free variable in the end, we will have a propositional function rather than a sentence; and a function, of course, is not either true or false. Thus, *you must make sure that each variable in your symbolization at the end is bound*, that is, that it falls within the scope of its own quantifier. You should get in the habit of quantifying over each new variable as it comes up; *do not* wait until the end and then try to put all your quantifiers at the front.

Once you have picked out the relational function for your subject or predicate phrase, remembering that you will need another quantifier, you still have to figure out *which* quantifier to use, whether universal or existential. Here look for the same clues as before. If the funtion is "*x* likes every girl," then obviously you should use a universal quantifier, and the function could be symbolized as $(y)(Gy \supset Lxy)$. If you have words such as "a" or "an," most often this will signal an existential quantifier.

The function we symbolized in the preceding paragraph was an example. Another example would be "x has a cat." This would be symbolized as $(\exists y)(Cy \cdot Hxy)$. In some cases there will be no indicating words or phrases at all, and then, as often happens, you will just have to use your common sense about what is meant.

One final word of caution on symbolizing functions: sometimes what looks like a relational statement between two individuals really isn't, because one of the terms cannot rightly be interpreted as naming an individual. In the function "x went for a swim," for example, it would be silly to translate it as $(\exists y)(Sy \cdot Wxy)$: "There is a y such that y is a swim and x went for y." It would be much better to symbolize this simply as a one-place function, Sx, meaning "x swam."

At this point, having seen many examples of functions in isolation, let us symbolize sentences containing such phrases. We will do the four listed earlier, for which we gave the paraphrases. We will give a step-by-step explanation of the first symbolization, and then will just list the symbolizations for the others and let you work out the rationale for yourself, given the abbreviations for the propositional functions.

> There is an x such that x is a dog and x has an owner who does not feed x, and x catches mice.
>
> $(\exists x)(x$ is a dog $\cdot$ x has an owner who doesn't feed x $\cdot$ x catches mice$)$
>
> $(\exists x)(Dx \cdot (\exists y)(y$ is the owner of x and y doesn't feed $x) \cdot x$ catches mice$)$
>
> $(\exists x)(Dx \cdot (\exists y)(Oyx \cdot \sim Fyx) \cdot x$ catches mice$)$
>
> $(\exists x)(Dx \cdot (\exists y)(Oyx \cdot \sim Fyx) \cdot (\exists z)(z$ is a mouse and x catches $z)$
>
> $(\exists x)(Dx \cdot (\exists y)(Oyx \cdot \sim Fyx) \cdot (\exists z)(Mz \cdot Cxz))$

1. $(\exists x)(Tx \cdot (\exists y)(Sy \cdot By \cdot Hxy) \cdot \sim (\exists z)(Fz \cdot Gxz))(Tx \equiv x$ is a teacher; $Sx \equiv x$ is a student; $Bx \equiv x$ is bright; $Hxy \equiv x$ has y; $Fx \equiv x$ is a failing grade; $Gxy \equiv x$ gives y)

2. $(x)((Bx \cdot Sx \cdot Gx) \supset (\exists y)(Cy \cdot Gxy))$ (Here we use Gx for x gets good grades; it would not be appropriate to symbolize this either as "all x's grades are good," or as "some of x's grades are good." The first is too strong and the second is too weak; $Cx \equiv x$ is a scholarship; $Gxy \equiv x$ gets y)

3. $\sim (\exists x)((Sx \cdot \sim Gx) \cdot ((\exists y)(Cy \cdot Gxy) \lor Lx))$ $(Lx \equiv x$ graduates cum laude$)$

5. Symbolizing English Sentences

We have already done a number of English sentences, but in the preceding section the emphasis was on how to symbolize various subject and predicate phrases. Here the emphasis will be on taking English sentences, some of which are rather complex, analyzing them into their components, and gradually working out their complete symbolizations. At the end of the section we will make a number of general comments, and cautions, on relational symbolizations.

Again, we follow the three-step procedure of first, determining the form of the sentence, second, identifying the subject and predicate of the English sentence, and third, symbolizing the subject and predicate phrases. Where the subject and predicate phrases are complex, we will symbolize them bit by bit, carefully analyzing each component. Let us take as an example "Any student who reads books will enjoy some of his or her courses and will learn something." Here the first thing to do is to analyze the form, and this is clearly an A proposition, a universal formula. Thus the overall structure will be $(x)(\_\_\_ \supset \_\_\_)$. The next thing is to pick out the subject and predicate of the sentence, keeping in mind that we mean subject *with all modifiers*. The subject phrase is *students who read books*, and the predicate phrase is *will enjoy some of his or her courses and will learn something*. So far this shoud be a piece of cake. The *paraphrase* of the sentence, which will tell us what we need to do to symbolize the various parts, is:

For any x, if x is a student and x reads books, then x will enjoy some of x's courses and x will learn something.

The subject phrase is a conjunction, with "x is a student" and "x reads books" as the two conjuncts. The first conjunct is simple, and can be symbolized as Sx. For the second conjunct we will need two other functions. We can use Bx for "x is a book" and Rxy for "x reads y." The second conjunct says, in effect, that there is at least one book that x reads; we can symbolize this as $(\exists y)(By \cdot Rxy)$. Putting the two conjuncts together, we have as the symbolization for our subject function $Sx \cdot (\exists y)(By \cdot Rxy)$.

The predicate phrase is also a conjunction, whose two conjuncts are "x will enjoy some of x's courses" and "x will learn something." The first conjunct will require the functions "x is a course," "x takes y," and "x enjoys y." We can abbreviate these as Cx, Txy, and Exy, respectively. We will again use an existential quantifier here, because of the word "some." We want to say that there is a course which x takes, and that x enjoys the course. For this we can use $(\exists z)(Cz \cdot Txz \cdot Exz)$. The second conjunct says that x learns something. This, too, is existential, and we will need another function Lxy for "x learns y." Notice, however, that it does not say *what* is learned, or what kind of thing is learned, so we can symbolize this phrase by saying simply that there is something which x learns, $(\exists w)Lxw$. Putting together the two conjuncts we have $(\exists z)(Cz \cdot Txz \cdot Exz) \cdot (\exists w)Lxw$ for our predicate phrase. Now, plugging the subject phrase and the predicate phrase into our form, we have the final symbolization:

$$(x)((Sx \cdot (\exists y)(By \cdot Rxy)) \supset ((\exists z)(Cz \cdot Txz \cdot Exz) \cdot (\exists w)Lxw)).$$

There are a number of things you should be aware of here. First, notice again that in our final symbolization *all variables are bound*. What guarantees this, aside from including the right quantifiers, is the proper use of *parentheses* to make sure the scope of the quantifier goes to the end of the function being symbolized. Notice that for each new quantifier we introduced, we included parentheses around the entire compound following the quantifier. In the end, we will also include parentheses around the whole thing to make sure the scope of the initial quantifier reaches clear to the end of the sentence. *Always be sure you have enough parentheses, and in the right places.*

Notice also that we have put the three additional quantifiers in the *middle* of the sentence, in their appropriate positions in the subject or predicate phrases. This may seem the obvious thing to do, but students are sometimes tempted to put all the quantifiers at the front of the formula, and this is a very risky procedure. The reason is that with complex formulas you are almost sure to be wrong; and the reason for this is that the rules that govern moving quantifiers from inside the formula to the outside are quite complex, and not what you would expect. It is true that there will always be an equivalent formula with all the quantifiers at the front, but the quantifiers will often be the opposite of what you would think. If you have all existentials, it won't hurt to put them all at the front of the sentence, but in general it is better just to put the quantifiers in the middle. This will be the natural thing to do if you symbolize the sentence bit by bit.

Notice also that we have the same pattern here as we have in one-variable logic: *universal* formulas have the *horseshoe* as the major operator of the function that follows them, and *existentials* are followed by functions with *conjunctions* as their major operators. This will not always be the case with these more complex formulas, but it is a good rule of thumb, and in any case, you can be almost certain that if you have a universal followed by a conjunction, or an existential followed by a conditional, it will be wrong.

So far we have been meticulously avoiding negations. The time has now come to plunge in to these more complicated symbolizations. Let us take a fairly simple example first: "Not all students enjoy all of their courses." We can use the same abbreviations as we did earlier. The *form* of the sentence is a *negated universal* and the *subject is students* and the *predicate is things that enjoy their courses*. The outer structure, then, will be $\sim (x)(\underline{\quad} \supset \underline{\quad})$, and the paraphrase, with subject and predicate italicized, will be "It is not the case that for every x if x *is a student* then x *enjoys all of x's courses*."

The subject phrase can be symbolized simply as Sx. The predicate phrase is here a *universal*, since it is saying something about *all* courses taken by the student. We want to say that for any course x takes, x enjoys that course, that is, for any y, if y is a course and x takes y, then x enjoys y. This could be symbolized as $(y)((Cy \cdot Txy) \supset Exy)$. Now, plugging the subject and predicate phrases into the form, we get the following symbolization:

$$\sim (x)(Sx \supset (y)((Cy \cdot Txy) \supset Exy))$$

Now, of course, any negated universal proposition will be equivalent to an existential. The equivalent English sentence here is "There are some students who do not enjoy all of their courses." This could be symbolized as:

$$(\exists x)(Sx \cdot \sim (y)((Cy \cdot Txy) \supset Exy))$$

This would also be equivalent to: $(\exists x)(Sx \cdot (\exists y)(Cy \cdot Txy \cdot \sim Exy))$, which says that there are some students who have some courses they take that they do not like. Any one of these symbolizations would be correct. Notice that the formulas are all equivalent just by applications of the C.Q.N. rules.

With negated quantifier sentences, especially if they are complex, you will have many possible different correct symbolizations. You may be able to apply C.Q.N. several times, and, of course, you can also apply your sentential replacement rules, such as Contraposition, DeMorgan's, or Exportation.

Let us take one more, rather complex example to illustrate both the symbolization process, and the fact that there can be many possible equivalent symbolizations: "No student who does not like any of his or her teachers will like any of his or her classes." Here the form is a negated existential, so the outer structure will be $\sim (\exists x)(\_\_\_ \cdot \_\_\_)$. The subject phrase is "student who does not like any of his or her teachers," and the predicate phrase is "will like any of his or her classes." We can paraphrase the sentence as follows: It is not the case that there exists an x such that *x is a student and x does not like any of x's teachers* and *x likes some of x's classes.*" The subject function is a conjunction, and the first conjunct can be symbolized simply as Sx. The second conjunct, *x does not like any of x's teachers*, will require abbreviations for "x is a teacher," "x likes y," and "x has y," for which we can use Tx, Lxy, and Hxy, respectively. The *form* of this second conjunct is also a negated existential; it is saying that there are no teachers x has that x likes. This could be paraphrased as "There is no y such that y is a teacher and x has y and x likes y," which could be symbolized as $\sim (\exists y)(Ty \cdot Hxy \cdot Lxy)$. The entire subject phrase, then, can be symbolized as $Sx \cdot \sim (\exists y)(Ty \cdot Hxy \cdot Lxy)$. Notice that the second conjunct could also be symbolized as $(y)((Ty \cdot Hxy) \supset \sim Lxy)$, which says that for any teacher x has, x does not like that teacher. The two forms are equivalent by C.Q.N.

The predicate phrase, "x likes some of x's classes," can be symbolized as $(\exists z)(Cz \cdot Hxz \cdot Lxz)$. The full symbolization for the sentence will then be:

1. $\sim (\exists x)(Sx \cdot \sim (\exists y)(Ty \cdot Hxy \cdot Lxy) \cdot (\exists z)(Cz \cdot Hxz \cdot Lxz))$

We could read this literally as "There is no x such that x is a student and x has no teachers x likes, and x does have classes x likes."

Some alternative symbolizations would be the following:

2. $(x)(Sx \supset \sim (\sim (\exists y)(Ty \cdot Hxy \cdot Lxy) \cdot (\exists z)(Cz \cdot Hxz \cdot Lxz)))$ (For all x, if x is a student, then it is not the case both that x has no teachers x likes and that x does have classes x likes.) This is equivalent to formula 1 by C.Q.N.; the placement of the negations and parentheses is essential.
3. $(x)(Sx \supset (\sim (\exists y)(Ty \cdot Hxy \cdot Lxy) \supset \sim (\exists z)(Cz \cdot Hxz \cdot Lxz)))$ (For all x, if x is a student, then if x has no teachers x likes, then x has no classes x likes.) This follows by an application of DeM. and C.E. to 2.
4. $(x)((Sx \cdot \sim (\exists y)(Ty \cdot Hxy \cdot Lxy)) \supset \sim (\exists z)(Cz \cdot Hxz \cdot Lxz))$ (For all x, if x is a student and x has no teachers x likes, then x has no classes x likes.) This is equivalent by an application of Exportation to 3.
5. $(x)(Sx \supset ((\exists z)(Cz \cdot Hxz \cdot Lxz) \supset (\exists y)(Ty \cdot Hxy \cdot Lxy)))$ (For every x, if x is a student, then if x has classes x likes, x has teachers x likes.) This is equivalent by Contraposition to 3. There are still other possible combinations, but this should give you the general idea.

We will take one more example in this section, and simply do a bit-by-bit analysis; this sentence has even greater complexity than the ones we have done so far. "No *dentist who treats patients who have no teeth will have either a large income or the respect of all the other dentists.*" (Negated existential.) Paraphrased, this reads "There is no x such that x *is a dentist* and x *treats patients who have no teeth*, and either x *has a large income* or x *has the respect of all the other dentists.*"

$\sim (\exists x)(x$ is a dentist $\cdot$ x treats patients who have no teeth $\cdot$ (x has a large income $\lor$ x has the respect of all the other dentists))

$\sim (\exists x)(Dx \cdot x$ treats patients who have no teeth $\cdot$ (x has a large income $\lor$ x has the respect of all the other dentists))

$\sim (\exists x)(Dx \cdot (\exists y)(y$ is a patient and x treats y and y has no teeth) . . . etc.)

$\sim (\exists x)(Dx \cdot (\exists y)(Py \cdot Txy \cdot y$ has no teeth) . . . etc.)

$\sim (\exists x)(Dx \cdot (\exists y)(Py \cdot Txy \cdot \sim (\exists z)(Tz \cdot Hyz)) \cdot (x$ has a large income $\lor$ x has the respect of all the other dentists)).

$\sim (\exists x)(Dx \cdot (\exists y)(Py \cdot Txy \cdot \sim (\exists z)(Tz \cdot Hyz)) \cdot ((\exists w)(Iw \cdot Hxw) \lor x$ has the respect of all the other dentists)).

Finally, we can paraphrase this last remaining function by saying that all the other dentists respect x; we can use Ox for "x is another dentist" and Rxy for "x respects y." Our final symbolization, then, will be:

$$\sim (\exists x)(Dx \cdot (\exists y)(Py \cdot Txy \cdot \sim (\exists z)(Tz \cdot Hyz)) \cdot ((\exists w)(Iw \cdot Hxw) \vee (u)(Ou \supset Rux)))$$

Read through the problem above until you understand the rationale for each part of the symbolization.

Notice that we used five different variables above. This was not strictly necessary, and we can point out now that *if the scopes of quantifiers do not overlap*, you may use the same letter for each. It would be correct, for instance, to symbolize "Any student with teeth visits a dentist" as $(x)((Sx \cdot (\exists y)(Ty \cdot Hxy)) \supset (\exists y)(Dy \cdot Vxy))$. There is no problem here because the scope of the first quantifier that uses y stops before we get to the second quantifier that uses y. Of course, if the scopes *do* overlap, as for x and y above, it is essential that you use different variables. One final reminder: *be sure all the variables are bound in your final symbolization*. You should now do Exercises 7, 8, and 9 at the end of the unit. If you think of the exercises as puzzles, you should find them an enjoyable challenge.

EXERCISES

1.   For each of the sentences below, pick out the propositional function and symbolize it, and then symbolize the sentence itself. These will all be simple singular sentences or their negations. They will not all be relational.

*a.   John is taller than Andrew.

 b.   Minnesota is colder than Texas.

*c.   Nixon wrote *Six Crises*.

 d.   Mary doesn't believe Richard.

*e.   Richard works for the CIA.

 f.   The CIA is a government agency.

*g.   The CIA spends more money than the EPA.

 h.   The EPA does not spend much.

*i.   Richard approves of the EPA.

 j.   Richard approves of Richard.

*k.   Richard told Mary about the EPA.

 l.   Mary hates the CIA because of Nixon.

*m.   The Freer Gallery is closer to the White House than the Washington Monument is.

 n.   The Freer Gallery is closer to the White House than the Washington Monument is to the Pentagon.

*o. The Freer Gallery is between the Washington Monument and the White House.

2. Symbolize the following, using lowercase letters for constants, *Bxyz* for "*x* is between *y* and *z*," *Mxy* for "*x* is the mother of *y*," *Oxy* for "*x* is older than *y*," *Lxy* for "*x* loves *y*," and *Txy* for "*x* is taller than *y*." Some sentences may be truth-functional compounds of simple sentences.

*a. Chicago is between New York and San Francisco.

b. Charles is not taller than Frank.

*c. Charles is older than Molly.

d. Rosalynn is Ann's mother.

*e. Stephen's mother is Josephine.

f. Minneapolis is not between New York and Chicago.

*g. John is taller than his mother Ann.

h. Peter is between John and Stephen in age.

*i. John is loved by Ann.

j. John does not love himself.

*k. Ann loves John but not Peter.

l. John loves Ann only if she is his mother.

*m. John is taller than neither Peter nor Stephen.

n. Stephen's mother is neither Ann nor Josephine.

*o. Ann is not the mother of both John and Stephen.

p. If John is older than Stephen and Ann, then John's mother is either Beatrice, Darlene, or Charlene, but not Ann or Martha.

3. Given the interpretations of the propositional functions, state the meanings of the following formulas in English. Restrict your domain to people.

Bxy ≡ *x* is a better bridge player than *y*; *Rxy* ≡ *x* respects *y*; *Lxyz* ≡ *x* likes *y* better than *x* likes *z*; *Txyz* ≡ *x* told stories about *y* to *z*; *a* = Anne; *b* = Bob; *c* = Charles; *d* = Dora.

*a. (∃ x)Bxa	*k. (∃ x)Lbxc
b. (x)Rbx	l. (x)Lxab
*c. (x)Rxc	*m. (∃ y)Lydc
d. (∃ y)Bdy	n. (y)Tacy
*e. (∃ x)(∃ y)Bxy	*o. (∃ x)(y)Txjy
f. (x)(∃ y)Byx	p. (∃ y)(x)Tjyx
*g. (x)(∃ y)Bxy	*q. (∃ y)(x)Tyjx
h. (∃ x)(y)Rxy	r. (x)(y)Txjy
*i. (∃ x)(y)Ryx	*s. (∃ x)(y)(z)Tyxz
j. (∃ x)(∃ y)Byx	t. (∃ x)(∃ y)(z)Tzxy

4. Symbolize the following English sentences, using the abbreviations indicated. These will all be quantifications over simple sentences; no negations will be required. Again, restrict your domain to human beings.

$Rxy \equiv x$ respects y; $Ixyz \equiv x$ introduced y to z.

*a.   Some people respect John.

 b.   Everyone respects Amy.

*c.   Some people are respected by someone.

 d.   John respects everyone.

*e.   Chris is respected by someone.

 f.   There is someone who respects everyone.

*g.   Someone introduced Amy to John.

 h.   Amy introduced John to everyone.

*i.   John introduced everyone to Amy.

 j.   There was someone who introduced John to everyone.

*k.   Everyone was introduced to Amy by someone (or other).

 l.   Everyone was introduced by Amy to someone (or other).

*m.   Everyone introduced someone to somebody.

 n.   There was someone who was introduced to some one person by everyone.

*o.   There was someone who introduced some one person to everyone.

 p.   Someone introduced himself to everyone.

5.   Apply Q.N. to the following formulas, starting from the outside and working your way in. See how many equivalent formulas you can get in each case.

*a. $\sim (x)(\exists y)(\exists z)Fxyz$

 b. $\sim (\exists x)(y)(z) \sim Fxyz$

*c. $\sim (x)(y)(z)(w) \sim Fxyzw$

 d. $\sim (\exists x) \sim (y) \sim (\exists z)Fxyz$

*e. $\sim (x) \sim (y) \sim (z) \sim (w)Fxyzw$

6.   Symbolize the following English sentences, using the indicated abbreviations. Where negations are involved, there may be more than one correct answer. Again, restrict your domain to people.

$Lxy \equiv x$ loves y; $Bxy \equiv x$ is the brother of y; $Sxy \equiv x$ is the sister of y; $Ixyz \equiv x$ introduced y to z.

*a.   Amy loves no one.

 b.   Not everyone loves John.

*c.   John does not love everyone.

 d.   Nobody loves Richard.

*e.   There is no one who loves everyone.

 f.   Not everyone loves somebody.

*g. There is no one who loves no one.

h. John has brothers.

*i. Amy has no sisters.

j. There is no one whom John does not love.

*k. Not everyone is not a brother of John.

l. Nobody introduced Andrew to Martha.

*m. Andrew didn't introduce anyone to Charlene.

n. John was not introduced to everyone by Amy.

*o. Not everybody failed to introduce Richard to Amy.

p. There was no one who didn't introduce someone to John.

*q. Not everyone introduced someone to Richard.

r. There are some people who introduced no one to John.

*s. Not everyone introduced one person to another.

t. No one introduced everybody to everybody.

7. Symbolize the following, which are basically complex categorical propositions, using the abbreviations indicated.

$Sx \equiv x$ is a student; $Bx \equiv x$ is a book; $Cx \equiv x$ is a comic; $Gx \equiv x$ gets good grades; $Rxy \equiv x$ reads y; $Lxy \equiv x$ listens to y; $Fx \equiv x$ is a professor; $Hxy \equiv x$ has y; $Wx \equiv x$ is well-rounded; $Px \equiv x$ is poetry; $Wxy \equiv x$ writes y; $Axyz \equiv x$ assigns y to z.

*a. Every student reads some books.

b. Some students read books and comics.

*c. No student reads all books.

d. No student reads only comics.

*e. Some students who read no books will still get good grades.

f. Some students listen to some of their professors.

*g. Not every student who reads no books will get good grades.

h. There is no student who listens to none of his professors.

*i. Not every student listens to all of his professors.

j. A student who reads neither books nor comics will not be well-rounded.

*k. A student who reads books but not comics will get good grades but will not be well-rounded.

l. Some students read some books assigned by some of their professors.

*m. Any student who reads all books assigned by all his professors will be well-rounded and will get good grades.

n. Some students write poetry.

*o. All students read some poetry.

p. Some well-rounded students who don't get good grades both read and write poetry.

*q.   No student who either gets good grades or writes poetry reads all comics.

r.   Not every student who reads all comics reads no books.

8.   Using the abbreviations above, decipher the following formulas, that is, write out the English sentence which is symbolized by the formula. Write an ordinary English sentence, not a literal logical reading.

*a. $\sim (x)(Sx \supset (\exists y)(Cy \cdot Rxy))$

b. $(\exists x)(\exists y)(Fx \cdot Sy \cdot Hyx \cdot Lxy)$

*c. $\sim (x)(Fx \supset (y)((Sy \cdot Hyx) \supset Lxy))$

d. $\sim (\exists x)(Fx \cdot \sim (\exists y)(Sy \cdot Hyx \cdot Lxy))$

*e. $(x)((Sx \cdot (y)((Fy \cdot Hxy) \supset Lxy)) \supset ((\exists z)(Bz \cdot Rxz) \cdot Gx))$

f. $(\exists x)(Fx \cdot (\exists y)(By \cdot Wxy) \cdot \sim (\exists z)(Pz \cdot Wxz))$

*g. $\sim (\exists x)(Bx \cdot (y)(Sy \supset Ryx))$

h. $(\exists x)(Px \cdot (y)(Sy \supset \sim Wyx))$

*i. $(x)(((Sx \lor Fx) \cdot (\exists y)(Py \cdot Wxy)) \supset (Wx \cdot \sim (\exists z)(Cz \cdot Rxz)))$

j. $(\exists x)(Fx \cdot (y)((Py \cdot \sim Wxy) \supset \sim Rxy))$

*k. $(x)((Sx \cdot (\exists y)(Cy \cdot Rxy) \cdot \sim (\exists z)((Bz \lor Pz) \cdot Rxz) \supset \sim (Gx \supset Wx))$

9.   Symbolize the following, using the abbreviations below.

$Dx \equiv x$ is a doctor; $Pxy \equiv x$ is a patient of y; $Lxy \equiv x$ likes y; $Rxy \equiv x$ respects y; $Txy \equiv x$ treats y; $Ox \equiv x$ comes to the office; $Mx \equiv x$ is money; $Hxy \equiv x$ has y; $Px \equiv x$ is a person; $Ax \equiv x$ is an ailment; $Nx \equiv x$ is a medicine; $Pxyz \equiv x$ prescribes y for z; $Exy \equiv x$ is a peer of y; $Sx \equiv x$ is a side effect; $Sxy \equiv x$ sues y; $Lx \equiv x$ is a lawyer; $Cx \equiv x$ is a large practice; $Bx \equiv x$ is a large bank account.

*a.   Some doctors don't like all of their patients.

b.   There are no doctors who don't like any of their patients.

*c.   Doctors who respect all their patients will like some of them.

d.   Not all doctors treat every patient who comes to the office.

*e.   Some doctors treat patients who have no money.

f.   Doctors who respect all of their patients will be liked and respected by them all.

*g.   There are some people with no ailments who are treated by doctors.

h.   No doctor treats all of her patients who have no ailments.

*i.   Some doctors prescribe medicine for all of their patients.

j.   All doctors prescribe medicine for some of their patients, unless they don't have a large practice.

*k.   No doctor who prescribes medicine for patients with no ailments will be respected by all his peers.

l.   Some doctors who prescribe medicine with side effects for some of their patients will be sued by some of their patients.

*m. There are no medicines with side effects that are prescribed by any doctor for patients with no ailments.

n. No doctor who prescribes medicine with side effects for a patient who has no ailment will be respected by any of his peers unless he has a large practice or a large bank account.

*o. Some medicines with side effects are prescribed by some doctors for any patient of theirs with no ailment who has a lawyer with a large bank account who sues doctors.

UNIT 18

Proofs and Invalidity for Relational Predicate Logic

A. INTRODUCTION

In this unit we discuss proof procedures and methods for demonstrating invalidity in relational predicate logic. The proofs will, of course, be more complex than what you did in Unit 15, since there will often be more than one quantifier in the premises and conclusion, and you will have to be much more careful in plotting out strategy. But there will be no new rules to learn—just more complexities in applying the ones you already have. And again, think of the proof problems as puzzles; they will be more complex, and thus more of a challenge, than the ones you had earlier, but for this very reason they will be more interesting, and will generate an even greater feeling of satisfaction when you come up with the right answer.

To show that arguments are invalid we will again be using the natural interpretation method and the model universe method; the only difference in practice from what was done in Unit 16 is the occurrence of multiple quantifiers and relational predicates. An important difference in principle, however, is the fact that in relational logic there is no algorithm for determining whether a counterexample exists; trial and error, plus a few rules of thumb, is the best we can do. In relational logic there may be

326

arguments for which we can find *neither* a proof nor a counterexample, so that we are left uncertain about the validity of the argument. There will always *be* either a proof or a counterexample, but there is no mechanical method for finding either; the status of the argument may thus remain unknown. This is perhaps an incentive to master thoroughly both the proof techniques and the semantic techniques: the more proficient we are, the more problems we will be able to solve. What you will need to know is listed below.

B. UNIT 18 OBJECTIVES

1. Learn to construct proofs for arguments containing relational formulas.
2. Learn to construct proofs of theorems in relational predicate logic.
3. Be able to apply both the natural interpretation method and the model universe method to demonstrate invalidity for relational arguments.

C. UNIT 18 TOPICS

1. Proofs in Relational Predicate Logic

We noted at the beginning of Unit 17 that there were some arguments that could not be proved given only the resources of one-variable predicate logic, and which needed relational logic for a demonstration of their validity. We cited as an example "Some people don't like any dogs, all puppies are dogs, so some people don't like puppies." You know now that this can be symbolized as follows:

$$(\exists x)(Px \cdot (y)(Dx \supset\; \sim Lxy)), (x)(Ux \supset Dx) \; / \therefore (\exists x)(Px \cdot (y)(Uy \supset\; \sim Lxy)).$$

Now, how can we construct a proof for this argument? It is really not difficult; simply apply the rules as you did in one-variable logic, observe the restrictions, be careful about strategy, and watch out for the few complications we will describe below. We will now construct a proof for that argument, explaining the steps as we go.

a. 1. $(\exists x)(Px \cdot (y)(Dy \supset\; \sim Lxy))$ Pr.
    2. $(x)(Ux \supset Dx)$ Pr. / $\therefore (\exists x)(Px \cdot (y)(Uy \supset\; \sim Lxy))$

We can apply E.I. here as before, noting in the justification that a is being flagged, and being sure to substitute a for *every* occurrence of x in the function of the premise. Thus we have the third step, and the next two follow by Simplification:

3. $Pa \cdot (y)(Dy \supset \sim Lay)$	E.I. 1, a/x, (flag a)
4. Pa	Simp. 3
5. $(y)(Dy \supset \sim Lay)$	Simp. 3

At this point we have to stop and think about strategy—what we need to do to derive the conclusion at which we are aiming. The conclusion itself is existential, and it will be no problem to derive this by E.G. The instance will obviously be $Pa \cdot (y)(Uy \supset \sim Lay)$, since we already have the Pa. Now, how will we derive the second conjunct? Since it is universal, it will most likely come by U.G., and we must remember that in order to use U.G. we must first set up a subproof flagged with the letter over which we will be quantifying. Thus we will need an instance of $(y)(Uy \supset \sim Lay)$ such that the letter which is substituted for y can be flagged. a can no longer be used, since it has already appeared in the proof. There is nothing to stop us from using b, however, so we might as well set up a b-flagged subproof, and then use U.G. on this letter at the end. We can then do our instantiations and arrive at an instance of the conclusion; we will use E.G. to get the last step.

6.	flag b	F.S. (U.G.)
7.	$Ub \supset Db$	U.I. 2, b/x
8.	$Db \supset \sim Lab$	U.I. 5, b/y
9.	$Ub \supset \sim Lab$	H.S. 7,8
10.	$(y)(Uy \supset \sim Lay)$	U.G. 9, b/y
11.	$Pa \cdot (y)(Uy \supset \sim Lay)$	Conj. 4,10
12.	$(\exists x)(Px \cdot (y)(Uy \supset \sim Lxy))$	E.G. 11, a/x

There are a few things about which you will need to be particularly careful in doing relational proofs. For one thing, as noted above, you must replace *every* occurrence of the variable whose quantifier you are instantiating with the instance letter. This is not always quite so automatic as it was in one-variable logic, since you will often have occurrences of the variable inside the scope of another quantifier; but if you are dropping an x quantifier, for instance, you must replace *every* occurrence of the x with the instance letter.

Another very important limitation on the use of these rules, which was emphasized in Unit 15, is that you may use the quantifier rules *only* on statements that are of quantifier form, that is, *only* on formulas that begin with a quantifier, whose scope extends to the end of the sentence. This is the same as in one-variable logic, but here there are more ways in which you might go wrong. For instance, since you may only use U.I.

or E.I. on the leftmost quantifier, *you may never use the instantiation rules on quantifiers in the middle of a formula.* In other words, the four quantifier rules U.I., E.I., U.G., and E.G., *may not be used on subformulas.* The following inferences would thus be *incorrect*:

$$(x)(Fx \supset (y)(Gy \supset Hxy)) \quad /\therefore \quad (x)(Fx \supset (Gb \supset Hxb)) \quad \text{U.I., } b/y \quad \textbf{Error}$$
$$(\exists x)(Fx \cdot (\exists y)(Hxy \cdot Jyx)) \quad /\therefore \quad (\exists x)(Fx \cdot Hxa \cdot Jax) \quad \text{E.I., } a/y \quad \textbf{Error}$$
$$Fa \supset (y)(Gy \supset Hay) \quad /\therefore \quad Fa \supset (Gb \supset Hab) \quad \text{U.I., } b/y \quad \textbf{Error}$$

The same limitation holds for the generalization rules. You may never infer, from an instance, a formula that puts the new quantifier in the middle; *the quantifier you are adding on must always be added to the very front (far left) of the formula.* It would be *incorrect*, for instance, to use E.G. in the following way:

$$(x)(Fx \supset (Gax \cdot Hax)) \quad /\therefore \quad (x)(Fx \supset (\exists y)(Gyx \cdot Hyx)) \quad \text{E.G., } a/y \quad \textbf{Error}$$

Another thing that was mentioned earlier, but which now becomes important in relational logic, is that *you may not use variables as instance letters.* Use only *a*, *b*, *c*, and so on in doing your instantiations, and not any letter we have used as a variable. The reason for this is that if you use a variable it may get "caught," or bound, by a quantifier appearing in another part of the formula. An example of this sort of incorrect inference would be the following, if we allowed ourselves to use *y* as an instance letter: $(x)(\exists y)Fyx$ /∴ $(\exists y)Fyy$. If this doesn't look particularly fallacious, take the following interpretation: let *Fxy* be "*x* is the biological father of *y*," and then restrict your domain to human beings. The premise then says something true, namely, that everyone has a biological father. The conclusion, however, says something ridiculous—that there is someone who is the biological father of himself. Thus, again, you should always use letters at the beginning of the alphabet for the instances.

Since you have more than one quantifier in many of these formulas, it may often happen that you have need for more than one flagged letter in a proof; you may want to use E.I. with one and U.G. with another. Be careful in these cases *never* to flag the same letter twice in a proof; *for every use of U.G. and E.I. you will need a separate flagged letter.* Thus if you use E.I. twice and U.G. three times in a proof, you will need five different flagged letters. Let us look at an example of a problem in which you need more than one flagged letter:

b. 1. $(x)(Px \supset (\exists y)Fyx)$ Pr.

   2. $(x)(y)(Fyx \supset Lyx)$ Pr. /∴ $(x)(Px \supset (\exists y)Lyx)$

Here our conclusion is universal, so we will need to use U.G. at the end. With this in mind, we may as well set up our flagged subproof:

3.   ┌─►flag a F.S. (U.G.)

Now, since a is the letter on which we will be using U.G. in the end, the instance we need will be $Pa \supset (\exists y)Lya$. This is a conditional, so we should probably assume the antecedent and try to derive the consequent. Thus we need to set up another subproof, this time for C.P.

4.   ┌─►Pa Assp. (C.P.)

We now need to derive $(\exists y)Lya$, and this should not be difficult. We will need, however, to use E.I., and we will have to be sure to use a letter other than a, since a has already appeared in the proof.

5.	$Pa \supset (\exists y)Fya$	U.I. 1, a/x
6.	$(\exists y)Fya$	M.P. 4,5
7.	Fba	E.I. 6, b/y (flag b)
8.	$(y)(Fya \supset Lya)$	U.I. 2, a/x
9.	$Fba \supset Lba$	U.I. 8, b/y
10.	Lba	M.P. 7,9
11.	$(\exists y)Lya$	E.G. 10, b/y
12.	$Pa \supset (\exists y)Lya$	C.P. 4–11
13.	$(x)(Px \supset (\exists y)Lyx)$	U.G. 12, a/x

One of the trickiest parts of constructing proofs in relational logic is to know what letter to use in your instantiations. To determine this, you will have to be careful in plotting your strategies, and also keep in mind the restrictions on the use of the quantifier rules. At this point, it might be a good idea to state these restrictions once again:

R$_1$ A letter being flagged must be new to the proof, that is, it may not appear, either in a formula or as a letter being flagged, previous to the step in which it gets flagged.

R$_2$ A flagged letter may not appear either in the premises or in the conclusion of a proof.

R$_3$ A flagged letter may not appear outside the subproof in which it gets flagged.

In plotting strategies for relational proofs, it is especially important to use the method of working backwards: determine the form of the conclusion, decide what you will need to get it, what preliminary steps you will need, and so on. Let us take one rather complex example to illustrate this procedure.

c. 1. $(x)((Fx \cdot \sim Gx) \supset \sim (\exists y)(Tyx \cdot Hxy))$ Pr.
    2. $\sim (\exists x)(Fx \cdot Gx)$ Pr.
    3. $(x)((y)(Hxy \supset Zxy) \supset Gx)$ Pr. $/\therefore (x)(Fx \supset (\exists y) \sim (Tyx \lor Zxy))$

Here the conclusion is again a universal statement, so we will set up an a-flagged subproof and then try to derive the instance $Fa \supset (\exists y) \sim (Tya \lor Zay)$. This is a conditional, so we will want to assume the antecedent and try to derive the consequent. The consequent is an existential statement, which we can get just by E.G. provided we have an instance. The instance would be equivalent by DeM. to $\sim Tba \cdot \sim Zab$, so this is what we need to aim for. We might first try to derive what we can from the premises, once we set up the subproofs for U.G. and C.P.

4.	flag a	F.S. (U.G.)
5.	Fa	Assp. (C.P.)
6.	$(x)(Fx \supset \sim Gx)$	C.Q.N. 2
7.	$Fa \supset \sim Ga$	U.I. 6, a/x
8.	$\sim Ga$	M.P. 5,7
9.	$Fa \cdot \sim Ga$	Conj. 5,8
10.	$(Fa \cdot \sim Ga) \supset \sim (\exists y)(Tya \cdot Hay)$	U.I. 1, a/x
11.	$\sim (\exists y)(Tya \cdot Hay)$	M.P. 9,10
12.	$(y)(Tya \supset \sim Hay)$	C.Q.N. 11

At this point we need to stop and take stock of the situation. Again, we need $\sim Tba$ and $\sim Zab$. We can eventually get $Tba \supset \sim Hab$ by U.I. from 12, and then we would need Hab and then $\sim \sim Hab$ so that we could derive $\sim Tba$ by M.T. The function Hxy appears in premises 1 and 3, but since we have already used 1, it is unlikely that we will be deriving Hab from there. Thus we need to look at premise 3. Notice here the placement of the parentheses, which determines that the antecedent is a universal statement and the consequent is simply Gx. We already have the negation of Ga, at step 8, so this looks like a possible application of M.T. Let us see what happens when we use U.I. and M.T.

13.	$(y)(Hay \supset Zay) \supset Ga$	U.I. 3, a/x
14.	$\sim (y)(Hay \supset Zay)$	M.T. 13,8
15.	$(\exists y)(Hay \cdot \sim Zay)$	C.Q.N. 14

At this point we do our anticipated E.I. step.

16. | | $Hab \cdot \sim Zab$ E.I. 15, b/y (flag b)

From now on it is a simple matter to derive our instance, $\sim Tba \cdot \sim Zab$, and then finish off the proof.

17.		Hab	Simp. 16
18.		$\sim Zab$	Simp. 16
19.		$Tba \supset \sim Hab$	U.I. 12, b/y
20.		$\sim \sim Hab$	D.N. 17
21.		$\sim Tba$	M.T. 19,20
22.		$\sim Tba \cdot \sim Zab$	Conj. 21,18
23.		$\sim (Tba \lor Zab)$	DeM. 22
24.		$(\exists y) \sim (Tya \lor Zay)$	E.G. 23, b/y
25.	$Fa \supset (\exists y) \sim (Tya \lor Zay)$		C.P. 5–24
26.	$(x)(Fx \supset (\exists y) \sim (Tyx \lor Zxy))$		U.G. 25, a/x

Remember that your rules of U.I. and E.G. are unrestricted, except that they must be used only on quantifier formulas. It is E.I. and U.G. that require the flagging restrictions.

One final reminder about the importance of parentheses. You must be very careful in dropping your quantifiers to be sure you really do have an instance of the propositional function. If you don't, and get your parentheses mixed up, you may scuttle your proof. If you had the following inference, for instance, you would use U.I. as shown: $(x)((\exists y)Fxy \supset Gx)$ /∴ $(\exists y)Fay \supset Ga$. Notice that there are no parentheses here, so that if you derived $\sim Ga$ you would then be able to use M.T. to get $\sim (\exists y)Fay$, which would then be equivalent by Q.N. to a universal proposition, on which you could use your *unrestricted* rule U.I. If you erroneously inferred $(\exists y)(Fay \supset Ga)$, however, sticking in an extra set of parentheses, you would be required to use E.I., and would need the restriction that the instance letter be flagged. This might make it impossible to carry out the proof. So do be very careful about the *form* of the instance you are inferring.

Finally, we will apply some of the things you have already learned to the proofs of *theorems* in relational predicate logic. Again, this is nothing different from what you did before. In proving a theorem, since you have no premises, you will always have to make an assumption at the first step. You will just have to be especially careful to analyze the form of the conclusion to see what assumption gets made first. If it is a conditional, assume the antecedent for C.P.; if you want to use I.P., assume the opposite of what you want to prove; and if it is a universal statement, you must set up a flagged subproof. The following, for example, is a *conditional*, and can be proved by using C.P.: $(x)(y)Fxy \supset (x)(\exists y)Fxy$. The proof follows.

e. 1. $(x)(y)Fxy$ Assp. (C.P.)
    2.   flag a F.S. (U.G.)
    3. $(y)Fay$ U.I. 1, a/x
    4. Fab U.I. 3, b/y
    5. $(\exists y)Fay$ E.G. 4, b/y
    6. $(x)(\exists y)Fxy$ U.G. 5, a/x
    7. $(x)(y)Fxy \supset (x)(\exists y)Fxy$ C.P. 1–6

The following is a *universal* formula, and for it we will need an outer flagged subproof so that we may use U.G. at the last step: $(x)((\exists y)Fxy \supset (\exists z)Fxz)$. Thus the first step in the proof is a flagging step. The *instance* of the conclusion is a *conditional*, so our next assumption must be the antecedent of the conditional, for C.P. We then proceed to derive the consequent of the conditional, use C.P., and then use U.G.

f. 1. flag a F.S. (U.G.)
    2. $(\exists y)Fay$ Assp. (C.P.)
    3. Fab E.I. 2, b/y (flag b)
    4. $(\exists z)Faz$ E.G. 3, b/z
    5. $(\exists y)Fay \supset (\exists z)Faz$ C.P. 2–4
    6. $(x)((\exists y)Fxy \supset (\exists z)Fxz)$ U.G. 5, a/x

These are rather simple examples, but there is nothing different in principle in doing proofs of theorems. You simply have to start with assumptions and plot strategy carefully. Exercise 2, at the end of the unit, will give you more practice in proving theorems.

2. Invalidity in Relational Predicate Logic

The methods for demonstrating invalidity in relational predicate logic are just an extention of what you learned in Unit 16; there is nothing new in principle. The only difference is the occurrence of relational predicates and multiple quantifiers. In the natural interpretation method we will still be looking for interpretations of the predicates and constants that result in true premises with a false conclusion; we will just have to be particularly careful in reading out the quantifier statements. Confusing $(x)(\exists y)Fxy$ with $(\exists y)(x)Fxy$, for instance, will lead to errors. In the model universe method we will still be rewriting quantifier statements as conjunctions and disjunctions, but with multiply quantified statements we will have to be careful to begin with the dominant quantifier, the one to the far left.

In the natural interpretation method we want to find meanings for the predicate letters and individual constants that yield a counterexample.

We must, of course, assign one-place predicates to one-place predicate letters, two-place predicates to two-place predicate letters, and so on. It is important to remember that the form of a multiply quantified formula is determined by its outermost, or leftmost quantifier; $(x)(\exists y)Fxy$, for instance, is a universal statement.

The following argument form is invalid; let us apply the natural interpretation method to demonstrate this fact. $(y)(\exists x)Fxy$ /∴ $(\exists x)(y)Fxy$. The first formula, the premise, says that for every y there is some x (not necessarily the same one) which is related to y in a certain way. The conclusion says that there is some *one* x which is related to *every* y in that way. We can show that this is invalid by here taking our domain as the set of human beings and letting Fxy stand for "x is the biological father of y." The premise then reads "For every y there is some x such that x is the biological father of y," that is, everyone has a biological father, and this is true for all the members of our domain. The conclusion, however, says that there is some *one* x such that for *every* y, x is the biological father of y, that is, that there is some x who is the biological father of all human beings, and this, of course, is false. Thus we have a counterexample. Again, the proper form for the answer to this problem would be the following:

1. Domain is human beings.
2. $Fxy \equiv x$ is the biological father of y.
3. Premise: Everyone has a biological father.
   Conclusion: There is someone who is the biological father of everyone.
4. The premise is true by biology.
   The conclusion is obviously false.

To show invalidity for arguments containing constants in the natural interpretation method, we must supply an interpretation for the constants as well as for the predicate letters. The following argument will illustrate this procedure: $(\exists x)Fxa, (x)Fxa \supset (\exists x)Gax$ /∴ $(\exists x)Gax$. If we restrict the domain to human beings and use $Fxy \equiv x$ is the biological father of y, $Gxy \equiv x$ is the biological mother of y, and $a \equiv$ Albert Einstein, then the premises are both true, since they say (tenselessly) "Albert Einstein has a biological father" and "If everyone is the biological father of Einstein, then Einstein is the biological mother of someone." (The latter is true simply because the antecedent of the conditional is false.) The conclusion, however, is false, since it says that Einstein is the biological mother of someone. Thus we have a counterexample.

We can also use the natural interpretation method for arguments containing categorical relational statements, though it may be harder to construct counterexamples because of the increased complexity. The following argument is invalid: $(x)(Fx \supset (\exists y)Gyx), (\exists x)(Fx \cdot (\exists y)Hyx)$ /∴ $(\exists x)(\exists y)(Gyx \cdot Hyx)$. We can get a counterexample by again taking as our domain the set of human beings, and using $Fx \equiv x$ is female, $Gxy \equiv x$ is the biological father of y, and $Hxy \equiv x$ is the biological mother of y. Then the first premise says "All females have a biological father," which is true. The second premise says "There is some female who has a biological mother," which is also true. The conclusion, however, says that there is someone who has a single individual y as both biological mother and biological father, and this, of course, is false.

The model universe method is also applied much as before. We select a domain and then rewrite the quantifier statements as conjunctions or disjunctions of instances from the domain. If there is more than one quantifier in a formula we take them one at a time, rewriting the leftmost quantifier first. In $(y)(\exists x)Fxy$ /∴ $(\exists x)(y)Fxy$ the outside quantifier for the premise is universal, so we would have a *conjunction* of instances. The question is, what instances? Well, the *function* following the universal quantifier is $(\exists x)Fxy$, and we need the instances of this function for every element in our domain. In this example a domain of one individual will not work, so we need a domain with at least two; we will use $\{a, b\}$. Reformulating $(y)(\exists x)Fxy$ in $\{a, b\}$, then, we get $(\exists x)Fxa \cdot (\exists x)Fxb$, where we substitute the a and b for the y in $(\exists x)Fxy$. Now we need to rewrite each of those existential statements in $\{a, b\}$. Each, of course will be a *disjunction*. $(\exists x)Fxa$ will be written as $(Faa \lor Fba)$, and $(\exists x)Fxb$ will be rewritten as $(Fab \lor Fbb)$. The full reformulation of our premise, then, will be $(Faa \lor Fba) \cdot (Fab \lor Fbb)$. Using the same approach on the conclusion, we note that this is an *existential* statement, so the overall form will be a disjunction. Rewriting our first quantifier we get $(y)Fay \lor (y)Fby$. We now have to rewrite each of the universal statements. The first will yield $(Faa \cdot Fab)$ and the second will yield $(Fba \cdot Fbb)$. The full reformulation of the conclusion, then, will be $(Faa \cdot Fab) \lor (Fba \cdot Fbb)$. Having rewritten our quantified formulas, we now need to come up with a counterexample, and this is not hard; we just make Faa and Fbb false and Fab and Fba true. The following diagram verifies that this is a counterexample to our argument form.

$$
\begin{array}{ccc}
\text{F} \quad \text{T} \quad \text{T} \quad \text{F} & & \text{F} \quad \text{T} \quad \text{T} \quad \text{F} \\
(Faa \lor Fba)\cdot(Fab \lor Fbb) & /\therefore & (Faa \cdot Fab)\lor(Fba \cdot Fbb) \\
\text{T} \qquad\qquad \text{T} & & \text{F} \qquad\qquad \text{F} \\
\qquad \text{T} & & \qquad \text{F}
\end{array}
$$

The proper form for the answer to this model universe problem would be the following:

1. Domain is $\{a, b\}$
2. $(Faa \lor Fba) \cdot (Fab \lor Fbb)$ /∴ $(Faa \cdot Fab) \lor (Fba \cdot Fbb)$

3. Faa = F, Fab = T, Fba = T, Fbb = F

4.   Since Fab and Fba are both true, both conjuncts in the premise are true, so the premise is true. Since Faa and Fbb are false, both disjuncts in the conclusion are false, so the conclusion is false.

For the previous argument containing constants, $(\exists x)Fxa$, $(x)Fxa \supset (\exists x)Gax$ /∴ $(\exists x)Gax$, we will need a domain with two individuals, and we will simply assign the individual a to the constant a. Rewriting the formulas in $\{a, b\}$ gives the following: $Faa \lor Fba$, $(Faa \cdot Fba) \supset (Gaa \lor Gab)$ /∴ $Gaa \lor Gab$. It is easy to get a counterexample; we simply let Faa be true and the other simple formulas be false. Our answer is then as follows:

1.   Domain is $\{a, b\}$

2, 3, 4. $Faa \lor Fba$, $(Faa \cdot Fba) \supset (Gaa \lor Gab)$ /∴ $Gaa \lor Gab$

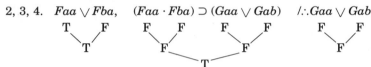

Arguments containing categorical statements are again a bit more complex, but the principle is the same. In the following problem we rewrite the formulas in two stages, doing the outermost ones first.

$(x)(Fx \supset (\exists y)Gxy), (\exists x)(Fx \cdot (\exists y)Hxy)$ /∴ $(\exists x)(\exists y)(Gxy \cdot Hxy)$

1.   Domain is $\{a, b\}$

2.   First premise: $(Fa \supset (\exists y)Gay) \cdot (Fb \supset (\exists y)Gby)$
                       $(Fa \supset (Gaa \lor Gab)) \cdot (Fb \supset (Gba \lor Gbb))$
      Second premise: $(Fa \cdot (\exists y)Hay) \lor (Fb \cdot (\exists y)Hby)$
                       $(Fa \cdot (Haa \lor Hab)) \lor (Fb \cdot (Hba \lor Hbb))$
      Conclusion: $(\exists y)(Gay \cdot Hay) \lor (\exists y)(Gby \cdot Hby)$
               $(Gaa \cdot Haa) \lor (Gab \cdot Hab) \lor (Gba \cdot Hba) \lor (Gbb \cdot Hbb)$

3. Fa = T, Fb = T, Gaa = F, Haa = T, Gab = T, Hab = F, Gba = F, Hba = T, Gbb = F, Hbb = T

4.   The conclusion is false because each disjunct is false; Premise 2 is true because the left disjunct is true; Premise 1 is true because both conjuncts are true.

You should now be able to do the invalidity exercises; you will have to be very careful in rewriting statements to get all the constants in the right place.

EXERCISES

1. Symbolize and construct proofs for the following arguments. Answers are provided for symbolizations at the back of the book.

*a. Some teachers are scholars. No scholar has time for either football or basketball. Therefore, some teachers do not have time for basketball.

*b. No one in his right mind would buy a used toothbrush. John buys used toothbrushes when they are on sale. Therefore, John is not in his right mind.

*c. Some boys like all girls, but no boys like any witches. So, no girls are witches.

*d. Anyone who likes John has a great deal of patience. Anything with a great deal of patience has steady nerves. No one here has steady nerves, so no one here likes John.

e. If anyone puts tabasco in the soup it will be spoiled, and if it is spoiled someone will be furious with John. No one is furious with John. Therefore, John did not put tabasco in the soup.

f. $(x)(Ex \supset (y)(Fy \supset Gxy))$, $(\exists x)(Ex \cdot (\exists y) \sim Gxy)$ /∴ $\sim (x)Fx$

g. $(x)(Kx \supset ((\exists y)Lxy \supset (\exists z)Lzx))$, $(x)((\exists z)Lzx \supset Lxx)$, $\sim (\exists x)Lxx$
   /∴ $(x)(Kx \supset (y) \sim Lxy)$

h. $(x)(Ox \supset (y)(Ry \supset \sim Lxy))$, $(x)(Ox \supset (\exists y)(Hy \cdot Lxy))$, $(\exists x)Ox$
   /∴ $(\exists x)(Hx \cdot \sim Rx)$

i. $\sim (x)(Fx \supset Gx)$, $(x)(Hx \supset Gx)$, $(x)(\sim (Hx \lor Gx) \supset (\exists y)(Axy \lor Byx))$,
   $\sim (\exists x)(\exists y)Axy$ /∴ $(\exists x)(\exists y)Byx$

*j. $(x)(\exists y)(Dy \cdot Kyx) \supset (z)(Az \supset Hxz))$, $(x)(y)(Fxy \supset \sim Hxy)$, $\sim (x)(Ax \supset \sim Dx)$,
   $(x)(Cx \supset (y)Kyx)$ /∴ $(x)(Cx \supset (\exists y)(Dy \cdot \sim Fxy))$

k. $(\exists x)(Px \cdot (y)((Sy \cdot (\exists z)(Pz \cdot Lyz)) \supset Lyx))$, $(x)(Sx \supset (\exists y)(Py \cdot Lxy))$
   /∴ $(\exists x)(Px \cdot (y)(Sy \supset Lyx))$

l. $(x)((Fx \cdot \sim (y)(Gy \supset Hxy)) \supset (z)(Iz \supset Tzx))$, $\sim (\exists x)(\exists y)(Wxy \cdot Hxy)$,
   $(x)(\exists y)(Fy \cdot Wyx) \supset Ix)$ /∴ $(x)(Fx \supset (y)((Gy \cdot Wxy) \supset Tyx))$

2. Construct proofs for the following theorems.

a. $(x)(y)Fxy \supset (\exists x)(\exists y)Fxy$

b. $(\exists x)(y)(z)Fxyz \supset (y)(z)(\exists x)Fxyz$

*c. $(x)((\exists y)Fxy \supset Gx) \supset (x)(y)(Fxy \supset Gx)$

*d. $(x)(\exists y)(Fxy \supset Gx) \supset (x)((y)Fxy \supset Gx)$

e. $(\exists y)(Fy \cdot (x)(Gx \supset Hxy)) \supset (x)(Gx \supset (\exists y)(Fy \cdot Hxy))$

3. Use *both* the natural interpretation method and the model universe method to show that the following argument forms are invalid.

*a. $(\exists x)(y)Fxy$ /∴ $(x)(y)Fxy$

b. $(x)(\exists y)Fxy$ /∴ $(x)(y)Fxy$

*c. $(x)(Fx \supset (\exists y)Gxy)$ /∴ $(\exists x)(y)(Fx \supset Gxy)$

d. $(x)(y)(\exists z)Fxyz$ /∴ $(x)(\exists z)(y)Fxyz$

*e. $(x)(y)(\exists z)Fxyz$ /∴ $(\exists z)(y)(x)Fxyz$

f. $(\exists x)Fxa$ /∴ Faa

g. $(\exists x)Fxa$, $(\exists x)Gxa$ /∴ $(\exists x)(Fxa \cdot Gxa)$

h. $(\exists x)(Fxa \cdot Gax)$ /∴ $(\exists x)(Fxa \cdot Gxa)$

i. $(x)Fxa \supset (\exists x)Fax$, $(\exists x)Fxa$ /∴ $(\exists x)Fax$

j. $(\exists x)(Fx \cdot (\exists y)Gxy)$, $(x)(y)(Gxy \supset Hx)$ /∴ $(x)(Fx \supset Hx)$

k. $(x)((y)Fyx \supset Lx)$ /∴ $(x)(y)(Fyx \supset Lx)$

l. $(x)(Fx \supset (\exists y)Gxy)$ /∴ $(\exists y)(x)(Fx \supset Gxy)$

m. $(x)((\exists y)Fxy \supset (\exists y)Gxy)$, $(x)(Hx \supset (\exists y)Fxy)$ /∴ $(x)(Hx \supset (y)Gxy)$

UNIT 19
Identity and Definite Descriptions

A. INTRODUCTION

In Unit 10 we talked about the three senses of "is," and the three kinds of subject-predicate propositions. Until now, however, we have been using only the first two, singular propositions and categorical propositions. In this unit, finally, you will learn about the third sense of "is," the "is" of identity. In the identity sentence both subject and predicate are individuals, and the sentence asserts that the two named individuals are one and the same. An example would be "Mark Twain is Samuel Clemens." You will learn here the basic symbolization for identity statements (which is very simple), how to distinguish identity sentences from other S–P propositions (most of which you already know), and then some interesting things that can be done with quantifier logic once we add the identity concept. We can, for instance, make *exceptive* statements, such as "Everyone except John liked the pie," and a closely related kind of sentence that uses "only," such as "John was the only one who got a raise." In relational logic we can, as you have seen, make comparative statements; at this point, given identity, we can make *superlative* statements, such as "Alaska is the biggest state in the union." Perhaps the most surprising result is that now, with the use of identity plus relational predicate logic, we can make *numerical statements* without using numbers! That is, for any finite number n, we can say "There are n people in the room." Finally, we can symbolize another kind of individual term, mentioned earlier: the *definite description*, such as "the person who ate the pie." A definite description, if you remember, is something like an individual constant in that it makes reference to a particular individual, but it does so by means of an identifying description rather than a proper name. In the next unit we introduce rules of proof for identity, but in this unit we confine ourselves to symbolization. What you need to know is listed below.

B. UNIT 19 OBJECTIVES

1. Learn how to recognize and symbolize identity statements.
2. Learn how to distinguish identity statements from singular statements and categorical propositions.
3. Learn how to symbolize exceptives, "only" statements, and superlatives.
4. Learn how to symbolize numerical statements.
5. Learn how to symbolize statements containing definite descriptions.

C. UNIT 19 TOPICS

1. Identity Statements and Their Negations

In Unit 10 you learned about the three different senses of "is," and the three different kinds of subject–predicate propositions that correspond to them: the singular sentence, the categorical proposition, and the identity sentence. Remember that the difference between these three kinds of sentences lies in the kinds of subjects and predicates they have. In the singular sentence the subject is an individual and the predicate is a class; in the categorical proposition, both subject and predicate are classes; and in the identity sentence both subject and predicate are individuals. A singular sentence states that an individual has a certain property, a categorical sentence states some kind of inclusion or exclusion relation between the subject and predicate classes, and an identity sentence states that the subject and predicate individuals are one and the same, or are identical.

The symbolization for identity sentences could hardly be easier. We simply place the identity symbol, ' $=$,' with which you are no doubt familiar, between the two individual constants that name the subject and predicate individuals. Keeping in mind that we generally use the first letter of a name as the individual constant which abbreviates it, we could symbolize

"Mark Twain is Samuel Clemens" simply as $m = s$. Other examples of identity statements, with their symbolizations, are below:

Superman is Clark Kent.	$s = c$
Dr. Jekyll is Mr. Hyde.	$j = h$
Mephistopheles is Satan.	$m = s$
Buffalo Bill was William Cody.	$b = w$

Symbolizing negative identity statements could be done simply by placing our negation sign in front of the identity statement to be negated. We could say that Lois is not Superwoman, for instance, by writing $\sim (i = w)$; but it is customary, and somehow easier, to use instead a slash through the identity sign: $\neq$. We could symbolize the above sentence, then, as $i \neq w$. Some further examples of negative identity statements, with their symbolizations, follow.

John Travolta is not Rocky.	$t \neq r$
The Bible is not the Koran.	$b \neq k$
Buffalo Bill was not Custer.	$b \neq c$
The Pentagon is not the White House.	$p \neq w$

Of course, you must be able to distinguish identity sentences from the other kinds of simple S–P propositions. This should not be too difficult in most cases; an identity sentence is simply one that names two individuals who are said to be the same. You should be careful, however, not to confuse identity statements with relational statements. In the sentence "Mary is John's sister," for instance, it looks as if we are asserting an identity between Mary and John's sister, but "John's sister" is *not* a proper name, so the sentence could not be an identity sentence. The best way to interpret this would be as a singular relational statement, which could be symbolized as Smj, where $Sxy \equiv x$ is the sister of y. We ought not to interpret it as a definite description, because there is no indication that Mary is *the* sister of John, that is, that John has only one sister. A pure identity statement, again, will contain two proper names, and it will be asserting that the two individuals are the same. Exercise 1, at the end of the unit, will help you become proficient in making the distinction between identity statements, monadic singular propositions, relational singular statements, and categorical propositions.

2. Exceptives and "Only" Statements

We often want to make something less than a universal claim; we may want to say, for instance, that everyone *except* John and Mary came to the party, or that everyone in the room *except* David has a cold. We could not do anything with propositions of this sort until now, however, because their symbolizations require the use of *negative identity* statements. When we make an exceptive statement, of the form "Every F except a is a G," what we are really saying is that *for every x, if x is an F but x is not a, then x is a G*, where the "is not" uses the "is" of identity. We would symbolize this as $(x)((Fx \cdot x \neq a) \supset Gx)$. In many, or most cases, we also mean to say that a is a counterexample to the universal claim, which means that it has the subject property but does not have the predicate property, so we will conjoin to the rest of the sentence the clause $Fa \cdot \sim Ga$. Thus the full symbolization for "Every F except a is a G" would, in most cases, be $(x)((Fx \cdot x \neq a) \supset Gx) \cdot Fa \cdot \sim Ga$. We would symbolize "Everyone in the room except David has a cold," for instance, as $(x)((Px \cdot Rx \cdot x \neq d) \supset Cx) \cdot Pd \cdot Rd \cdot \sim Cd$.

In some cases there may not be a subject term proper, as in the sentence "Everything except God was created by God." This would be symbolized just as $(x)(x \neq g \supset Cx) \cdot \sim Cg$, where Cx means "x was created by God." Notice that we do include the statement that God was not created by God.

In sentences that do have a subject term F, you should always conjoin to the universal formula the claim that a is an F: Fa. In most cases, as noted, you will also want to say that a is not a G, $\sim Ga$, since a is being *excepted* from the universal claim. However, there are cases where it is inappropriate to include $\sim Ga$; if, for instance, we just don't know whether a is a G or not, as in "Everyone except George thinks the plan will work, but I don't know about George." This would be symbolized as $(x)((Px \cdot x \neq g) \supset Tx) \cdot Pg$, where Tx means "x thinks that the plan will work." We do include the claim that George is a person.

The basic pattern for exceptives, then, will be the following: a *universal proposition, with a negative identity statement conjoined to the subject function* (if there is one), to indicate that the named individual (or individuals) is (or are) excluded from the universal claim. As noted, we should also include a claim to the effect that *the named individual has the subject property*, and we will usually want to say as well that *the individual does not have the predicate property*. Some further exceptive statements, with their symbolizations, are as follows:

1.   Every Democratic incumbent except for George will be reelected.

     $(x)((Dx \cdot Ix \cdot x \neq g) \supset Rx) \cdot Dg \cdot Ig \cdot \sim Rg$

2.   Everyone except John understood the explanation.

     $(x)((Px \cdot x \neq j) \supset Ux) \cdot Pj \cdot \sim Uj$

3. Every planet except the Earth is too hot or too cold to sustain life.

$(x)((Px \cdot x \neq e) \supset (Hx \vee Cx)) \cdot Pe \cdot \sim (He \vee Ce)$

We can also except more than one individual from our universal claim; in such cases, we simply conjoin the other negative identity statements to the subject function, and add the other singular statements at the end. Examples follow

4. Everyone at the party except for Bob and Marion drank a lot.

$(x)((Px \cdot Ax \cdot x \neq b \cdot x \neq m) \supset Dx) \cdot Pb \cdot Ab \cdot \sim Db \cdot Pm \cdot Am \cdot \sim Dm$

We can also, of course, move the exceptive clause around in the English sentence, for stylistic variation. Some examples would be the following:

5. Everyone here has a Toyota except for John and Andrew.

$(x)((Px \cdot Hx \cdot x \neq j \cdot x \neq a) \supset Tx) \cdot Pj \cdot Hj \cdot \sim Tj \cdot Pa \cdot Ha \cdot \sim Ta$

6. Everyone on the flight thought it would crash, except for Bob, Mary, and Evelyn.

$(x)((Px \cdot Fx \cdot x \neq b \cdot x \neq m \cdot x \neq e) \supset Cx) \cdot Pb \cdot Fb \cdot \sim Cb \cdot Pm \cdot Fm \cdot \sim Cm \cdot Pe \cdot Fe \cdot \sim Ce$

It should be pointed out that we are here making statements which except *individuals*, and it is this for which we need identity. We could earlier have made statements like the following, which except certain *classes* from a universal claim: "Everything in this store except for the cat food is of poor quality." This could have been symbolized, without identity, simply as $(x)((Sx \cdot \sim Cx) \supset Px)$. Some sentences of this sort will be included in the exercises, to test your perspicacity.

So far we have been discussing only what we might call *exceptive A* statements, of the form "*All F* except *a* are *G*." We can also make exceptive *E* statements, of the form "*No F* except *a* are *G*." Such sentences can be symbolized exactly like the *A* forms, except for having a negated predicate. An example would be "No player except John got a home run." This would be symbolized as $(x)(Px \cdot x \neq j) \supset \sim Hx) \cdot Pj \cdot Hj$. Here the negations are just reversed from the way they are in the *A* statement. For comparison, we will symbolize the *A* claim "*All* players except John got a home run": $(x)((Px \cdot x \neq j) \supset Hx) \cdot Pj \cdot \sim Hj$. Some other examples of exceptive *E* statements, with their symbolizations, are the following; note that "but" may be used instead of "except."

1. No one at the party except Alfred got sick.

    $(x)((Px \cdot Ax \cdot x \neq a) \supset \sim Sx) \cdot Pa \cdot Aa \cdot Sa$

2. No player except Jennifer got a standing ovation.

    $(x)((Lx \cdot x \neq j) \supset \sim Ox) \cdot Lj \cdot Oj$

3. No one but Andrew likes to cook.

    $(x)((Px \cdot x \neq a) \supset \sim Cx) \cdot Pa \cdot Ca$

Another kind of sentence, which is very close in meaning and structure to the exceptive sentence, is one containing phrases such as "only John." An example would be "Only John got a raise." This again will be a universal proposition, conjoined to appropriate singular sentences, because we are saying that *everyone else did not* get a raise, but that *John did.* We could paraphrase the above sentence as "For all x, if x is a person and x is not John, then x did not get a raise, but John is a person who did get a raise." This would be symbolized as $(x)((Px \cdot x \neq j) \supset \sim Rx) \cdot Pj \cdot Rj$. Notice that this is *exactly* the same pattern as we had for the exceptive E statements! In fact, the English sentence above could be rephrased as "No one except John got a raise." Thus, this kind of "only" statement really turns out to be an E exceptive. Some further examples of "only" statements are given below, with their symbolizations. Notice the stylistic variations.

1. The only great U.S. president was Lincoln.

    $(x)((Ux \cdot x \neq l) \supset \sim Gx) \cdot Ul \cdot Gl$

2. The only native of Dry Flats, Texas, who is a millionaire is Slim.

    $(x)((Nx \cdot x \neq s) \supset \sim Mx) \cdot Ns \cdot Ms$

3. The only one who respects Richard is Richard.

    $(x)((Px \cdot x \neq r) \supset \sim Rxr) \cdot Pr \cdot Rrr$

4. Of all the people in the class, only Mary got an A in Computer Science.

    $(x)((Px \cdot Cx \cdot x \neq m) \supset \sim Ax) \cdot Pm \cdot Cm \cdot Am$

Notice that in all these cases, as in the exceptive E statements which are equivalent to them, we *do* conjoin the singular sentence which says that the individual has the predicate property. This is because, unlike the A exceptives, the implication seems to be clear in these cases that the named individual *does* have the predicate property. To say *"Only* John got a raise" certainly implies that John did get a raise. Notice also that there are a number of other equivalent constructions in English. We could say, instead of the above, "John got a raise, but no one else did" or "No one else besides John got a raise." Keep an eye out for these various English constructions. Exercise 3, at the end of the unit, will test your ability to sort all this out.

3. Superlatives

As noted earlier, in relational logic we were able to make *comparative statements*, statements that use words such as "better," "higher," "smarter," "poorer," and so on. Superlatives, in case you have forgotten your English grammar, are those in which we use words with an "est," or "st" ending—"best," "highest," "smartest," "poorest," are examples. Sentences containing such words can be symbolized by combining the comparative terms with a negative identity clause, and they will end up looking much like our exceptive statements—they are, in fact, just a special kind of exceptive. When we say, for instance, that Billy is the fastest runner in his class, what we mean is that *he is faster than anyone else*. We cannot say simply that he is faster than *anyone* in the class, because this would imply that he is faster than himself, which is impossible. Thus we cannot symbolize the sentence simply as $(x)(Cx \supset Fbx)$. What we are saying is that *he is faster than anyone in the class except himself*, so this turns out to be another exceptive. We could paraphrase the sentence as "For any x, if x is in the class and x is not Billy, then Billy is faster than x (and Billy is in the class)" and we could symbolize it as follows: $(x)((Cx \cdot x \neq b) \supset Fbx) \cdot Cb$. This will be the standard pattern for superlatives: a universal statement, with a negative identity clause conjoined to the subject function, and a *comparative* statement as the predicate function, and the whole conjoined to a singular statement asserting that the named individual is in the subject class. Let us take several such examples and symbolize them:

1. Lincoln was the best U.S. president.

   $(x)((Ux \cdot x \neq l) \supset Blx) \cdot Ul$ ($Ux \equiv x$ was a U.S. president; $Bxy \equiv x$ is better than y; l = Lincoln)

2. Mary is the brightest person in Computer Science.

   $(x)((Px \cdot Cx \cdot x \neq m) \supset Bmx) \cdot Pm \cdot Cm$ ($Px \equiv x$ is a person; $Cx \equiv x$ is in Computer Science; $Bxy \equiv x$ is brighter than y; m = Mary)

3. The Exxon Building is the highest point in Houston, Texas.

   $(x)((Tx \cdot x \neq e) \supset Hex) \cdot Te$ ($Tx \equiv x$ is in Houston, Texas; $Hxy \equiv x$ is higher than y; e = the Exxon Building)

4. International Falls, Minnesota is the coldest place in the contiguous
    United States.

    $(x)((Ux \cdot x \neq i) \supset Cix) \cdot Ui$ ($Ux \equiv x$ is in the contiguous United
    States; $Cxy \equiv x$ is colder than y; i = International Falls, Minnesota)

Exercise 3, at the end of the unit, will give you practice in symbolizing
superlatives. (It may get rather tedious at this point, because all the
sentences look alike!)

4. Numerical Statements

As noted earlier, given relational logic with identity, it is possible
to make numerical statements without using numbers, statements to the
effect that there are a certain number of things of a certain sort. Examples
of such numerical statements, which we could symbolize (theoretically)
by using our quantifier logic with identity, would be the following: "There
are two senators from each state." "There are nine planets," and "There
are 237,845 ants in my ant colony." Where n is any finite positive whole
number, we can symbolize the statement that there are n things of a
certain sort by saying first that there are *at least n* of those things, and
then adding that there are *at most n* of those things. Thus, there must
be *exactly n* such things. If there are at least three women in the class,
and at most three women in the class, then there must be exactly three
women in the class, or simply, there are three women in the class. Thus,
what we need to do to symbolize numerical statements is learn how to
symbolize "at least" and "at most," and then just put them together.

For symbolizing "at least" clauses, you must keep in mind that the
existential quantifier means "There is at least one." Thus, if we want to
say just that there is *at least one* millionaire in the crowd, all we have to
say is $(\exists x)(Mx \cdot Cx)$. No use of identity is required. Suppose we want
to say there are *at least two* millionaires in the crowd. Well, we will certainly need two existential quantifiers. We will have to begin by saying
$(\exists x)(\exists y)(Mx \cdot Cx \cdot My \cdot Cy\text{—})$. But this is not enough. There is nothing
in that formula to rule out the possibility that x and y are the same
individual, in which case there would be only *one* millionaire around.
So to *guarantee* that there are at least two, we must add a clause which
says that x and y are *not* the same individual; that is, conjoin to the
above $x \neq y$. The correct symbolization for that sentence would then be
$(\exists x)(\exists y)(Mx \cdot Cx \cdot My \cdot Cy \cdot x \neq y)$.

If you want to say that there are *at least three* millionaires in the
crowd, you will have to use three existential quantifiers, and then add
statements to the effect that no one of the individuals is the same as any
other. Hence you will need: $(\exists x)(\exists y)(\exists z)(Mx \cdot Cx \cdot My \cdot Cy \cdot Mz \cdot Cz \cdot$
$x \neq y \cdot y \neq z \cdot x \neq z)$. The pattern will be the same for any number in an

"at least" statement: to say that there are at least n ϕ's, use n different existential quantifiers, state that the individuals have the property ϕ, and then use as many negative identity statements as you need to guarantee that no individual is the same as any other. (As an exercise, you might try to figure out the formula which tells you how many negative identity statements you will need to make a claim that there are at least n individuals.) We will do one example of four and then quit, because the formulas get too long. The number of negative identity statements needed goes up rather fast.

$$(\exists x)(\exists y)(\exists z)(\exists w)(Mx \cdot Cx \cdot My \cdot Cy \cdot Mz \cdot Cz \cdot Mw \cdot Cw \cdot x \neq y \cdot x \neq z \cdot$$
$$x \neq w \cdot y \neq z \cdot y \neq w \cdot z \neq w).$$

To say there are *at most n* things of a certain sort, we use *universal* quantifiers rather than existential. We don't need any existential quantifiers, because an "at most" statement does *not* imply that *there are* any of those things. We might argue in theology, for instance, that there is at most one god, because given the properties that we ascribe to a divine being (such as omnipotence) there could be *only* one. If there were more than one god, they couldn't both be omnipotent, and so one could not be a god. But this kind of argument, that there can be *at most* one god, *does not imply* that *there is* a god. Different sorts of arguments are needed to establish existence. (Whether there are any sound arguments for the *existence* of God has been, and still is, a very lively topic of debate among philosophers.) So what do we do to say there are *at most n* things? Suppose, for example, that we wanted to say that there is at most one god. Well, we might reason in this way: If there seems to be more than one, then one must be the same as the other; must be *identical* to the other. In other words, for any *two* things that have the property of being god, the two are the same, so they "collapse down" into one after all. We could symbolize this, which is rather clumsy in English, in the following way:

$$(x)(y)((Gx \cdot Gy) \supset x = y)$$

Suppose we wanted to say that there are *at most two* things of a certain sort, such as millionaires in the crowd. (Remember that this does not imply that there are any.) We begin by using *three* quantifiers, and then say, in effect, that if there seem to be three, then two of them must be the same, so that the three collapse down into two. We could paraphrase this by saying "For any three things, x, y, and z, if x is a millionaire in the crowd, and so is y and so is z, then *either* x is the same as y, or y is the same as z, or x is the same as z." In symbols:

$$(x)(y)(z)((Mx \cdot Cx \cdot My \cdot Cy \cdot Mz \cdot Cz) \supset (x = y \lor y = z \lor x = z)).$$

The general pattern for saying that there are *at most n things* with a certain property will be to begin with *n + 1 universal quantifiers*, and then say if all those things have the property, then two of them must be the same. This latter phrase will always be a disjunction, since you only know that *some two* must be the same, but you don't know which. Thus *either x = y* or *x = z* or . . . . (Again, we leave it as an exercise to figure out how many disjuncts there must be.) We will do one example of a statement that says there are at most three things of a certain sort, this time Arizona congressmen:

$$(x)(y)(z)(w)((Cx \cdot Ax \cdot Cy \cdot Ay \cdot Cz \cdot Az \cdot Cw \cdot Aw) \supset (x = y \lor x = z \lor x = w \\ \lor y = z \lor y = w \lor z = w))$$

Again, limitations of space preclude the illustration of very high numbers, but you should see the general pattern, and also that, *in principle*, it is possible to say that there are at most *n* things, for any finite *n*.

Finally, to say that there are *exactly n* things is just to say that *there are at least n and there are at most n* of those things. Thus we could symbolize that there are *n* things just by conjoining the "at least" and the "at most" statements. This would result in a very long formula in most cases, however (though it would be perfectly correct), so instead we combine them in a more efficient way. We start out by saying that there are *at least n* of those things, and then conjoin to that formula one which says that for any *other* thing that has that property, it must be one of the "at least *n*" bunch. To say that there is exactly one god, for instance, we begin by saying that *there is at least one* god, which is simply $(\exists x)Gx$. We then say "For any y, if y is a god, then y must be the same as x," $(y)(Gy \supset y = x)$. Putting the two together, we have our symbolization for "There is (exactly) one god": $(\exists x)(Gx \cdot (y)(Gy \supset y = x))$. Notice that the scope of the initial x quantifier must extend to the end of the formula to pick up the x at the end of the identity statement. Notice also that since identity is commutative (as we will see in the next unit), it doesn't really matter whether we say $x = y$ or $y = x$.

We say that there are (exactly) *two* things of a certain sort by first saying that there are at least two, and then, in effect, saying "that's all"; that whatever other things there seem to be with that property, they must be identical to one of the two. We can say that there are exactly two West Virginia U.S. Senators, for instance, in the following way:

$$(\exists x)(\exists y)(Wx \cdot Sx \cdot Wy \cdot Sy \cdot x \neq y \cdot (z)((Wz \cdot Sz) \supset (z = x \lor z = y)))$$

Another thing to notice about these longer formulas is that it really doesn't make any difference in what order we list the conjuncts; it would have been perfectly correct, in the above sentence, to use $Wx \cdot Wy \cdot Sx \cdot Sy \cdot x \neq y$ . . . . In general, again, to say that there are exactly n things, say first that there are *at least n*, and then *conjoin* a universal statement which says that, for any other thing which has that property, it is identical to one of the initial n things. This means, by the way, that there will always be just n *disjuncts of identity statements* at the end. Let us do the formulas for three and four, to fix the pattern:

There are (exactly) three serious presidential candidates.

$$(\exists x)(\exists y)(\exists z)(Px \cdot Sx \cdot Py \cdot Sy \cdot Pz \cdot Sz \cdot x \neq y \cdot y \neq z \cdot x \neq z \cdot (w)((Pw \cdot Sw) \supset (w = x \lor w = y \lor w = z)))$$

There are (exactly) four cats in this room.

$$(\exists x)(\exists y)(\exists z)(\exists w)(Cx \cdot Rx \cdot Cy \cdot Ry \cdot Cz \cdot Rz \cdot Cw \cdot Rw \cdot x \neq y \cdot x \neq z \cdot x \neq w \cdot y \neq z \cdot y \neq w \cdot z \neq w \cdot (u)((Cu \cdot Ru) \supset (u = x \lor u = y \lor u = z \lor u = w)))$$

In principle, then, it is possible to make numerical statements for any finite n, but as you can see, the formulas become very large very fast, so for practical purposes it is not of much use. Also, of course, for *very* large numbers, such as the number of ants in my ant colony, you would never know whether you had the right formula unless you could count in the first place, since you would need to know whether you had the right number of existential quantifiers! (And you would need to be a numerical genius, or at least have a computer, to figure out how many negative identity statements you would need for that formula.) Despite the lack of practical applications, however, it is still of considerable theoretical interest that, in principle, it is possible to make numerical statements without numbers, using only quantifier logic and identity.

To say that there are *at most n* is to put an upper bound on the number of objects of a certain sort. It means that there are *not more than n*, which is really equivalent to saying that there are *not at least n + 1* of those things. Thus the "at most n" statement will be equivalent to the "not at least $n + 1$" statement. Conversely, to say that there are *at least n* is to say that the number of objects does not stop at $n - 1$, that there are *not at most n − 1*. Thus, again, there will be an equivalence between the "at least n" statement and the "not at most $n - 1$" statement. This is just an exercise in applying Q.N. and C.Q.N. and your other equivalences. You should try to carry out a couple of these transformations.

Another equivalence, mentioned earlier, is between the simply conjoined statements that there are at least n things and at most n things, and our more efficient symbolization. You should be able to prove this equivalence using the quantifier rules already given (though it is a rather long proof).

5. Definite Descriptions

Definite descriptions are expressions which uniquely identify an individual by describing it in some detail. They are like proper names in that they refer to a particular individual, but they do so not by means of a name but by means of a description which (purportedly) fits only one thing. Examples would be "the first president of the United States," "the first book ever printed on a printing press," "the capital of Yugoslavia," "the residence of the Prime Minister of Great Britain," and so on. These are not *names*, but they do all point to a single individual. Such expressions have given rise to some knotty philosophical problems which were finally solved in this century by the use of quantifier logic. What do you do, for example, with a sentence such as the following, in which there is no referent for the description: "The king of the United States of America treats his subjects well"? Is it true? False? Neither? It seems to be a perfectly grammatical sentence, so there is no particular reason why it should not have a truth value. And it would be very odd to call it true. On the other hand, it doesn't seem quite right to call it false either, as if what we were saying about the king of the United States was not accurate. There have been various solutions, but the only one we will discuss here is the one which is generally accepted as correct, Bertrand Russell's "theory of descriptions."[1]

What Russell said is that definite descriptions are really not grammatically just like names; they do not refer in the same simple way that names do. Rather, he said, a definite description makes sense only *in context*, when it is used as a part of a sentence. It cannot be analyzed in isolation, by itself. (Such terms, which cannot be analyzed out of context, are called "syncategorematic.") However, we can analyze the definite description when it appears within a sentence. When we say *the* ϕ has property ψ, we are really saying *three* things: first, *there is* a ϕ, second, *there is no more than one* such ϕ, and third, *the ϕ does have the property ψ*. Thus, to take our previous sentence, we would have to analyze it as "There is a king of the United States, and there is only one king of the United States, and that king treats his subjects well." Thus paraphrased, it is not hard to symbolize. Using the obvious abbreviations, we could symbolize it as follows: $(\exists x)(Kx \cdot (y)(Ky \supset x = y) \cdot Tx)$. Notice here that we have a part of the sentence which says simply that there is exactly one king of the United States; this is what is meant by a definite description: there is one and only one of those things ("The one and only . . .").

---

[1] Russell was one of the most influential philosophers and logicians in the first half of the twentieth century. He and A. N. Whitehead first brought symbolic logic to the attention of philosophers and logicians in *Principia Mathematica*, in 1910.

Analyzed in this way, a sentence containing a definite description will turn out to be *false* either if *there isn't any such thing* (like the king of the United States), or if *there is more than one such thing*, or if there is just one such thing, but *it doesn't have the property* ascribed to it. Examples of such false propositions are below, with their symbolizations. They are all false for different reasons.

1. The woman who walked on the moon left footprints.

   $(\exists x)(Wx \cdot Mx \cdot (y)((Wy \cdot My) \supset x = y) \cdot Fx)$

2. The book in the library is a philosophy text.

   $(\exists x)(Bx \cdot Lx \cdot (y)((By \cdot Ly) \supset x = y) \cdot Px)$

3. The person who wrote *War and Peace* was an American.

   $(\exists x)(Px \cdot Wx \cdot (y)((Py \cdot Wy) \supset x = y) \cdot Ax)$

Examples of true sentences containing definite descriptions would be the following:

4. The person who wrote *War and Peace* was a Russian landowner.

   $(\exists x)(Px \cdot Wx \cdot (y)((Py \cdot Wy) \supset x = y) \cdot Rx \cdot Lx)$

5. The president in office during the Civil War was a Republican.

   $(\exists x)(Px \cdot Cx \cdot (y)((Py \cdot Cy) \supset x = y) \cdot Rx)$

Exercise 5 will give you practice in symbolizing sentences containing definite descriptions.

EXERCISES

1. For each of the following sentences, state whether it is an identity sentence, a monadic singular sentence, a relational singular sentence, or a categorical proposition.

*a. David is John's friend.

b. John is cuter than David.

*c. David is "Peaches."

d. David is cute.

*e. Little boys are cute.

f. David is older than John.

*g. Buffalo Bill is William Cody.

h. Kit Carson is a scout.

*i.   Scouts are fearless.

 j.   Kit Carson is not Custer.

*k.   Custer is crazier than Wild Bill.

 l.   Custer is reckless.

*m.   Reckless people die young.

 n.   Wild Bill did not die young.

*o.   Geronimo is not Indian George.

 p.   Geronimo is not Navajo.

*q.   Navajos are not aggressive.

 r.   Geronimo is braver than Kit Carson.

*s.   Custer is Custer.

 t.   Custer died young.

2. Identify the forms of the following, which are fairly simple examples of the three kinds of S–P propositions, and then symbolize, using the indicated abbreviations. Note that some of the singular sentences may be relational, or quantifications over relational predicates.

a = Ahab; d = the devil; g = God; i = Ishmael; m = Moby Dick; Ax ≡ x is albino; Fx ≡ x is a fish; Mx ≡ x is a mammal; Px ≡ x is a person; Wx ≡ x is a whale; Dxy ≡ x defeats y; Exy ≡ x is an enemy of y; Fxy ≡ x is a friend of y; Mxy ≡ x is meaner than y; Sxy ≡ x is swifter than y.

*a.   Moby Dick is a whale.

 b.   A whale is a mammal.

*c.   Moby Dick is Ahab's enemy.

 d.   Moby Dick is not the devil.

*e.   Ishmael is not albino.

 f.   Ishmael is Moby Dick's friend.

*g.   Ishmael is not God.

 h.   Moby Dick is not a fish.

*i.   Moby Dick defeats Ahab.

 j.   Ahab is meaner than the devil.

*k.   Ishmael is everyone's friend.

 l.   Ahab is not Ishmael's enemy.

*m.   Moby Dick is albino.

 n.   No one is meaner than Ahab.

*o.   Ahab is meaner than anyone.

 p.   God is not a whale.

*q.   The devil is no one's friend.

 r.   Ahab has no friends.

*s.   Ishmael does not defeat Ahab.

  t.   God is not the devil.

3.   Symbolize the following exceptives and superlatives, using the abbreviations above and the additional ones given here.

Zxy = x fears y; Nx = x is an animal; Sx = x has a spout; Bxy = x is a better sailor than y; Qx = x is on the Pequod.

*a.   Everyone except Ahab fears Moby Dick.

  b.   No one except Ahab can defeat Moby Dick.

*c.   Moby Dick is the only albino whale.

  d.   Only Ishmael is a friend of Moby Dick.

*e.   No one is meaner than Ahab except the devil.

  f.   Every whale except Moby Dick is a friend of Ahab.

*g.   There are other whales besides Moby Dick which are feared by Ishmael.

  h.   Moby Dick is the only whale that is swifter than Ahab.

*i.   Moby Dick is the swiftest whale.

  j.   Ahab is the meanest person.

*k.   No animals except whales have spouts.

  l.   Ishmael is the best sailor on the Pequod.

4.   Symbolize the following numerical statements, using the abbreviations indicated, and some from above.

*a.   Ahab has at least two enemies.

  b.   There is one God. (Gx = x is a god)

*c.   Moby Dick has exactly two friends.

  d.   There are at least two albino whales.

*e.   There are at least three West Virginia congressmen. (Cx = x is a congressman; Wx = x is from West Virginia)

  f.   There are at most three buildings on this lot. (Bx, Lx)

*g.   There are exactly three forms of the deity. (Dx = x is a form of the deity)

  h.   There are at least five planets. (Px = x is a planet)

5.   Symbolize the following sentences, which contain definite descriptions, using the abbreviations indicated.

Bx = x is a book; Tx = x is on the table; Nx = x is a novel; Cx = x is a U.S. capitol building; Ixy = x is in y; Px = x was a Pony Express rider; Bxy = x was braver than y; Ox = x was an orphan; Fxy = x was faster than y; Fx = x was famous; Mx = x is a man; Sx = x started the Pony Express; Bx = x was a businessman; Lx = x lost a lot of money.

*a.   The book on the table is a novel.

  b.   The book on the table is *War and Peace*.

*c.   The U.S. Capitol building is in Washington, D.C.

d. The bravest Pony Express rider was Billy Tate, who was an orphan.

*e. The fastest Pony Express rider was Bob Haslam.

f. The only famous Pony Express rider was William Cody.

*g. The man who started the Pony Express was a businessman, and lost a lot of money.

UNIT 20
Proofs Involving Identity

A. INTRODUCTION

This unit will be mercifully short; the only new things you will be learning are the rules for identity, and since these are very simple and very natural, this should not be difficult. This introduction is also mercifully short. What you will need to learn in the unit is listed below.

B. UNIT 20 OBJECTIVES

1. Learn the three rules for identity.
2. Learn to construct proofs for arguments containing identity statements.

C. UNIT 20 TOPICS

1. Rules for Identity

The identity relation is a very special relation; it is what is sometimes called an *equivalence relation*. This means simply that it has three very important properties, which we can state symbolically as follows:

$(x) x = x$ Everything is equal to itself. That is, the iden-
                                     tity relation is *reflexive*.

$(x)(y)(x = y \supset y = x)$ For any two things, if the first is equal to the
                                     second, then the second is equal to the first.
                                     This means that identity is a *symmetric rela-
                                     tion*.

$(x)(y)(z)((x = y \cdot y = z) \supset x = z)$ For any three things, if the first is equal to the
                                     second, and the second is equal to the third,
                                     then the first is equal to the third. This means
                                     that identity is a *transitive* relation.

There are many other relations that have these properties as well
and which are thus equivalence relations, such as *being the same age as*
or *living in the same family as*. But they all tend to make use of the concept
of *sameness*, and thus identity is a very central relation.

The rules for identity will parallel very closely these three properties:
in fact, we could just throw the above formulas in as axioms, but since
this is a system of logic in which we use rules instead of axioms, we will
put the properties in the form of rules. The first rule just restates the
content of the reflexivity property in a slightly different form; it says that
given any statement whatsoever, we may infer for any instance letter 'a'
that $a = a$. This rule will be called, for ease of reference, *Identity Reflexivity*,
or *I Ref.* for short, and will be stated as follows:

$$\frac{p}{/\therefore a = a}$$

The second identity rule will reflect the symmetry property, and so
will be called *Identity Symmetry*, abbreviated as *I Sym*. This rule really
just tells us that identity is commutative, that we can switch the order
of the terms in an identity statement. The rule can be stated as follows:

$$a = b :: b = a$$

Notice that this second identity rule is stated as a *replacement* rule,
so that you may use it on subformulas as well as on whole formulas. This
will make many of the proofs that require this rule considerably easier.

The third identity rule is closely related to the third property, transitivity, but it is more general. The transitivity property, in fact, can be
derived from this third rule. The rule itself is very natural, and states a
substitution property for identity statements. It says, roughly, that if one
thing is the same as the other, then if the first thing has a certain property,
then the second must have it as well. We can state this rule, which we
will call *Identity Substitution*, or *I Sub.*, as follows:

$$a = b$$
$$\phi a$$
$$/\therefore \phi b$$

This means that in constructing a proof, whenever you have a statement that asserts something about a (that is, just any singular statement containing 'a'), then given the identity $a = b$, you may infer that same statement with regard to b. That is, you may take the formula ϕa and replace all the a's with b's. The reason is obvious: if two things really are one and the same, then whatever property the one has, the other must have as well.

Some examples of inferences using these three identity rules are below.

1. $(\exists x)(Fx \cdot (Gax \supset Hxa)) /\therefore a = a$ I Ref.
2. $(Fab \cdot Gba) \supset a = b /\therefore (Fab \cdot Gba) \supset b = a$ I Sym.
3. $(\exists x)(Fx \equiv Gxa), a = c /\therefore (\exists x)(Fx \equiv Gxc)$ I Sub.
4. $Bcra, r = s /\therefore Bcsa$ I Sub.
5. $(x)((\exists y)Fya \supset (\exists z)Fxaz), a = d /\therefore (x)((\exists y)Fyd \supset (\exists z)Fxdz)$ I Sub.

We can derive the *transitivity* property, which we will now state simply as "$a = b, b = c /\therefore a = c$," from the rule of *I Sub.* in the following way. We will take the fundamental identity statement to be the second premise of transitivity, $b = c$. We can interpret the first premise as making a statement *about* b, namely, that a is equal to b; it really has the form ϕb, where the ϕ says "a is equal to." Now since $b = c$, we can make the *same statement* about c, that is, we can assert ϕc, which would say that a is equal to c. Thus the argument, which is really just an application of the substitution rule I Sub., would look like the following; the part of the formula that is taken as ϕ has been outlined, so it is easier to see that the substitution rule is being applied.

$$\boxed{a =}\, b$$
$$b = c$$
$$/\therefore \boxed{a =}\, c$$

Thus, if you ever need to use the transitivity property of identity in a proof, you may simply cite the rule of *I Sub.*, since transitivity is just a special case of substitution. You will frequently find proofs in which transitivity is needed, so be sure to keep this procedure in mind.

Another thing you will often have to do in identity proofs is to show that one individual is *not* the same as another, that $a \neq b$. There is an easy way to do this, which also makes use of the rule of *I. Sub.* Since I Sub. says, in effect, that two individuals that are identical have all their properties in common, it follows that if there are two individuals, of which one has a property and the other doesn't, then the two individuals cannot be the same; they are not equal. This means that given as premises ϕa and $\sim\phi b$, you may infer that a and b are *not* the same, that $a \neq b$. We will not have a *rule* that allows you to do this directly, but you can always do it by the use of I.P. Given ϕa and $\sim\phi b$ (for any property ϕ), you can derive $a \neq b$ by using I.P. as follows:

1. ϕa Pr.
2. $\sim \phi b$ Pr. /$\therefore a \neq b$
3. $\quad a = b$ Assp. (I.P.)
4. $\quad\quad \phi b$ I Sub. 1,3
5. $\quad\quad \phi b \cdot \sim \phi b$ Conj. 4,2
6. $a \neq b$ I.P. 3–5

Thus, whenever you need a negative identity statement in a proof, simply find some property such that one individual has that property and the other does not. By using I.P. the way we have done above, you can show that the two individuals are not the same.

At this point, having learned the rules for identity and some strategies for proving certain formulas by means of these rules, you should be ready for constructing proofs that require the identity rules.

2. Proofs Containing Identity Statements

Constructing proofs for arguments containing identity statements is no different in principle from constructing any other kind of proof. You just have to be especially careful to watch for cases in which you need to use the symmetry and transitivity properties, or when you need reflexive statements, like $a = a$, or negative identity statements like $a \neq b$. This may require a little practice and experience. One thing that will be slightly different is that proofs involving identity often get rather long, since the typical formulas using identity (in particular, numerical statements and definite descriptions) are often rather long formulas. To make the proofs a little easier, we will at this point allow you to *combine* some of the simple rules such as Conjunction, Simplification, Commutation, and Association, so that in practice you can infer, say, Ga *directly* by Simp. from $(Fa \cdot (Ha \vee Ia) \cdot Ga \cdot Ta \cdot Wa)$, without doing all the Com. and Assoc. steps that would, strictly speaking, be required. With this in mind, we will now construct a proof for the following argument, which requires the use of *I Sub.* "The man standing by the door is a doctor. John knows the man standing by the door. Therefore, John knows a doctor."

a.   1. $(\exists x)(Mx \cdot Sx \cdot (y)((My \cdot Sy) \supset x = y) \cdot Dx)$ Pr.

     2. $(\exists x)(Mx \cdot Sx \cdot (y)((My \cdot Sy) \supset x = y) \cdot Kjx)$ Pr. /∴ $(\exists x)(Dx \cdot Kjx)$

     3. $Ma \cdot Sa \cdot (y)((My \cdot Sy) \supset a = y) \cdot Da$ E.I. 1, a/x (flag a)

     4. $Mb \cdot Sb \cdot (y)((My \cdot Sy) \supset b = y) \cdot Kjb$ E.I. 2, a/x (flag b)

     5. $Ma \cdot Sa$ Simp. 3

     6. Da Simp. 3

     7. $(y)((My \cdot Sy) \supset a = y)$ Simp. 3

     8. $(Mb \cdot Sb) \supset a = b$ U.I. 7, b/y

     9. $Mb \cdot Sb$ Simp. 4

   10. Kjb Simp. 4

   11. $a = b$ M.P. 8,9

   12. Kja I. Sub. 10,11

   13. $Da \cdot Kja$ Conj. 6,12

   14. $(\exists x)(Dx \cdot Kjx)$ E.G. 13, a/x

In this proof, the conclusion is $(\exists x)(Dx \cdot Kjx)$, so we need an *instance* of that formula, which could be either $Da \cdot Kja$ or $Db \cdot Kjb$ (or any other instance, for that matter). Da obviously will come from premise 1, and Kja will obviously come from premise 2. It is extremely important to remember in doing these proofs, however, that *you may not use the same instance letter for two different existential propositions.* Thus we could *not* have done this proof simply by instantiating both premises with the letter 'a.' We must use *different* instance letters, and then derive the *identity* statement $a = b$, so that we can infer Kj**a** from Kj**b**, our original instance. *Do not be tempted* in these proofs to take the shortcut of using the same letter for different existential premises; your proof will simply be wrong.

The strategy of working backwards will be especially helpful in proofs invoving identity, since the proofs do tend to get rather long, and it is easy to lose track of where you are going and what instance letter you will be needing. You should carefully analyze your conclusion to see exactly what you will need and what means can be used to get it. Remember that for any occurrence of a universal quantifier in the conclusion you will need a flagged subproof, and remember also the three flagging restrictions. A letter may not be flagged if it has already appeared in the proof, no flagged letter may occur in either the premises or the conclusion of a proof, and if a letter gets flagged within a subproof (as U.G. letters will always be), that letter may not then appear outside that subproof. We will do one more example of a rather long, but typical, proof involving identities, explaining our strategy as we go along.

b.   1. $(x)(Ax \supset Bx)$ Pr.

     2. $(x)(Bx \supset Cx)$ Pr.

3. $(x)(y)(z)((Cx \cdot Cy \cdot Cz) \supset (x = y \lor x = z \lor y = z))$ Pr.
4. $(\exists x)(\exists y)(Ax \cdot Ay \cdot x \neq y)$ Pr.

/ $\therefore$ $(\exists x)(\exists y)(Bx \cdot By \cdot x \neq y \cdot (z)(Bz \supset (z = x \lor z = y)))$

Here, since our conclusion is a doubly existentially quantified formula, we will need as an instance $(Ba \cdot Bb \cdot a \neq b \cdot (z)(Bz \supset (z = a \lor z = b)))$. We will have to use U.G. to derive the universal statement, and, of course, for this we will need a letter other than a or b. Presumably the instance can be $(Bc \supset (c = a \lor c = b))$, and this we will have to derive by C.P. The most promising source for this latter formula is premise 3, since we there have a formula with three disjuncts, and if we could get the negation of one, we would be left, by D.S., with the other two disjuncts. Having thought this through, we can begin our instantiations.

5. $(\exists y)(Aa \cdot Ay \cdot a \neq y)$ E.I. 4, a/x (flag a)
6. $Aa \cdot Ab \cdot a \neq b$ E.I. 5, b/y (flag b)
7. $(y)(z)((Ca \cdot Cy \cdot Cz) \supset (a = y \lor a = z \lor y = z))$ U.I. 3, a/x
8. $(z)((Ca \cdot Cb \cdot Cz) \supset (a = b \lor a = z \lor b = z))$ U.I. 7, b/y

Since we will be using U.G. on the letter c in the end, we must set up a c-flagged subproof before we do any instantiations using 'c.' We can do this now, and then complete the instantiation.

9.   ┌──▶ flag c F.S. (U.G.)
10. │ $(Ca \cdot Cb \cdot Cc) \supset (a = b \lor a = c \lor b = c)$ U.I. 8, c/z

Now, how can we get $(Ca \cdot Cb \cdot Cc)$ so that we can derive the disjunction by M.P.? Given that we have Aa and Ab (by Simp. from 6), we can get Ca and Cb easily enough by using the appropriate instantiations from premises 1 and 2.

11. │ Aa Simp. 6
12. │ Ab Simp. 6
13. │ $Aa \supset Ba$ U.I. 1, a/x
14. │ $Ab \supset Bb$ U.I. 1, b/x
15. │ $Ba \supset Ca$ U.I. 2, a/x
16. │ $Bb \supset Cb$ U.I. 2, b/x
17. │ Ba M.P. 11,13
18. │ Ca M.P. 15,17
19. │ Bb M.P. 12,14
20. │ Cb M.P. 16,19
21. │ $Ca \cdot Cb$ Conj. 18,20

Now, how do we get Cc, our last remaining conjunct? If we remember what we were aiming at, the instance $(Bc \supset (c = a \vee c = b))$, it is fairly obvious. We will assume Bc, for C.P., and given Bc it will be easy to get Cc. The rest of the proof is easy; we will be able to get the disjunction $(a = b \vee a = c \vee b = c)$, and we also will have $a \neq b$, so we can get $(a = c \vee b = c)$ by D.S. We will then just apply the appropriate versions of C.P., U.G., and E.G.

22.	$\rightarrow Bc$	Assp. (C.P.)
23.	$Bc \supset Cc$	U.I. 2, c/x
24.	Cc	M.P. 22,23
25.	$Ca \cdot Cb \cdot Cc$	Conj. 21,24
26.	$a = b \vee a = c \vee b = c$	M.P. 10,25
27.	$a \neq b$	Simp. 6
28.	$a = c \vee b = c$	D.S. 26,27
29.	$Bc \supset (a = c \vee b = c)$	C.P. 22–28
30.	$Bc \supset (c = a \vee c = b)$	I Sym. 29
31.	$(z)(Bz \supset (z = a \vee z = b))$	U.G. 30, c/z
32.	$Ba \cdot Bb \cdot a \neq b \cdot (z)(Bz \supset (z = a \vee z = b))$	Conj. 17,19,27,31
33.	$(\exists y)(Ba \cdot By \cdot a \neq y \cdot (z)(Bz \supset (z = a \vee z = y)))$	E.G. 32, b/y
34.	$(\exists x)(\exists y)(Bx \cdot By \cdot x \neq y \cdot (z)(Bz \supset (z = x \vee z = y)))$	E.G. 33, a/x

Finally, we need to say a few words about proving theorems involving identity. Again, these are simply formulas that can be proved without initially given premises, so that our first step must be an assumption. An example of such a theorem is $Fa \equiv (x)(x = a \supset Fx)$. Since the form is a biconditional, we will first prove the two conditionals by C.P., then join them by Conj., and then apply B.E.

c.	1.	$\rightarrow Fa$	Assp. (C.P.)
	2.	$\rightarrow$ flag b	F.S. (U.G.)
	3.	$\rightarrow b = a$	Assp. (C.P.)
	4.	Fb	I. Sub. 1,3
	5.	$b = a \supset Fb$	C.P. 3–4
	6.	$(x)(x = a \supset Fx)$	U.G. 5, b/x
	7.	$Fa \supset (x)(x = a \supset Fx)$	C.P. 1–6
	8.	$\rightarrow (x)(x = a \supset Fx)$	Assp. (C.P.)
	9.	$a = a \supset Fa$	U.I. 8, a/x
	10.	$a = a$	I Ref., 9
	11.	Fa	M.P. 9,10
	12.	$(x)(x = a \supset Fx) \supset Fa$	C.P. 8–11
	13.	$(Fa \supset (x)(x = a \supset Fx)) \cdot ((x)(x = a \supset Fx) \supset Fa)$	Conj. 7,12
	14.	$Fa \equiv (x)(x = a \supset Fx)$	B.E. 13

We have not yet demonstrated, in an actual proof problem, the technique for deriving *negative* identity statements. The following theorem requires this procedure, as well as two uses of U.G.: $(x)(y)(\sim (Fx \supset Fy) \supset x \neq y)$. There are a few theorems for you to prove in the exercises.

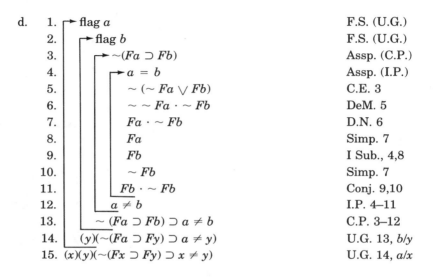

d.	1.	flag a	F.S. (U.G.)
	2.	flag b	F.S. (U.G.)
	3.	$\sim(Fa \supset Fb)$	Assp. (C.P.)
	4.	$a = b$	Assp. (I.P.)
	5.	$\sim (\sim Fa \lor Fb)$	C.E. 3
	6.	$\sim \sim Fa \cdot \sim Fb$	DeM. 5
	7.	$Fa \cdot \sim Fb$	D.N. 6
	8.	Fa	Simp. 7
	9.	Fb	I Sub., 4,8
	10.	$\sim Fb$	Simp. 7
	11.	$Fb \cdot \sim Fb$	Conj. 9,10
	12.	$a \neq b$	I.P. 4–11
	13.	$\sim (Fa \supset Fb) \supset a \neq b$	C.P. 3–12
	14.	$(y)(\sim(Fa \supset Fy) \supset a \neq y)$	U.G. 13, b/y
	15.	$(x)(y)(\sim(Fx \supset Fy) \supset x \neq y)$	U.G. 14, a/x

At this point, we have reached the end of our presentation of what is sometimes called "first-order" predicate logic, the logic that is usually covered in a first symbolic logic course. As we have indicated from time to time, this is by no means all there is to the study of logic. You could go on from here to study the *properties* of first order logic; sometimes called "metalogic," branch into *alternative* logics, such as those that make use of a stronger "if-then" operator, or begin the study of "higher-order" logics, such as those that deal with sets. This is a natural stopping place, however, because first-order predicate logic with identity is the *last* system in our series for which it is possible to prove consistency and completeness.[1] When we reach the higher-order logics we lose completeness (the systems are probably *in*complete), and we can no longer *prove* consistency, though we have good reason to believe the systems are consistent. In any case, we will stop here, and hope that you are motivated enough to continue with logic in more advanced courses. Some of the extra credit units will give you a glimpse into these more advanced topics.

[1] These concepts were defined in Unit 9, Section 7.

SUMMARY OF IDENTITY RULES

IDENTITY REFLEXIVITY (I. REF.)	IDENTITY SYMMETRY (I. SYM.)	IDENTITY SUBSTITUTION (I. SUB.)

$$\frac{p}{/\!\!\therefore\ a = a}$$

$$a = b \ ::\ b = a$$

$$\frac{\begin{array}{c} a = b \\ \phi a \end{array}}{/\!\!\therefore\ \phi b}$$

EXERCISES

1. Symbolize and prove the following, using only the abbreviations given. Answers are provided in the back of the book for all symbolizations. Proofs are provided in the back of the book for starred exercises.

a. The sports car buff who owns a Maserati is unemployed. Mary is a sports car buff and owns a Maserati. So, Mary is unemployed. ($Bx \equiv x$ is a sports car buff; $Mx \equiv x$ owns a Maserati; $Ex \equiv x$ is unemployed; $m = $ Mary)

*b. The person in town who owns a Ferrari is a wealthy Mafioso. John is in town, but he is a person who is not a Mafioso. Therefore, John does not own a Ferrari. ($Px \equiv x$ is a person; $Tx \equiv x$ is in town; $Fx \equiv x$ owns a Ferrari; $Wx \equiv x$ is wealthy; $Mx \equiv x$ is a Mafioso; $j = $ John.)

c. The person who owns a Ferrari is illegally parked. The person who is illegally parked will be towed. So someone who owns a Ferrari will be towed. (Px, Fx, Ix, Tx)

d. The person who owns a Ferrari is John. The person who owns a Volkswagen is Mr. Smith. No one who owns a Volkswagen owns a Ferrari. So Mr. Smith is not John. (Px, Fx, Vx, j, s)

*e. John is the only person who owns a Ferrari. Mr. Capone is the only person who owns a Maserati. There is someone who owns both a Ferrari and a Maserati. So John is Mr. Capone. (Px, Fx, Mx, j, c)

*f. Some council members support the mayor. No Republican council member supports the mayor, but some are in favor of lowering the city sales tax. Some council members who don't support the mayor are not in favor of lowering the city sales tax. Therefore, there are at least three members of the council. ($Cx \equiv x$ is a council member; $Mx \equiv x$ supports the Mayor; $Rx \equiv x$ is a Republican; $Lx \equiv x$ is in favor of lowering the city sales tax)

g. There is exactly one current president of the United States. One is Commander-in-Chief of the U.S. armed forces if and only if one is the current president of the United States. Therefore, there is exactly one Commander-in-Chief of the U.S. armed forces. ($Px \equiv x$ is current president of the United States; $Cx \equiv x$ is Commander-in-Chief of the U.S. armed forces)

h. All John's pets are Siamese cats. John has at least one pet and at most one Siamese cat. Therefore, John has exactly one pet, which is a Siamese cat. ($Mx \equiv x$ is a pet; $Sx \equiv x$ is a Siamese cat; $Hxy \equiv x$ has y; $j = $ John)

*i. There is one person in the front seat of my car. There is one person in the back seat of my car. No one is in both the front seat and the back seat of my car. Therefore, there are two people in my car. ($Px \equiv x$ is a person; $Fx \equiv x$ is in the front seat of my car; $Bx \equiv x$ is in the back seat of my car) (Note: you can use ($Fx \lor Bx$) for "x is in my car.")

j. Adam and Eve were the only people in the Garden of Eden, and they were tempted. Anyone in the Garden of Eden who was tempted succumbed to temptation. Anyone who succumbed to temptation was kicked out of the Garden of Eden. Therefore, everyone in the Garden of Eden was kicked out. ($Px \equiv x$ is a person; $Ex \equiv x$ was in the Garden of Eden; $Tx \equiv x$ was tempted; $Sx \equiv x$ succumbed to temptation; $Kx \equiv x$ was kicked out of the Garden of Eden; $a =$ Adam; $e =$ Eve)

k. There are at least two pianists in the room. All the pianists in the room are composers. There are at most two composers in the room. Therefore, there are exactly two pianists in the room. ($Px \equiv x$ is a pianist; $Rx \equiv x$ is in the room; $Cx \equiv x$ is a composer)

l. There are exactly three composers in the room. Exactly one of the composers in the room is a pianist. Any composer in the room who is not a pianist is an opera singer. Therefore, there are at least two opera singers in the room. ($Cx \equiv x$ is a composer; $Rx \equiv x$ is in the room; $Px \equiv x$ is a pianist; $Ox \equiv x$ is an opera singer)

m. The fastest animal on the track is a dog. Therefore, any animal on the track that isn't a dog can be outrun by some dog. ($Ax \equiv x$ is an animal; $Tx \equiv x$ is on the track; $Dx \equiv x$ is a dog; $Fxy \equiv x$ is faster than y)

2. Construct proofs for each of the following.

a. $(x)((Fx \cdot (\exists y)(Fy \cdot x \neq y)) \supset (Axb \lor Abx))$
$Fa \cdot Fb, Ga \cdot \sim Gb \mathbin{/\therefore} Aab \lor Aba$

*b. $(\exists x)(Ax \cdot (y)(Ay \supset x = y) \cdot x = a)$
$Ab \lor Ac \mathbin{/\therefore} a = b \lor a = c$

c. $(\exists x)(Ax \cdot (y)(Ay \supset x = y) \cdot (\exists z)(Bz \cdot (w)(Bw \supset z = w) \cdot z = x))$
$Ba \mathbin{/\therefore} (x)(Ax \supset x = a)$

d. $(\exists x)(\exists y)(\exists z)(Fx \cdot Fy \cdot Fz \cdot x \neq y \cdot y \neq z \cdot x \neq z),$
$(\exists x)(Fx \cdot Gx \cdot (y)((Fy \cdot Gy) \supset x = y)), (x)(\sim Gx \supset Hx)$
$\mathbin{/\therefore} (\exists x)(\exists y)(Hx \cdot Hy \cdot x \neq y)$

*e. $(\exists x)(\exists y)(\exists z)(Fx \cdot Fy \cdot Fz \cdot x \neq y \cdot y \neq z \cdot x \neq z), (x)(Fx \supset Gx),$
$(x)(y)(z)(w)((Gx \cdot Gy \cdot Gz \cdot Gw) \supset (x = y \lor x = z \lor x = w \lor y = z \lor$
$y = w \lor z = w)) \mathbin{/\therefore} (\exists x)(\exists y)(\exists z)(Fx \cdot Fy \cdot Fz \cdot x \neq y \cdot y \neq z \cdot x \neq z \cdot$
$(w)(Fw \supset (w = x \lor w = y \lor w = z)))$

3. Prove the following theorems.

*a. $Fa \equiv (\exists x)(Fx \cdot x = a)$

*b. $(\exists x)(y)(Fy \equiv x = y) \equiv (\exists x)(Fx \cdot (y)(Fy \equiv x = y))$

c. $(x)(y)(z)((x = y \cdot y = z) \supset x = z)$

d. $(x)(y)(z)(w)((x = y \cdot y = z \cdot z = w) \supset x = w)$

e. $((\exists x)Fx \cdot (x)(y)((Fx \cdot Fy) \supset x = y)) \equiv (\exists x)(Fx \cdot (y)(Fy \supset x = y))$

UNIT 21

Well-Formed Formulas for Sentential Logic

There are some strings of symbols which are obviously "good" formulas, and others which make no sense at all. $((A \cdot B) \vee C)$ would be an example of the former, and $\supset BC \cdot\cdot AA$ would be an example of the latter. Sometimes, however, it is not so easy to tell at first glance whether a string of symbols is a proper, or "well-formed," formula. For instance, you would probably have a hard time determining, without some figuring, whether the following is a meaningful formula or not: $((((A \cdot A) \supset \sim (B \vee C)) \equiv ((C \vee B) \cdot B) \vee (F \cdot G))))$. It has no obvious defects, but it is hard to tell whether everything "fits together" in the right way. We cannot really be sure about formulas like this unless we have a strict definition of what is to count as a meaningful formula, or, as we will say, a "well-formed-formula," or "Wff." The definition of "Well-formed formula" is as follows, and really only makes explicit what was said in Unit 2 about the way formulas are built up.

RECURSIVE DEFINITION OF WELL-FORMED FORMULA (WFF)

1. (Base clause)   Any statement constant is a Wff.
2. (Recursion clause)   If $\mathcal{A}$ and $\mathcal{B}$ are any Wffs, then all of the following are also Wffs: (a) $(\mathcal{A} \cdot \mathcal{B})$, (b) $(\mathcal{A} \vee \mathcal{B})$, (c) $(\mathcal{A} \supset \mathcal{B})$, (d) $(\mathcal{A} \equiv \mathcal{B})$, and (e) $\sim \mathcal{A}$.
3. (Closure clause)   Nothing will count as a Wff unless it can be constructed according to clauses 1 and 2.

This definition may look rather odd to you. It certainly isn't the sort of thing you would find in a dictionary, but it is very well suited for our purposes. This kind of definition is called a *recursive*, or *generative*, definition, because it tells us exactly how to generate instances of the things we are trying to define. Notice that the definition first tells us how we get started; this part is aptly called the *base clause*. It then goes on to tell us how to build up more complex formulas out of those we already have; this is called the *recursion*, *or generative*, *clause*. The third part "cuts off" the definition, and is aptly termed the *closure clause*, since it tells us that there is nothing else that will count as one of our defined entities; that is, nothing will be a Wff that cannot be constructed according to the first two clauses. Every recursive definition has these three parts: a base clause, which gives the starting points; a recursion clause, which says how to generate more of the entities; and a closure clause, which says "that's all."

Let us look in more detail at each of the three clauses which make up our definition of "Wff." The base clause tells us that any statement constant is a Wff. This makes sense, since these letters are supposed to stand for the simple sentences, and we certainly want to consider simple sentences as proper parts of our symbolic language, that is, as formulas.

We need to pay particular attention to the recursion clause. Note first that it has the form of an "if-then" sentence: *if* certain formulas are Wffs, then we can construct others out of them. Fortunately, we do have the statement constants to start with, from clause number one. Second, note that for the four two-place operators (those that have two components) the formula *must be enclosed in parentheses*. This is to ensure that no ambiguity arises when we go on to use a formula as a component of a larger formula. If we built up formulas without parentheses we would have something like $A \lor B \cdot C \equiv D \supset F$, which would be entirely useless, since we couldn't tell whether it was a disjunction, conjunction, etc. If we use parentheses at every point, no such ambiguity arises. Note also that negations do *not* require parentheses; this is because if we start out with something that is already a Wff, then if it is compound, it will already be enclosed in parentheses and thus will not give rise to ambiguity. (In practice, if a formula is not to be used as a component of another formula, the outside parentheses are sometimes dropped for the sake of convenience; this means you may often see in the book, and on the board, formulas such as $(A \lor B) \supset C$, which *strictly speaking* are not Wffs but which do not give rise to any problems, since there is no ambiguity involved. Here, however, we are concerned only with strictly proper Wffs, and so outside parentheses are necessary.)

Another thing that needs special mention is the use of Gothic letters in the definition rather than capital sentence letters. Capital letters, of course, are supposed to be used only for sententially simple sentences, while the Gothic letters are higher-order variables which refer to *any* Wff, simple or complex. It would not do to use the capital letters in our definition of "Wff," because the definition would then allow only the simplest sorts of formulas, those composed of only one or two sentence letters plus one operator. Our recursion clause would read, in effect, "Given two statement constants *A* and *B*, the following are all Wffs: $(A \cdot B)$, $(A \lor B)$, $(A \supset B)$, $(A \equiv B)$ and $\sim A$." There would be no provision made for building up more complex formulas out of those which are already complex. Thus we need in our definition the more general Gothic letters, which can stand for any Wff at all, simple or complex. This ensures that given *any* two Wffs, not just simple ones, we can construct another by joining them with the dot, the wedge, or some other symbol.

Given the first two clauses of the definition we are able to say what things *are* Wffs; that is, we know how to construct Wffs out of elementary components. But how do we know that $))AA \lor (B($ is *not* a Wff? This is the function of the closure clause. By inspection, it can be seen that there is no way to build up the above formula according to the first two clauses; the third clause, then, tells us that it is *not* a Wff, since *nothing* that cannot be constructed using only clauses 1 and 2 can be a Wff. The three clauses together, then, give us a complete definition of Well-formed formula: the first two tell us what things *are* Wffs, and the third tells us what things are not, that is, that there is nothing that is a Wff except for what is contained in clauses 1 and 2, thereby "closing off" the definition.

Finally, we must note that the definition here has been given in terms of statement constants, and so is a definition of *statement* Wffs. (Remember that statements are formulas that have constants as their smallest units.) We could give an exactly parallel definition of Wff for statement forms simply by using the following base clause instead: "Any statement variable is a Wff." The recursion and closure clauses remain the same. The two types of formulas must be kept separate, however; it is not correct to combine sentence constants and variables in a single formula. $((A \lor p) \supset (q \cdot B))$, for instance, is not a Wff.

EXERCISES

On the basis of the definition of Wff given in this unit, decide which of the following are Wffs. For those which are not, indicate why not.

*1. *A*

*2. $A \lor B$

*3. *ABC*

*4. $\sim (A \cdot B)$

*5. $((A \cdot A) \supset A)$

*6. $((A \cdot B))$

*7. $(\sim \sim A \cdot \sim \sim B)$
*8. $\sim \sim \sim \sim \sim \sim A$
*9. $\sim (\sim A)$
*10. $\sim ((A \cdot B \cdot C) \vee (B \cdot C \cdot A))$
*11. $\sim ((A \vee B) \vee (C \vee D))$
*12 $((A \cdot A) \supset (A \vee A))$
*13. $((((A \supset B) \supset C) \supset D)$
*14. $(A \vee (B \vee (C \vee D)))$
*15. $\sim (\sim (A \vee B) \supset ((B \vee C) \equiv (\sim B \cdot C)))$
*16. $\sim (\sim (\sim (\sim (A \vee B) \vee C) \vee D) \vee E)$
*17. $(((\sim A \supset \sim B) \supset ((B \vee C)) \equiv (\sim B \vee C) \vee D))$
*18. $(((A \cdot B) \supset (B \cdot C)) \supset ((\sim C \supset (D \supset E)) \supset F))$
*19. $\sim (A \cdot B) \vee (((B \supset C) \cdot (D \vee E)) \cdot F)$
*20. $((A \supset B) \equiv (\sim (\sim A \vee B) \supset (\sim B \supset C)))$

UNIT 22
Polish Notation for Sentential Logic

The symbolism we have been using for sentential logic is an *infix* system; that is, the operators (except for negation) are written *between* the formulas to be compounded. Another possibility, however, is to use a *prefix* notation, where the operators are all written in front of the formulas (as negation and the quantifiers are). This is sometimes called *Polish notation*, since it was developed by Polish logicians. This prefix notation is often what is used in calculators or computers, so it is useful to know. (Sometimes reverse Polish notation is also used.)

Another difference in the two symbol systems is that in Polish notation we use *capital letters* to stand for the operators, rather than the dot, horsehoe, tilde, wedge, and triple bar, which we have been using so far. The letters are the following:

C for "if-then"	(the **c**onditional)
A for "or"	(**a**lternation)
K for "and"	(**k**onjunction, in German)
E for "if and only if"	(**e**quivalence)
N for "not"	(**n**egation)

Thus,

for $(p \supset q)$ we write Cpq

for $(p \lor q)$ we write Apq

for $(p \cdot q)$ we write Kpq

for $(p \equiv q)$ we write Epq

for $\sim p$ we write Np

For complex formulas we write down the *major operator at the beginning*, followed by the components of that operator. Each complex component, similarly, begins with the operator followed by the components. Thus we would write $(p \cdot q) \supset (r \lor s)$ as $C(p \cdot q)(r \lor s)$, which becomes *CKpqArs*. We would write $((p \supset q) \lor (q \supset p)) \supset \sim r$ as *CACpqCqpNr*. More examples, with answers, will be given below. One of the very interesting things about this notation is that it *requires no parentheses*. The formulas are unambiguous simply because of the order in which the components are arranged in the formula. For instance, the formulas $((p \lor q) \cdot r)$ and $(p \lor (q \cdot r))$ can be distinguished in Polish notation without parentheses. The first would be written as *KApqr* and the second as *ApKqr*, since in the first the major operator is the conjunction and in the second the major operator is the disjunction.

When doing the exercises, don't try to write down everything at once; you will be very likely to make mistakes. Do them bit by bit; major operator followed by components. You might *underline* the various subformulas, in both the infix (ours) and the prefix (Polish) notation, to make sure you are keeping things straight. We will work out two examples below, working bit by bit until we get to the final Polish formula.

(1) $\sim (p \cdot q) \lor ((r \supset s) \cdot (s \lor t))$ (2) $(p \equiv q) \supset ((p \supset q) \cdot (q \supset p))$

    $A \sim (p \cdot q), ((r \supset s) \cdot (s \lor t))$ $C (p \equiv q), ((p \supset q) \cdot (q \supset p))$

    $ANKpq ((r \supset s) \cdot (s \lor t))$ $CEpq ((p \supset q) \cdot (q \supset p))$

    $ANKpqK (r \supset s), (s \lor t)$ $CEpqK (p \supset q), (q \supset p)$

    $ANKpqKCrsAst$ $CEpqKCpqCqp$

You might also want to work from the *inside out* rather than from the outside in, as we have done above. In this second approach, you start with the *smallest* components, translate them, and then go on to the larger components. We will do the same examples, using this method, below. Obviously, the results will be the same.

(1) $\sim (p \cdot q) \lor ((r \supset s) \cdot (s \lor t))$ (2) $(p \equiv q) \supset ((p \supset q) \cdot (q \supset p))$

    $\sim Kpq \lor (Crs \cdot Ast)$ $Epq \supset (Cpq \cdot Cqp)$

    $NKpq \lor KCrsAst$ $Epq \supset KCpqCqp$

    $ANKpqKCrsAst$ $CEpqKCpqCqp$

Use whichever method seems clearest to you; some prefer working from the outside in and some from the inside out. For very complex cases, it might also help to underline or circle the subformulas.

In reversing this process, going from Polish notation into the standard notation, it is probably easiest to work from the inside out. First identify the simplest compound formulas, circle or underline them, and then see how they fit together with other operators to form larger complexes. The next-to-last step will have just one operator letter, followed by a single Wff if it is an N, and by two Wffs if it is C, E, A, or K. We will do one example to illustrate.

$$CN \; \boxed{Apq} \; K \; \boxed{Np} \; \boxed{Nq}$$
$$CNApqK \sim p, \sim q$$
$$CNApq \, (\sim p \cdot \sim q)$$
$$CN \, (p \lor q), (\sim p \cdot \sim q)$$
$$C \sim (p \lor q), (\sim p \cdot \sim q)$$
$$\sim (p \lor q) \supset (\sim p \cdot \sim q)$$

The translation process back and forth between Polish notation and standard notation is obviously a mechanical process; with a little practice, and if you pay attention to details, you should have no difficulty with the exercises.

Not all strings of letters will be proper formulas, just as not all strings of symbols in our original notation are proper formulas. $KApq$ is not a proper, or "well-formed," formula because there is only one formula to conjoin, namely Apq, $(p \lor q)$, but conjunction requires two formulas. Nor is $KAprqs$, since there are too many variables; Apr is one formula, which could be conjoined with q, but then we have s left over. You should learn to identify well-formed formulas, as well as translate back and forth.

EXERCISES

*1. Decide which of the following are proper formulas. For those that are not, explain why not.

a. $KpCqr$

b. $KKpqCCqr$

c. $ACNpNqKNpq$

d. $NNNNNNNCpq$

e. $CKNpNqNApq$

f. $CCCCpqrst$

g. $NpKpCpq$

h. $KpqKrs$

i. $CCpCqrCCpqCpr$

*2. Write the following in the standard (our) notation.

a. All the formulas in Exercise 1 that are well-formed

b. $CrEst$

c. $CrANKsup$

d. $KNApsACKssss$

e. $CArtErs$

f. $KNEEpANtsss$

g. $KEEpout$

h. $KEEptrACKssss$

*3. Write the following in Polish notation.

a. $(p \cdot q) \supset p$

b. $((p \supset q) \cdot p) \supset q$

c. $(p \cdot (q \cdot r)) \equiv ((p \cdot q) \cdot r)$

d. $\sim (p \cdot q) \equiv (\sim p \vee \sim q)$

e. $p \supset \sim \sim p$

f. $((p \vee q) \supset r) \supset (p \supset r)$

g. $(p \cdot (q \vee r)) \equiv ((p \cdot q) \vee (p \cdot r))$

h. $(p \vee (q \cdot r)) \equiv ((p \vee q) \cdot (p \vee r))$

i. $\sim (p \supset (q \cdot r)) \supset (p \cdot (\sim q \vee \sim r))$

j. $p \equiv ((p \cdot q) \vee (p \cdot \sim q))$

Proof Trees for Sentential Logic

In doing proofs for sentential logic you learned (or were supposed to have learned) some twenty rules, and a proof consisted of a linear sequence of statements, leading from the premises to the conclusion. In the *proof tree* method we have a *branching* structure, which looks like an upside-down tree, rather than a linear sequence, and we need only nine rules; rules which, in effect, show us how to *break down complex formulas into simpler components*. Many of these nine rules are strongly analogous to the rules you already know or to the truth tables for the operators, so they shouldn't be hard to learn.

In constructing proof trees, rules are applied to compound formulas and the consequences are written below the formulas. For some formulas, such as $\sim (A \lor B)$, the conclusions, in this case $\sim A$ and $\sim B$, *both* follow directly, so they are written *linearly* underneath the premise. For other formulas, such as $A \lor B$, we know only that *one or the other* of the formulas, A or B, follows, and in these cases the proof *branches*, with one formula to the left and one to the right. The branches represent alternate ways in which the premise could be true; if $A \lor B$ is true, then either A is true or B is true. (Proof trees are actually sometimes called truth trees.) The compound formula $\sim (A \lor B) \lor (\sim (B \lor C) \lor \sim (A \lor C))$ would break down as follows:

$$\sim (A \vee B) \vee (\sim (B \vee C) \vee \sim (A \vee C))$$

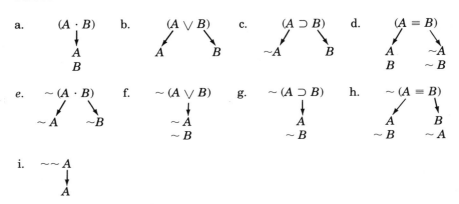

The three branches tell us that for the compound formula to be true, either $\sim A$ and $\sim B$ are true, or $\sim B$ and $\sim C$ are true, or $\sim A$ and $\sim C$ are true.

For each type of compound formula, except for negated letters, there is a rule saying how that formula breaks down. The set of rules is given below.

Notice that for rules a, f, and g, the two conclusions are listed one below the other in linear order. This is because *both* follow from the premise. In g, for instance, if the negated conditional is true, so that the conditional is false, then the antecedent A is true and the consequent B is false, so that $\sim B$ is true. For most of the other rules, the conclusions *branch*; this is because all we know, if the premise is true, is that *one or the other* of the conclusions is true, and the branching represents these alternative possibilities. In rule c, for instance, if the conditional is true, then *either* the antecedent A is false, so that $\sim A$ is true, or the consequent B is true.

In using the tree method to test arguments for validity, we first list the premises and the *negation* of the conclusion. We then construct the proof tree for this set of formulas by applying the tree rules to compound formulas, breaking them down one by one. The result is a tree, branching downward, in which the branches represent all possible ways in which the formulas could be true simultaneously. Unless contradictions appear on a branch, in which case the branch *closes*, we continue breaking down formulas until we are left with just sentence letters or their negations at the bottom of the branch. We will first illustrate this method with an example and then explain how we can determine validity or invalidity from the tree.

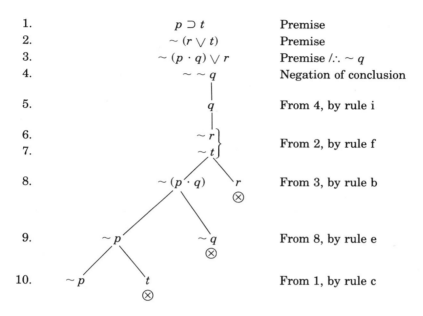

1.	$p \supset t$	Premise
2.	$\sim (r \lor t)$	Premise
3.	$\sim (p \cdot q) \lor r$	Premise /∴ $\sim q$
4.	$\sim \sim q$	Negation of conclusion
5.	q	From 4, by rule i
6.	$\sim r$	From 2, by rule f
7.	$\sim t$	
8.	$\sim (p \cdot q)$ r	From 3, by rule b
9.	$\sim p$ $\sim q$	From 8, by rule e
10.	$\sim p$ t	From 1, by rule c

Each direct line, from top to bottom, is a branch of the tree, so the above tree has four branches, ending, from left to right, with $\sim p$, t, $\sim q$, and r. If both a formula and its negation appear on a single branch, that is, if the branch contains a *contradiction*, then we say that the branch *closes*. We draw a heavy X beneath the branch as soon as the contradiction appears. In the tree above, three of the four branches are closed. If *all* branches in a completed tree close, then the argument (or form) is valid; if at least one branch remains open, then the argument is invalid. The argument above is invalid since the leftmost branch remains open. We explain below how it is possible to determine validity or invalidity by looking at the completed tree.

The tree method of testing for validity is analogous to the short truth table method: we are really checking to see whether there is any possible way to have all the premises true with a false conclusion, which means *any way to simultaneously make all the premises and the negation of the conclusion true*. Thus, our initial formulas are the premises and the negation of the conclusion. It is important that all our rules are *truth-preserving* in the sense that if the premise of the inference is true then *both* conclusions are true in the nonbranching rules and *at least one* of the conclusions is true in the branching rules. Because of the nature of the branching, *all the initial formulas can be made true if and only if at least one branch in the tree remains open.*

What does this mean for validity? Well, if a branch remains open, then all the initial formulas—premises plus negation of conclusion—can be true, which means that it is possible for premises to be true and conclusion false, which means that the argument (form) is invalid. Indeed, any open branch will generate a counterexample: if you make the un-negated letters on that branch true and the negated letters false (if $\sim p$ appears, for instance, make p false) you will have an instance with true premises and a false conclusion. In the example above, for instance, the first branch is open, on which we have $\sim p$, q, $\sim r$, and $\sim t$. If q is true and p, r, and t are all false, then clearly $p \supset t$, $\sim (r \lor t)$, and $\sim (p \cdot q) \lor r$ are all true, while the conclusion, $\sim q$, is false.

Why can we conclude that the argument is valid if all branches close? If all branches close, then it cannot be the case that all the initial formulas are true, which means that it is impossible for the premises and the *negation* of the conclusion to be true, which means that if the premises are true, then the negation of the conclusion must be false, so that the conclusion is true. Thus, if all branches close, then if all the premises are true, then the conclusion must be true as well, so that the argument is valid.

Two more examples are given below, the first valid and the second invalid.

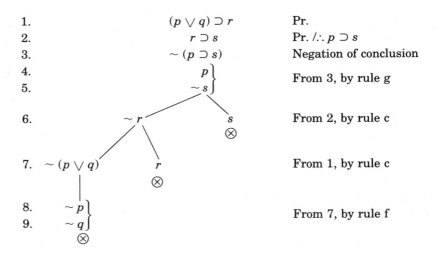

1.	$(p \lor q) \supset r$	Pr.
2.	$r \supset s$	Pr. /∴ $p \supset s$
3.	$\sim (p \supset s)$	Negation of conclusion
4.	p	From 3, by rule g
5.	$\sim s$	
6.	$\sim r$ s	From 2, by rule c
7.	$\sim (p \lor q)$ r	From 1, by rule c
8.	$\sim p$	From 7, by rule f
9.	$\sim q$	

This argument is valid since all branches close.

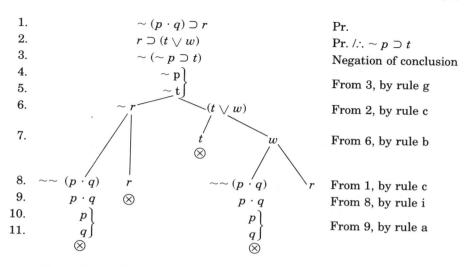

1.	$\sim(p \cdot q) \supset r$	Pr.
2.	$r \supset (t \vee w)$	Pr. /∴ $\sim p \supset t$
3.	$\sim(\sim p \supset t)$	Negation of conclusion
4.	$\sim p$	
5.	$\sim t$	From 3, by rule g
6.	$\sim r$ $\qquad (t \vee w)$	From 2, by rule c
7.	t $\qquad w$	From 6, by rule b
8.	$\sim\sim(p \cdot q)$ $\quad r$ $\qquad \sim\sim(p \cdot q)$ $\quad r$	From 1, by rule c
9.	$p \cdot q$ $\qquad p \cdot q$	From 8, by rule i
10.	p $\qquad p$	
11.	q $\qquad q$	From 9, by rule a

Since the right-hand branch, which ends in r, does *not* close, the argument is invalid.

In constructing proof trees certain procedures must be observed or the tree will be incorrect. In the first place, if there is more than one open branch on a tree being constructed, then in applying a rule to a formula you must write the conclusion at the bottom of *each open branch directly below that formula*. In our last example, for instance, in steps 6 and 7 we have two open branches, so we write the conclusion from step 1 at the bottom of each of those open branches. However, you *must not* append the conclusions from a particular formula to any branch on which that formula does *not* appear; that is, you do not "leap" branches. Also, you must not append *any* conclusions to a branch once that branch is closed. In the example above, for instance, we write the conclusions from $(t \vee w)$ *only* beneath that formula, and *not* beneath $\sim r$, which is on another branch. Also, we do not append the conclusions from step 1 to the branch ending in t, since that branch is closed.

It is best, though not essential, to apply your nonbranching rules a, f, and g first, since this will make the resulting tree simpler. Note that, as in using the rule of Indirect Proof, a contradiction need not be just a single letter and its negation; a compound formula and its negation also counts as a contradiction. Note also that the two formulas that contradict one another may appear *anywhere* on a branch, not necessarily next to each other. This means that you must inspect the branches carefully to make sure that you have not overlooked any contradictions; otherwise you may think that an argument is invalid, that it has an open branch, when in fact all branches close, so that it is valid.

A summary of the method is as follows:

1. List premises and the *negation* of the conclusion.

2. Pick one of the complex sentences and apply the appropriate rule of inference to break it down into its components. (It doesn't matter where you begin, but it is *easier* to begin with formulas that do *not* branch, do all of them, and then go on to the ones that do branch.) Once you have used a formula, check it off; it will *not* be used again.

3. Continue to break down formulas, keeping in mind the following:
   a. If a formula and also its negation appear on any branch, *close* that branch by writing a large × underneath it. Once a branch is closed, you need not do anything more to it; you do *not* need to break down any remaining complex formulas in it.
   b. When breaking down a formula, append its conclusions to *every* open branch which appears below it, in some direct line. (Do not "cross" lines.)

4. Continue until *either* every branch closes *or* until all complex formulas on unclosed lines are broken down into single letters or their negations.

5. If every branch closes, the argument is valid.

6. If some branch is left open after all formulas are broken down, then the argument is invalid.

Once you catch on to this method it is very easy and very fast. Why, then, didn't we teach it to begin with? Mostly because it is really *not* the way we reason in ordinary discourse; we generally reason linearly and make use of the standard inference patterns you have learned. But it is an interesting method, and one which has applications in the more advanced stages of logic.

This method can also be used to show that formulas are tautologies: simply assume the *negation* of the formula and apply the rules. If all branches close, the formula *is* a tautology (there is no way for it to be false, that is, for its negation to be true); if some branch remains open it is *not* a tautology (since there *is* a way for the negation to be true, that is, for it to be false).

One disadvantage of this method is that the tree structures can get very large and complicated with complex arguments. You can minimize this problem if you always do first the formulas that do *not* branch, and if you are very neat and systematic.

EXERCISES

*1. Apply this method to the exercises in Unit 5, 1e, f, i, 2a, c, 3a, b, c, e.

*2. Apply this method to the exercises in Unit 5, 1g, h, j, 2b, 3d.

3. Apply this method to Exercise 7 in Unit 9 (all of which are invalid).

Answers will be supplied to only a few of these, since they require a lot of paper and space. But since all exercises in 1 are valid, and all exercises in 2 and 3 are invalid (and thus should have an open line), this will give you some idea of whether you are doing them right.

Using Venn Diagrams to Prove Validity

In Unit 12, on one-variable predicate logic, you learned how to use Venn diagrams to symbolize the four kinds of categorical propositions. This method can be extended to prove validity for a very limited set of arguments in predicate logic: those in which there are just two premises, both of which are categorical propositions, and in which the conclusion is a categorical proposition. An example of this sort of argument would be "All professors are intelligent people, and some professors are conservative, so some intelligent people are conservative." Arguments of this sort are called *categorical syllogisms*. In addition to containing just categorical propositions, they must also meet one other very important requirement: *they must contain only three class terms*. In the example cited, for instance, the three terms are "professors," "intelligent people," and "conservatives."

In order to test these arguments for validity using the method of Venn diagrams, we draw *three* interlocking circles, representing the three class terms. In order to keep things uniform, let us stipulate that the two top circles will represent the subject term (on the left) and the predicate term (on the right) of the *first premise*. The remaining term, which will appear in the second premise and in the conclusion, will be drawn underneath these, but intersecting both of them. The class terms for the above argument could be represented as follows:

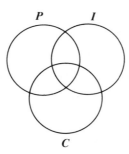

In diagramming the sentences we will use the same procedure as in Unit 12: to show that there is nothing in a certain portion of the class, that is, that that part of the class is empty, we shade it out, and to show that there is something in a part of a class, we draw in an ×. As we will explain below, however, we need to modify somewhat the procedure for showing existence.

The three interlocking circles represent eight different regions, as indicated below. Region 1 represents things that are *P* but not *I* or *C*. Region 2 represents things that are *P* and *I* but not *C*; region 3, things that are *I*, but not *P* or *C*, etc. Region 5 represents things that are *P*, *I*, *and C*; and region 8, things that are *neither P* nor *I* nor *C*.

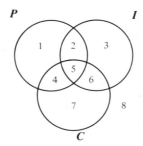

Notice that the intersection between *P* and *I* contains the *two* regions 2 and 5; that of *I* and *C*, regions 5 and 6; and that of *P* and *C*, regions 4 and 5. Similarly, the part that is *P* but *not I* contains regions 1 and 4; *I* but not *C* contains regions 2 and 3; and *P* but not *C* contains regions 1 and 2. This means that in diagramming the *A* proposition "All *P* are *I*" we must shade out *both* sections 1 and 4. If we diagram "All *P* are *C*" (There are no *P* that are not *C*) we must shade out both regions 1 and 2, and so on.

What happens if we want to say *there is* some *P* that is *C*? We have two possibilities: it could be in either region 4 or region 5. What we will do is to draw ×'s in each section, with a bar connecting them. What this means, it is essential to remember, is that there is an x *in one or the other. It does not mean there is something in both*. It simply indicates that *we do not know which* region it falls into. If our other premise is a universal proposition, which allows us to shade out one of those regions, then we can conclude that the x is in the other (by the rule of D.S.). This is, in fact, what we will do in proving the argument below.

Let us now diagram the premises. Since the first premise is an A proposition, which says that all *P* are *I*, the result of the diagramming would be as follows, with regions 1 and 4 (those that are *P* but not *I*), shaded out.

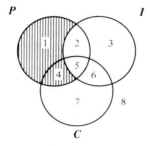

The second premise is an I proposition, which says that some *P* are *C*. If we were to diagram it separately, we would have the following result:

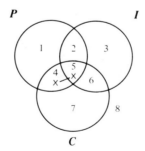

Here we have an × in both regions 4 and 5, joined by a bar, which indicates that there is something *either* in region 4 *or* 5 (not both).

Now, if we combine the results of the two premises, which jointly tell us that there is something either in region 4 or 5, but not in 1 and not in 4, we get the following, which indicates that there *is* something in region 5.

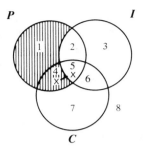

At this point, all we need to do is look at the result and see whether the conclusion is "contained" in the diagram for the premises. In this case it is; there is an x in the overlapping portion between conservatives and intelligent people, so we can *read off* from the diagram that some conservatives are intelligent, which is what the conclusion says. The conclusion, then, does follow from the premises, and the argument is valid.

In valid arguments, as indicated above, we can read off the conclusion once we have diagrammed the premises. In *invalid* arguments, on the other hand, the conclusion cannot be seen from the diagramming of the premises. It might be useful, at this point, to look at an example of an invalid argument.

The argument "No cats are dogs, and no dogs are tigers, so no cats are tigers" is obviously invalid, and can be shown to be so by the Venn diagram method.

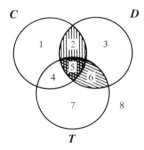

In digramming the two premises, we shade out regions 2, 5, and 6. The conclusion would require that *both* regions 4 and 5 be shaded out, but they are not, so the conclusion does *not* follow from the premises.

Another example of an invalid argument would be "Some doctors are wealthy and some lawyers are wealthy, so some doctors are lawyers." This would be diagrammed as below. Keep in mind that the ×'s joined by a bar indicate only that there is something in *one or the other* of the two regions.

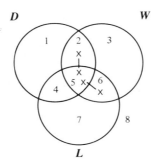

For the conclusion to follow from the premises, that is, for the argument to be valid, we would have to have an × in either region 4 or 5, but we cannot conclude that there is an × in either one of these, since all we know is that there is one in 2 or 5 and one in 5 or 6. It *might* be that the ×'s are in 2 and 6, so we cannot conclude that there is one in 5; that is, it does not follow that there are doctors who are lawyers.

It should be pointed out that some arguments which seem to have more than three terms may turn out to have only three if the premises or conclusion are restated. The argument "All lawyers are honest, and all crooks are dishonest, so no lawyer is a crook," in which we seem to have the four terms, "lawyers," "honest people," "dishonest people," and "crooks," can be restated so that it has only three terms. All we need to do is to replace the second premise with its equivalent form: "No crooks are honest." Similarly, negated categorical statements, such as "Not all lawyers are honest," should be replaced by their equivalent forms, in this case by "Some lawyers are dishonest." If an argument *cannot* be reduced to three terms, it cannot be tested by Venn diagrams.

EXERCISES

Use the method of Venn diagrams to test the following syllogisms for validity. If they contain more than three terms, restate some of the premises or the conclusion so that there are only three terms in all.

*1. All trespassers will be prosecuted, and some will be shot, so some of the prosecuted will be shot.

*2. Some dogs have fleas, and no creature with fleas makes a good pet, so no dog makes a good pet.

*3. There are politicians who are dishonest, and no dishonest person is trustworthy, so some politician is untrustworthy.

*4. Some students make good teachers, and all good teachers go to heaven, so some students will go to heaven.

*5.   No nuclear plant is safe, and no unsafe thing should be funded by the government, so no nuclear plant should be funded by the government.

*6.   Everybody loves a lover, and all lovers are happy, so everybody loves a happy person.

*7.   All coffee drinkers are nervous, and no coffee drinkers have large savings accounts, so no nervous people have large savings accounts.

*8.   No one with a large savings account is nervous, and all coffee drinkers are nervous, so no coffee drinker has a large savings account.

*9.   Not all criminals are nervous, but all coffee drinkers are nervous, so some coffee drinkers are not criminals.

*10.   All hijackings are crimes, and some hijackings are political acts, so some political acts are crimes.

Stroke (nand) and Dagger (nor) Operators

In Unit 8 you learned that $p \equiv q$ can be replaced by $(p \supset q) \cdot (q \supset p)$, and that $p \supset q$ can be replaced by $\sim p \vee q$. Furthermore, a bit of thought shows that a conjunction $p \cdot q$ is equivalent to and replacable by $\sim (\sim p \vee \sim q)$, by DeMorgan's and D.N. Thus, the three operators $\equiv$, $\supset$, and $\cdot$, are really not needed, but are just convenient abbreviations. They can be replaced by equivalent formulas using only $\vee$ and $\sim$, so we can reduce our set of operators from $\{\equiv, \supset, \cdot, \vee, \sim\}$ to $\{\vee, \sim\}$. Some examples of formulas written with only these two operators follow. You can use the truth table method to verify that in each case the pairs are logically equivalent.

$p \cdot \sim q$	can be written as	$\sim (\sim p \vee q)$
$\sim p \supset \sim q$	can be written as	$p \vee \sim q$
$p \supset (q \cdot r)$	can be written as	$\sim p \vee \sim (\sim q \vee \sim r)$
$(q \cdot r) \supset p$	can be written as	$(\sim q \vee \sim r) \vee p$
$p \equiv q$	can be written as	$\sim (\sim p \vee \sim q) \vee \sim (p \vee q)$
$\sim (p \equiv q)$	can be written as	$\sim (\sim p \vee q) \vee \sim (\sim q \vee p)$

It should be clear that whatever formula can be written with our initial five operators has an equivalent formula written with just $\vee$ and

~. In other words, the set $\{\vee, \sim\}$ can do whatever the set $\{\equiv, \supset, \cdot, \vee, \sim\}$ can do, though in most cases, less efficiently. But what can the set $\{\equiv, \supset, \cdot, \vee, \sim\}$ do? This set has an interesting property called "expressive completeness." This means that whatever *truth table* can be written down, of no matter how many variables, there will be a *formula* written with just those five operators that "expresses," or has, that truth table.

The four-row table to the right can be expressed by the formula $\sim (p \equiv q)$. (We assume the usual base column arrangement.)	F T T F

The eight-row truth table below it can be expressed by the longer formula $(p \supset \sim q) \cdot (p \vee (q \equiv r))$.	F T F T T F F T

Logicians were not content to observe that our set of five operators could be reduced to a set of two. They had to ask whether it was possible to use only *one* operator! This would mean that just one operator could do the job of five, that a single operator could be expressively complete. It turns out that this is possible, but the one operator cannot be one of the five we have been using. Disjunction alone is not expressively complete, for instance, because formulas using only disjunctions can never express truth tables with an F in the top row. There are two other operators, however, the stroke / and the dagger $\downarrow$, that are in themselves expressively complete. Their truth tables follow, and from these tables it is clear why they are sometimes called the "nand" and "nor" operators. They have, respectively, the same truth tables as the negated conjunction and negated disjunction.

p	q	$p\,/\,q$	$p \downarrow q$
T	T	F	F
T	F	T	F
F	T	T	F
F	F	T	T

How is it that a single operator can do the job of five? In the same way that the two, $\vee$ and $\sim$, can do the job of five. We can *define* the others in terms of / (or of $\downarrow$), that is, find logically equivalent formulas that use only / (or $\downarrow$). These definitions are given below. The biconditional is omitted because the formula is quite long, and it can obviously be defined in terms of $\supset$ and $\cdot$, once we have those operators.

FORMULA	DEFINED BY STROKE	DEFINED BY DAGGER
$p \vee q$	$(p / p) / (q / q)$	$(p \downarrow q) \downarrow (p \downarrow q)$
$p \cdot q$	$(p / q) / (p / q)$	$(p \downarrow p) \downarrow (q \downarrow q)$
$p \supset q$	$p / (q / q)$	$((p \downarrow p) \downarrow q) \downarrow ((p \downarrow p) \downarrow q)$
$\sim p$	p / p	$p \downarrow p$

Note the symmetry between conjunction and disjunction, and that negation is defined the same for both stroke and dagger. One thing to keep in mind while doing translations between the different kinds of notations is that $(p / p) / (p / p)$ will be equivalent to just p, since it represents the double negation of p. This is also true for the dagger. We do examples below of translating a formula into the stroke and dagger notation. Note that we proceed step by step from smaller subformulas to larger ones, and that the correct placement of parentheses is essential.

a. $\sim p \supset (\sim q \vee r)$ becomes in stroke notation
     $(p / p) \supset (q / q \vee r)$ which becomes
     $(p / p) \supset (((q / q) / (q / q)) / (r / r))$ which becomes
     $(p / p) \supset (q / (r / r))$ which becomes
     $(p / p) / ((q / (r / r)) / (q / (r / r)))$

b. $\sim p \supset (\sim q \vee r)$ becomes in the dagger notation
     $(p \downarrow p) \supset ((q \downarrow q) \supset r)$ which becomes
     $(p \downarrow p) \supset ((q \downarrow q) \downarrow r) \downarrow ((q \downarrow q) \downarrow r)$ which becomes
     $(((p \downarrow p) \downarrow (p \downarrow p)) \downarrow (((q \downarrow q) \downarrow r) \downarrow ((q \downarrow q) \downarrow r))) \downarrow$
     $(((p \downarrow p) \downarrow (p \downarrow p)) \downarrow (((q \downarrow q) \downarrow r) \downarrow ((q \downarrow q) \downarrow r)))$ which
     becomes
$(p \downarrow (((q \downarrow q) \downarrow r) \downarrow ((q \downarrow q) \downarrow r))) \downarrow (p \downarrow (((q \downarrow q) \downarrow r) \downarrow ((q \downarrow q) \downarrow r)))$

In going the other way, from stroke or dagger notation into standard notation, you may want to simplify the formulas by applying replacement rules. One example of this sort is done below.

     $\{((p / p) / (q / q)) / ((p / p) / (q / q))\} / \{(r / r) / (q / q)\}$ becomes
     $((p / p) \cdot (q / q)) / ((r / r) / (q / q))$ which becomes
     $(\sim p \cdot \sim q) / (r \vee q)$ which becomes
     $\sim (\sim p \cdot \sim q) \vee \sim (r \vee q)$ which is equivalent to
     $(p \vee q) \vee \sim (r \vee q)$

It is unlikely that you would want to use a system with only one operator, since the formulas are long and quite unintuitive. However, it is of considerable theoretical importance that logic can be "reduced" to a single operator. The following exercises will give you some familiarity with the stroke and dagger operators.

EXERCISES

1. Express the following in stroke notation.

*a. $\sim p \vee \sim q$ *g. $\sim (p \vee q) \supset r$

 b. $\sim p \supset \sim q$ h. $(p \supset q) \supset (r \supset s)$

*c. $\sim p \cdot \sim q$ *i. $(\sim p \supset \sim q) \supset q$

 d. $\sim (p \cdot q) \supset r$ j. $p \vee \sim p$

*e. $p \supset \sim (q \cdot r)$ *k. $p \supset (q \supset (p \cdot q))$

 f. $\sim (p \supset \sim (q \cdot r))$ l. $(p \cdot q) \supset (q \cdot p)$

*2. Express the formulas in Exercise 1 in dagger notation.

*3. Translate the following into standard notation.

a. $((p \mid q) \mid (p \mid q)) \mid ((r \mid s) \mid (r \mid s))$
b. $(((p \mid p) \mid q) \mid ((p \mid p) \mid q)) \mid (q \mid (p \mid p))$
c. $((p \mid q) \mid (r \mid s)) \mid (((p \mid p) \mid (q \mid q)) \mid ((r \mid r) \mid (s \mid s)))$
d. $(((p \mid q) \mid (r \mid s)) \mid ((p \mid q) \mid (r \mid s))) \mid (((p \mid p) \mid (q \mid q)) \mid ((r \mid r) \mid (s \mid s)))$

*4. Do Exercise 3, replacing the stroke with the dagger.

5. Find formulas that express the following truth tables.

*a.		*c.		*e.		f.		*g.		h.	
F		F		T		F		F		T	
T		F		F		F		F		F	
F		T		T		F		T		F	
F		T		F		T		T		F	
				F		F		T		F	
b.	F	d.	T	T		T		F		F	
	F		T	F		F		F		T	
	T		F	T		T		T		T	
	F		T								

Partial Answers to Exercises

Note: answers are given to only those exercises starred in text.

Unit 1

1.
   a. Valid
   b. Invalid
   c. Invalid
   d. Invalid
   e. Valid
   f. Invalid
   g. Valid
   h. Invalid
   i. Valid
   j. Valid

2.
   a. Invalid
   b. Valid
   c. Invalid
   d. Invalid
   e. Invalid
   f. Invalid
   g. Valid
   h. Invalid
   i. Valid
   j. Invalid

Unit 2

1.
   a. Compound; Mary hit a home run; Mary hit a triple.
   c. Simple
   e. Compound; John will get cancer; John will have a heart attack; John stops eating fatty beef.
   g. Compound; Some people can survive for a long time on junk food.
   i. Simple
   k. Simple
   m. Compound; Human beings will die out; Human beings will be mutated; There is an atomic war.

o. Compound; Mike will have to clean up the kitchen after dinner; John will have to clean up the kitchen after dinner.

q. Compound; John will lose weight; Mary quits teasing John.

s. Compound; Everyone likes John.

2. a. Dot (conjunction)   i. Dot
   c. Triple bar (biconditional)   k. Wedge
   e. Second horseshoe   m. First horseshoe
   g. Wedge (Disjunction)   o. First horseshoe

Unit 3

1. a. F   c. F   e. T   g. T   i. F   k. T   m. F
   o. T   q. F   s. T

2. a. F   (because one conjunct, $\sim A$, is false)
   c. T   (since one disjunct, $\sim \sim A$, is true)
   e. T   (since the antecedent, $\sim (A \vee B)$ is false)
   g. T   (since the antecedent, $\sim (X \vee \sim Y)$ is false)
   i. T   (because the antecedent is false. Since A is true, the disjunction is true, so the negated disjunction, which is the antecedent, is false)

3. a. No.   Since A and B are true, the truth value of $(A \cdot G)$ and $(B \cdot H)$ would depend on G and H, which we are not given.
   c. Yes.   It must be true, no matter what the values of H and G. If H is true, then $G \supset H$, the consequent, must be true, so there cannot be a true antecedent with a false consequent.
   e. Yes.   Since A is true, $A \vee G$ is true, so $\sim (A \vee G)$ is false.
   g. Yes. $A \equiv B$ is true since both A and B are true, so $(H \equiv G) \supset (A \equiv B)$ is true because the consequent is true.
   i. No.   Since A is true but we don't know the value of G, we can't compute the value of $(A \cdot G)$ or $\sim (A \cdot G)$.

Unit 4

1. a. $M \equiv$ Man is descended from monkeys. $\sim M$
   c. $A \equiv$ Mary is the most athletic girl in her class.
      $S \equiv$ Mary is the smartest girl in her class. $\sim A \cdot S$
   e. Not truth-functional; must be symbolized with a single letter.
   g. $A \equiv$ Americans will stop driving big cars.
      $C \equiv$ There are comfortable small cars on the market. $A \supset C$
   i. Not truth-functional; must be symbolized with a single letter.
   k. $V \equiv$ John drives his van.
      $S \equiv$ John needs the space $\sim S \supset \sim V$, or $V \supset S$
   m. $J \equiv$ John is married.
      $M \equiv$ Mary is married.
      $E \equiv$ John and Mary are married to each other. $(J \cdot M) \cdot \sim E$
   o. $J \equiv$ John thinks that Mary married the wrong person.
      $M \equiv$ Mary knows she married the right person. $(J \cdot M)$

q. $J \equiv$ John will bring a pie to the picnic.
$M \equiv$ Mary will bring a pie to the picnic.
$T \equiv$ Ted is going to bring a birthday cake.
$T \supset \sim (J \cdot M)$ or $T \supset (\sim J \lor \sim M)$

s. $T \equiv$ Ted will bring the dessert.
$J \equiv$ John will bring the dessert. $(T \lor J) \cdot \sim (T \cdot J)$

2. a. $J \supset W$ or $\sim W \supset \sim J$
c. $\sim H \supset \sim J$ or $J \supset H$ or $H \lor \sim J$.
e. $\sim (J \cdot M)$ or $(\sim J \lor \sim M)$
g. $(L \cdot W) \supset \sim 0$
i. $0 \supset \sim (J \lor M)$
k. $M \equiv \sim J$

3. a. $\sim (D \lor E) \supset G$ or $(\sim D \cdot \sim E) \supset G$
c. $\sim M \supset \sim (D \cdot E)$ or $(D \cdot E) \supset M$
e. $\sim (W \cdot \sim D) \supset ((S \lor R) \supset \sim G)$ or $((S \lor R) \supset \sim G) \lor (W \cdot \sim D)$
g. $(G \equiv \sim (D \lor E)) \supset ((S \lor R) \cdot \sim W)$
i. $(D \cdot E) \supset (M \cdot \sim T)$

4. a. $G \supset (L \cdot M)$ or $\sim (L \cdot M) \supset \sim G$, or $(\sim L \lor \sim M) \supset \sim G$
c. $(\sim N \supset \sim R) \cdot (N \supset F)$ or $(R \supset N) \cdot (\sim F \supset \sim N)$
e. $\sim (G \lor C) \supset S$ or $(\sim G \cdot \sim C) \supset S$
g. $O \supset \sim (S \cdot G)$ or $O \supset (\sim S \lor \sim G)$
i. $\sim R \supset ((L \cdot \sim C) \supset (S \cdot G))$ or $(\sim R \cdot L \cdot \sim C) \supset (S \cdot G)$
k. $(O \lor R) \supset (\sim G \cdot \sim S)$
m. $\sim R \supset (\sim (G \lor C) \supset S)$ or $R \lor ((\sim G \cdot \sim C) \supset S)$
o. $(A \lor V) \cdot (\sim (G \lor Z) \supset \sim(A \cdot V))$ or $(A \lor V) \cdot ((A \cdot V) \supset (G \lor Z))$

Unit 5

Study Question 8:

a. T   b. F   c. F   d. F   e. F
f. F   g. F

1. a. Valid   g. Invalid (second row)
c. Valid   i. Valid
e. Valid

2. a. Valid   d. Valid
c. Valid   f. Invalid (where p = F, q = F, r = T, and s = F)

3. a. Valid   d. Invalid (let p = T, t = F, s = F, r = T)
b. Valid   e. Valid
c. Valid

4. a. $(T \supset F) \cdot (F \supset \sim G), \sim G \supset \sim E$ /∴ $\sim E$
Invalid : let E = T, G = T, F = F and T = F.
c. $(R \supset P) \cdot (P \supset D), D \supset F, H \supset \sim F$ /∴ $R \supset \sim H$
Valid
e. $C \supset (W \cdot \sim (H \lor R)), B \supset (R \equiv W)$ /∴ $\sim (C \cdot B)$ Valid
g. $\sim (D \cdot E) \supset G, E \supset \sim (T \lor L), \sim V \supset (D \supset T), \sim V$ /∴ G Valid

   i. $(A \supset L) \cdot (L \equiv S), (D \lor W) \supset S, E \supset W /\therefore E \supset A$ Invalid
   k. $(G \equiv D) \cdot (D \supset (T \cdot I)), \sim (F \lor V) \supset \sim I, M \supset G /\therefore \sim F \supset \sim M$ Invalid
   m. $\sim W \supset \sim (R \cdot P), W \supset (F \lor H), (H \supset E) \cdot (E \supset \sim W),$
      $(F \supset D) \cdot (D \supset \sim W) /\therefore \sim R$ Invalid

Unit 6

1. a. Contradiction   g. Tautology
    c. Tautology   i. Contingent
    e. Tautology   k. Can't tell; could be anything.
2. a. Tautology   g. Tautology
    c. Contradiction   i. Tautology
    e. Contingent   k. Tautology
3. The following pairs are logically equivalent: a, c, f, h, i, j
4. a. 1 logically implies 2
    c. Both (both are contradictions)
    e. Both
5. The following sets are consistent:
        a. (all T in third row),   c. (all T in fourth row)
        d. (all T in second row),   e. (all T in last row)
6. a. $(\sim R \supset P) \cdot (R \supset \sim P)$ Contingent   c. $(F \lor \sim F) \supset S$ Contingent
    e. $(J \cdot M) \lor (F \cdot (\sim J \lor \sim M)) \lor (\sim F \cdot \sim (J \cdot M))$ Tautologous
    g. $(J \equiv R) \cdot (\sim J \supset R)$ Contingent
    i. $(E \supset \sim T) \cdot (E \supset S) \cdot (S \supset T)$ Contingent
7. a. (1) $(F \lor \sim F) \supset R$ (2) R The two forms are logically equivalent.
    c. (1) $M \equiv \sim G$ (2) $G \lor M$ Not equivalent, but 1 logically implies 2
    e. (1) $J \supset (T \supset S)$ (2) $\sim S \supset (T \supset \sim J)$ Logically equivalent
    g. (1) $(J \supset S) \supset \sim S$ (2) $\sim J$ Neither implies the other
8. a. (1) $W \supset A$, (2) $W \supset \sim A$, (3) A, (4) $\sim W$
        Consistent; true if A = T, W = F
    c. (1) $J \supset (S \cdot \sim T)$ (2) $J \equiv \sim M$ (3) $M \supset (\sim T \cdot S)$ Consistent

Unit 7

1. The following *are* instances of the given form.
    (1) a, b, d, f   (2) a, b, d, f
    (3) b, d, e   (4) a, c, d, e, f
    (5) b, c, e   (6) a, b, e, f
2. The following are correct applications:
    a, e, f, g, k, o, p, r, s, y
3. a. D.S.   f. Dil.   k. D.S.
    b. M.P.   g. M.T.   l. Conj.
    c. M.T.   h. M.P.   m. M.P.
    d. Simp.   i. H.S.   n. Add.
    e. H.S.   j. M.T.   o. M.T.

4. a. 5. Simp. 3
    6. D.S. 2,5
    7. Simp. 3
    8. Simp. 7
    9. M.T. 4,8
    10. Add. 6
    11. Simp. 7
    12. M.P. 1,10
    13. Add. 11

  b. 5. Simp. 3
    6. M.T. 2,5
    7. Simp. 3
    8. Add. 6
    9. D.S. 6,7
    10. M.P. 4,8
    11. Add. 10
    12. M.P. 1,11
    13. Conj. 9,12

  c. 5. Simp. 1
    6. Simp. 5
    7. Simp. 5
    8. Simp. 1
    9. Add. 7
    10. M.P. 6,8
    11. M.P. 2,9
    12. M.P. 6,11
    13. Simp. 3
    14. Simp. 3
    15. D.S. 14,10
    16. Conj. 15,6
    17. M.P. 4,16
    18. M.P. 10,17
    19. M.P. 15,18
    20. M.P. 12,19
    21. Add. 20

  d. 6. Simp. 3
    7. Simp. 6
    8. Simp. 3
    9. Simp. 6
    10. Add. 7
    11. M.P. 4,9
    12. H.S. 4,5
    13. Add. 9

    14. M.P. 5,11
    15. M.P. 2,14
    16. M.T. 7,8
    17. D.S. 15,16
    18. Add. 9
    19. D.S. 17,18
    20. Conj. 14,19
    21. M.P. 1,20

5. a. 1. $\sim A \vee \sim B$
    2. $(E \supset A) \supset D$
    3. $\sim \sim (A \vee B)$
    4. $G \supset R$
    5. $\sim (B \equiv C)$
    6. $(A \supset D) \vee C$
    7. $\sim (A \vee \sim B)$
    8. $\sim \sim (A \vee (B \supset \sim C))$
    9. $(D \supset (D \supset F)) \supset (D \supset F)$
    10. $(D \vee \sim F) \vee (\sim D \vee F)$

  b. 1. $\sim (\sim D \cdot F)$
    2. $(A \supset B) \supset (A \supset (A \supset B))$
    3. $(E \supset \sim F) \supset (\sim F \supset E)$
    4. $\sim F \supset (A \equiv \sim F)$
    5. $((A \cdot B) \supset (B \vee C)) \supset (A \supset (B \vee C))$
    6. $(\sim B \vee \sim C) \supset \sim A$
    7. $(F \supset (G \vee H)) \vee \sim (A \vee B)$
    8. $\sim (\sim A \supset B) \supset (\sim B \vee (A \supset C))$
    9. $(B \supset \sim A) \supset (B \supset (A \supset B))$
    10. $(\sim A \supset B) \vee (\sim B \supset \sim A)$

  c. 1. $\sim (C \vee \sim D)$ M.P.
    2. $\sim (\sim B \vee C) \supset (\sim B \vee C)$ H.S.

        3. $\sim A \supset \sim B$ D.S.

        4. $\sim (\sim (C \vee \sim D) \vee \sim C)$ M.T.

        5. $F \equiv (\sim E \equiv \sim F)$ D.S.

        6. $(B \vee \sim C) \vee (\sim A \vee (C \vee \sim B))$ Dil.

        7. $\sim (\sim A \supset (\sim B \vee \sim A))$ M.T.

        8. $(C \supset (A \supset C)) \supset (A \supset (C \supset A))$ H.S.

        9. $\sim (A \supset (C \supset A))$ M.P.

     10. $\sim (C \vee (A \vee C))$ D.S.

6.   e.   1. $\sim A \supset \sim B$ Pr.

            2. $A \supset C$ Pr.

            3. $Z \supset W$ Pr.

            4. $\sim C \cdot \sim W$ Pr. /∴ $\sim B \vee W$

            5. $\sim C$ Simp. 4

            6. $\sim W$ Simp. 4

            7. $\sim Z$ M.T. 3,6

            8. $\sim A$ M.T. 2,5

            9. $\sim B$ M.P. 1,8

          10. $\sim B \vee W$ Add. 9

      g.   1. $(A \cdot B) \supset \sim C$ Pr.

            2. $C \vee \sim D$ Pr.

            3. $A \supset B$ Pr.

            4. $E \cdot A$ Pr. /∴ $\sim D$

            5. E Simp. 4

            6. A Simp. 4

            7. B M.P. 3,6

            8. $A \cdot B$ Conj. 6,7

            9. $\sim C$ M.P. 1,8

          10. $\sim D$ D.S. 2,9

      i.   1. $F \supset (G \supset \sim H)$ Pr.

            2. $(F \cdot \sim W) \supset (G \vee T)$ Pr.

            3. $F \cdot \sim T$ Pr.

            4. $W \supset T$ Pr. /∴ $\sim H$

            5. F Simp. 3

            6. $\sim T$ Simp. 3

            7. $\sim W$ M.T. 4,6

            8. $F \cdot \sim W$ Conj. 5,7

            9. $G \vee T$ M.P. 2,8

          10. G D.S. 6,9

          11. $G \supset \sim H$ M.P. 1,5

          12. $\sim H$ M.P. 10,11

      j.   1. $P \supset (Q \supset (R \vee S))$ Pr.

            2. $P \cdot Q$ Pr.

            3. $S \supset T$ Pr.

            4. $\sim T \vee \sim W$ Pr.

            5. $\sim \sim W$ Pr. /∴ R

6. $\sim T$	D.S. 4,5
7. $\sim S$	M.T. 3,6
8. P	Simp. 2
9. Q	Simp. 2
10. $Q \supset (R \vee S)$	M.P. 1,8
11. $R \vee S$	M.P. 9,10
12. R	D.S. 7,11

7. e.
1. $A \supset ((C \vee D) \supset B)$	Pr.
2. $(\sim W \vee \sim T) \supset (A \cdot C)$	Pr.
3. $W \supset (S \vee P)$	Pr
4. $\sim H \vee \sim (S \vee P)$	Pr.
5. $\sim H \supset Z$	Pr.
6. $\sim Z \cdot \sim Y$	Pr. /∴ $B \cdot \sim Y$
7. $\sim Z$	Simp. 6
8. $\sim Y$	Simp. 6
9. $\sim \sim H$	M.T. 5,7
10. $\sim (S \vee P)$	D.S. 4,9
11. $\sim W$	M.T. 3,10
12. $\sim W \vee \sim T$	Add. 11
13. $A \cdot C$	M.P. 2,12
14. A	Simp. 13
15. $(C \vee D) \supset B$	M.P. 1,14
16. C	Simp. 13
17. $C \vee D$	Add. 16
18. B	M.P. 15,17
19. $B \cdot \sim Y$	Conj. 18,8

   f.
1. $(\sim A \cdot \sim B) \supset (\sim C \vee \sim D)$	Pr.
2. $(E \vee \sim F) \supset \sim A$	Pr.
3. $\sim H \supset (B \supset J)$	Pr.
4. $(\sim F \cdot \sim H) \supset (\sim \sim C \cdot \sim J)$	Pr.
5. $\sim H \cdot (F \supset H)$	Pr. /∴ $\sim D$
6. $\sim H$	Simp. 5
7. $F \supset H$	Simp. 5
8. $\sim F$	M.T. 6,7
9. $\sim F \cdot \sim H$	Conj. 8,6
10. $\sim \sim C \cdot \sim J$	M.P. 4.9
11. $B \supset J$	M.P. 3,6
12. $\sim J$	Simp. 10
13. $\sim B$	M.T. 11,12
14. $E \vee \sim F$	Add. 8
15. $\sim A$	M.P. 2,14
16. $\sim A \cdot \sim B$	Conj. 15,13
17. $\sim C \vee \sim D$	M.P. 1,16
18. $\sim \sim C$	Simp. 10
19. $\sim D$	D.S. 17,18

8. a. 1. $P \vee D$ Pr.
        2. $(P \supset \sim A) \cdot (B \supset \sim D)$ Pr.
        3. B Pr. /∴ $\sim A$
        4. $B \supset \sim D$ Simp. 2
        5. $\sim D$ M.P. 3,4
        6. P D.S. 1,5
        7. $P \supset \sim A$ Simp. 2
        8. $\sim A$ M.P. 6,7

    b. 1. $((S \vee D) \supset \sim W) \cdot ((\sim W \vee \sim E) \supset R)$ Pr.
        2. $R \supset (\sim X \cdot \sim T)$ Pr.
        3. S Pr. /∴ $\sim T$
        4. $(S \vee D) \supset \sim W$ Simp. 1
        5. $S \vee D$ Add. 3
        6. $\sim W$ M.P. 4,5
        7. $(\sim W \vee \sim E) \supset R$ Simp. 1
        8. $\sim W \vee \sim E$ Add. 6
        9. R M.P. 7,8
       10. $\sim X \cdot \sim T$ M.P. 2,9
       11. $\sim T$ Simp. 10

    c. 1. $B \supset E$ Pr.
        2. $E \supset (C \cdot S)$ Pr.
        3. $\sim S$ Pr.
        4. B Pr. /∴ $\sim G$
        5. E M.P. 1,4
        6. $C \cdot S$ M.P. 2,5
        7. S Simp. 6
        8. $S \vee \sim G$ Add. 7
        9. $\sim G$ D.S. 3,8

    d. 1. $N \supset (A \vee W)$ Pr.
        2. $(W \cdot I) \supset C$ Pr.
        3. $A \supset S$ Pr.
        4. $(N \cdot I) \cdot \sim S$ Pr. /∴ C
        5. $N \cdot I$ Simp. 4
        6. $\sim S$ Simp. 4
        7. N Simp. 5
        8. I Simp. 5
        9. $A \vee W$ M.P. 1,7
       10. $\sim A$ M.T. 3,6
       11. W D.S. 9,10
       12. $W \cdot I$ Conj. 11,8
       13. C M.P. 2,12

    e. Symbolization only:
        1. $\sim S \vee (B \cdot E)$ Pr.
        2. $\sim S \supset M$ Pr.
        3. $E \supset (H \vee D)$ Pr.
        4. $\sim M \cdot \sim D$ Pr. /∴ H

Unit 8

1. The following are correct: b, c, e, g, i, j, l, m, q, r.

2.
a. Contrap.	f. DeM.	k. Dist.	p. C.E.	u. Exp.
b. Exp.	g. Dist.	l. DeM.	q. Com.	v. Dup.
c. D.N.	h. DeM.	m. Contrap.	r. DeM.	w. Contrap.
d. C.E.	i. Dup.	n. D.N.	s. B.E.	x. Dist.
e. Assoc.	j. C.E.	o. Dup.	t. Assoc.	y. C.E.

3. a.
5. DeM. 4
6. Simp. 5
7. Simp. 5
8. Add. 7
9. C.E. 8
10. Contrap. 3
11. M.P. 9,10
12. D.N. 11
13. D.S. 2,12
14. DeM. 13
15. Simp. 14

16. Conj. 12,15
17. DeM. 16
18. C.E. 17
19. Add. 18
20. DeM. 19
21. B.E. 20
22. B.E. 1
23. Simp. 2
24. M.P. 21,23
25. Conj. 24,6
26. DeM. 25

b.
3. C.E. 1
4. Dist. 3
5. Simp. 4
6. Com. 5
7. DeM. 6
8. Dist. 7
9. Simp. 8

10. C.E. 2
11. Dist. 10
12. Simp. 11
13. Com. 9
14. C.E. 13
15. C.E. 12
16. H.S. 14,15

c.
4. DeM. 1
5. DeM. 4
6. Com. 5
7. Dist. 6
8. DeM. 7
9. Simp. 8 (or 5–7)
10. Com. 3
11. Simp. 4 (or 5)
12. DeM. 11
13. Simp. 12
14. D.S. 10,13

15. D.N. 14
16. D.S. 9,15
17. Exp. 2
18. Contrap. 17
19. DeM. 18
20. C.E. 19
21. M.P. 16,20
22. Exp. 21
23. DeM. 22
24. Contrap. 23

d.
5. Dist. 2
6. DeM. 3
7. Dup. 4
8. Com. 5
9. Simp. 6
10. D.S. 8,9

11. Conj. 10,7
12. D.N. 11
13. DeM. 12
14. C.E. 13
15. D.S. 1,14
16. M.P. 10,15

4. e.
1. $\sim A \supset \sim B$ Pr.
2. B Pr. /∴ A
3. $\sim \sim B$ D.N. 2

       4. $\sim \sim A$ M.T. 1,3
       5. A D.N. 4

f.   1. $A \supset B$ Pr.
      2. $A \supset C$ Pr. /∴ $A \supset (B \cdot C)$
      3. $\sim A \vee B$ C.E. 1
      4. $\sim A \vee C$ C.E. 2
      5. $(\sim A \vee B) \cdot (\sim A \vee C)$ Conj. 3,4
      6. $\sim A \vee (B \cdot C)$ Dist. 5
      7. $A \supset (B \cdot C)$ C.E. 6

g. 1. $A \supset B$ Pr.
      2. $C \supset B$ Pr. /∴ $(A \vee C) \supset B$
      3. $\sim A \vee B$ C.E. 1
      4. $\sim C \vee B$ C.E. 2
      5. $B \vee \sim A$ Com. 3
      6. $B \vee \sim C$ Com. 4
      7. $(B \vee \sim A) \cdot (B \vee \sim C)$ Conj. 5,6
      8. $B \vee (\sim A \cdot \sim C)$ Dist. 7
      9. $(\sim A \cdot \sim C) \vee B$ Com. 8
    10. $\sim (A \vee C) \vee B$ DeM. 9
    11. $(A \vee C) \supset B$ C.E. 10

h. 1. $\sim (A \cdot B)$ Pr.
      2. A Pr. /∴ $\sim B$
      3. $\sim A \vee \sim B$ DeM. 1
      4. $\sim \sim A$ D.N. 2
      5. $\sim B$ D.S. 3,4

i.   1. $\sim (A \cdot B)$ Pr. /∴ $B \supset \sim A$
      2. $\sim A \vee \sim B$ DeM. 1
      3. $\sim B \vee \sim A$ Com. 2
      4. $B \supset \sim A$ C.E. 3

l.   1. $(A \supset B) \vee (A \supset C)$ /∴ $A \supset (B \vee C)$ Pr.
      2. $(\sim A \vee B) \vee (\sim A \vee C)$ C.E. 1
      3. $\sim A \vee (B \vee (\sim A \vee C))$ Assoc. 2
      4. $\sim A \vee ((\sim A \vee C) \vee B)$ Com. 3
      5. $\sim A \vee (\sim A \vee (C \vee B))$ Assoc. 4
      6. $(\sim A \vee \sim A) \vee (C \vee B)$ Assoc. 5
      7. $\sim A \vee (C \vee B)$ Dup. 6
      8. $A \supset (C \vee B)$ C.E. 7
      9. $A \supset (B \vee C)$ Com. 8

q. 1. $\sim A \supset A$ Pr. /∴ A
      2. $\sim \sim A \vee A$ C.E. 1
      3. $A \vee A$ D.N. 2
      4. A Dup. 3

5. b. 1. $(A \vee B) \supset \sim (C \vee D)$ Pr.
         2. $(A \cdot E) \vee \sim F$ Pr.
         3. F Pr. /∴ $\sim C$
         4. $\sim \sim F$ D.N. 3

    5. $A \cdot E$ D.S. 2,4
    6. A Simp. 5
    7. $A \lor B$ Add. 6
    8. $\sim (C \lor D)$ M.P. 1,7
    9. $\sim C \cdot \sim D$ DeM. 8
    10. $\sim C$ Simp. 9

e. 1. $(A \equiv B) \supset C$ Pr.
    2. $\sim (C \lor A)$ Pr. /∴ B
    3. $\sim C \cdot \sim A$ DeM. 2
    4. $\sim C$ Simp. 3
    5. $\sim A$ Simp. 3
    6. $\sim (A \equiv B)$ M.T. 1,4
    7. $\sim ((A \supset B) \lor \sim (B \supset A))$ B.E. 6
    8. $\sim (A \supset B) \lor \sim (B \supset A)$ DeM. 7
    9. $\sim A \lor B$ Add. 5
    10. $A \supset B$ C.E. 9
    11. $\sim \sim (A \supset B)$ D.N. 10
    12. $\sim (B \supset A)$ D.S. 8,11
    13. $\sim (\sim B \lor A)$ C.E. 12
    14. $\sim \sim B \cdot \sim A$ DeM. 13
    15. $\sim \sim B$ Simp. 14
    16. B D.N. 15

g. 1. $X \equiv \sim Y$ Pr.
    2. $(Y \lor Z) \supset T$ Pr.
    3. $\sim (T \lor W)$ Pr. /∴ $P \supset X$
    4. $\sim T \cdot \sim W$ DeM. 3
    5. $\sim T$ Simp. 4
    6. $\sim (Y \lor Z)$ M.T. 2,5
    7. $\sim Y \cdot \sim Z$ DeM. 6
    8. $\sim Y$ Simp. 7
    9. $(X \supset \sim Y) \cdot (\sim Y \supset X)$ B.E. 1
    10. $\sim Y \supset X$ Simp. 9
    11. X M.P. 8,10
    12. $\sim P \lor X$ Add. 11
    13. $P \supset X$ C.E. 12

h. 1. $(F \cdot \sim G) \lor (T \cdot \sim W)$ Pr.
    2. $W \cdot H$ Pr.
    3. $\sim (F \supset G) \supset (H \supset \sim S)$ Pr. /∴ $\sim S$
    4. W Simp. 2
    5. H Simp. 2
    6. $\sim \sim W$ D.N. 4
    7. $\sim T \lor \sim \sim W$ Add. 6
    8. $\sim (T \cdot \sim W)$ DeM. 7
    9. $F \cdot \sim G$ D.S. 1,8
    10. $\sim (\sim F \lor G) \supset (H \supset \sim S)$ C.E. 3
    11. $(\sim \sim F \cdot \sim G) \supset (H \supset \sim S)$ DeM. 10
    12. $(F \cdot \sim G) \supset (H \supset \sim S)$ D.N. 11

13. $H \supset \sim S$ M.P. 12,9
14. $\sim S$ M.P. 5,13

6. Symbolizations only for b, c, d, and e.
   a. 1. $I \supset (B \equiv M)$ Pr.
       2. $M \supset T$ Pr.
       3. $\sim T \cdot B$ Pr. /∴ $\sim I$
       4. $\sim T$ Simp. 3
       5. B Simp. 3
       6. $\sim M$ M.T. 2,4
       7. $B \cdot \sim M$ Conj. 5,6
       8. $\sim \sim B \cdot \sim M$ D.N. 7
       9. $\sim (\sim B \vee M)$ DeM. 8
      10. $\sim (B \supset M)$ C.E. 9
      11. $\sim (B \supset M) \vee \sim (M \supset B)$ Add. 10
      12. $\sim ((B \supset M) \cdot (M \supset B))$ DeM. 11
      13. $\sim (B \equiv M)$ B.E. 12
      14. $\sim I$ M.T. 1,13

   b. $(J \supset W) \cdot (W \supset ((R \cdot I) \cdot E)), ((R \cdot I) \equiv H) \cdot (H \supset \sim T) \cdot (\sim T \supset \sim E)$
      /∴ $\sim J$
   c. $I \supset (F \vee M), (F \supset A) \cdot (A \supset (S \vee B)), (S \supset U) \cdot (B \supset W),$
      $\sim (W \vee U) \cdot I$ /∴ M
   d. $O \supset (\sim S \cdot \sim P), \sim E \supset (\sim A \cdot \sim T), O \vee \sim E$ /∴ $\sim T \vee \sim P$
   e. $M \supset (I \equiv S), K \supset (I \equiv D), L \supset (I \equiv F), (S \supset T) \cdot (F \supset T),$
      $(\sim D \cdot \sim T) \cdot I$ /∴ $\sim (M \vee (K \vee L))$

Unit 9

1. a. Assume $((A \vee B) \supset C)$.
   b. First assume $(A \supset B)$, then $(A \supset C)$.
   c. First assume $((A \supset (C \supset A)) \supset (B \supset C))$, then A.
   d. First assume $(A \supset B)$, then A, then $(B \supset C)$, then A.
   e. Assume $((A \supset B) \supset A)$.
   f. First assume $(A \supset A)$, then $(B \supset (C \supset B))$, then B.
   g. First assume A, then B, then C, then $(A \vee B)$.
   h. Assume only $((A \supset B) \supset C) \supset (A \vee B)$.
   i. First assume $((A \supset A) \supset (B \supset (A \supset (A \supset B)))))$, then A, then B.
   j. First assume $((A \supset (B \supset A)) \supset B)$, then $(B \supset A)$, then $((C \supset B) \supset B)$, then A.

2. c, e, h, and j are contradictions. The others are not.
3. a. $\sim A \supset \sim B$ can be the conclusion of C.P., not of I.P., since it is a conditional.
   b. $\sim (A \supset B)$ can be the conclusion of I.P. since it is a negation. It can also be the conclusion of M.T. and M.P. Since it is not a conditional, it cannot be the conclusion of C.P.
   c. $\sim A \vee B$ cannot be the conclusion of either I.P. or M.T., since it is not a negation. It can be the conclusion of: M.P., Add., D.S., Simp., Dil., C.E., D.N., Com., and Dup.

   d. $\sim A \lor B$ can be a premise for Add. and D.S., but not H.S. It could also be one premise for M.P., Conj., Dil., D.N., Com., Dup., and C.E.

   e. The last step in a subproof for C.P. is the consequent of the conditional being proved. The last step in a subproof for I.P. is a contradiction.

4. b.

1.	$(D \cdot E) \supset \sim F$	Pr.
2.	$F \lor (G \cdot W)$	Pr.
3.	$D \supset E$	Pr. /$\therefore D \supset G$
4.	$\rightarrow D$	Assp. (C.P.)
5.	E	M.P. 3,4
6.	$D \cdot E$	Conj. 4,5
7.	$\sim F$	M.P. 1,6
8.	$G \cdot W$	D.S. 2,7
9.	G	Simp. 8
10.	$D \supset G$	C.P. 4–9

  f.

1.	$(A \lor B) \supset \sim C$	Pr.
2.	$D \supset (\sim F \cdot \sim G)$	Pr. /$\therefore (A \lor D) \supset \sim (C \cdot F)$
3.	$\rightarrow A \lor D$	Assp. (C.P.)
4.	$\rightarrow C \cdot F$	Assp. (I.P.)
5.	C	Simp. 4
6.	$\sim \sim C$	D.N. 5
7.	$\sim (A \lor B)$	M.T. 1,6
8.	$\sim A \cdot \sim B$	DeM. 7
9.	$\sim A$	Simp. 8
10.	D	D.S. 3,9
11.	$\sim F \cdot \sim G$	M.P. 2,10
12.	$\sim F$	Simp. 11
13.	F	Simp. 4
14.	$F \cdot \sim F$	Conj. 12,13
15.	$\sim (C \cdot F)$	I.P. 4–14
16.	$(A \lor D) \supset \sim (C \cdot F)$	C.P. 3–15

  h.

1.	$(A \lor B) \supset (A \cdot B)$	Pr. /$\therefore A \equiv B$
2.	$\rightarrow A$	Assp. (C.P.)
3.	$A \lor B$	Add. 2
4.	$A \cdot B$	M.P. 1,3
5.	B	Simp. 4
6.	$A \supset B$	C.P. 2–5
7.	$\rightarrow B$	Assp. (C.P.)
8.	$A \lor B$	Add. 7
9.	$A \cdot B$	M.P. 1,8
10.	A	Simp. 9
11.	$B \supset A$	C.P. 7–10
12.	$(A \supset B) \cdot (B \supset A)$	Conj. 6,11
13.	$A \equiv B$	B.E. 12

  i.

1.	$A \lor B$	Pr.
2.	$\sim A \lor \sim B$	Pr. /$\therefore \sim (A \equiv B)$
3.	$\rightarrow A \equiv B$	Assp. (I.P.)
4.	$(A \supset B) \cdot (B \supset A)$	B.E. 3

5.	A	Assp. (I.P.)
6.	$A \supset B$	Simp. 4
7.	B	M.P. 5,6
8.	$\sim \sim A$	D.N. 5
9.	$\sim B$	D.S. 2,8
10.	$B \cdot \sim B$	Conj. 7,9
11.	$\sim A$	I.P. 5–10
12.	$B \supset A$	Simp. 4
13.	$\sim B$	M.T. 11,12
14.	A	D.S. 1,13
15.	$A \cdot \sim A$	Conj. 14,11
16.	$\sim (A \equiv B)$	I.P. 3–15

q.
1.	$(A \lor B) \supset \sim (F \cdot D)$	Pr.
2.	$\sim (A \cdot \sim D)$	Pr.
3.	$\sim F \supset \sim (C \cdot D)$	Pr. /∴ $\sim (A \cdot C)$
4.	$A \cdot C$	Assp. (I.P.)
5.	A	Simp. 4
6.	C	Simp. 4
7.	$A \lor B$	Add. 5
8.	$\sim (F \cdot D)$	M.P. 1,7
9.	$\sim A \lor \sim \sim D$	DeM. 2
10.	$\sim \sim A$	D.N. 5
11.	$\sim \sim D$	D.S. 9,10
12.	$\sim F \lor \sim D$	DeM. 8
13.	$\sim F$	D.S. 11,12
14.	$\sim (C \cdot D)$	M.P. 3,13
15.	D	D.N. 11
16.	$C \cdot D$	Conj. 6,15
17.	$(C \cdot D) \cdot \sim (C \cdot D)$	Conj. 16,14
18.	$\sim (A \cdot C)$	I.P. 4–17

5. e.
1.	$(X \cdot Y) \lor \sim (Z \lor W)$	Pr.
2.	$(Z \cdot X) \supset (Y \supset \sim B)$	Pr.
3.	$C \supset \sim C$	Pr.
4.	$\sim (B \lor C) \supset \sim Y$	Pr. /∴ $\sim Z$
5.	Z	Assp. (I.P.)
6.	$Z \lor W$	Add. 5
7.	$\sim \sim (Z \lor W)$	D.N. 6
8.	$X \cdot Y$	D.S. 1,7
9.	X	Simp. 8
10.	Y	Simp. 8
11.	$Z \cdot X$	Conj. 5,9
12.	$Y \supset \sim B$	M.P. 2,11
13.	$\sim B$	M.P. 10,12
14.	$\sim C \lor \sim C$	C.E. 3
15.	$\sim C$	Dup. 14
16.	$\sim B \cdot \sim C$	Conj. 13,15
17.	$\sim (B \lor C)$	DeM. 16

18.	$\sim Y$	M.P. 4,17
19.	$Y \cdot \sim Y$	Conj. 10,18
20.	$\sim Z$	I.P. 5–19

6. a.

1.	$\sim p \supset (\sim q \supset \sim r)$	Assp. (C.P.)
2.	r	Assp. (C.P.)
3.	$(\sim p \cdot \sim q) \supset \sim r$	Exp. 1
4.	$\sim \sim r$	D.N. 2
5.	$\sim (\sim p \cdot \sim q)$	M.T. 4,3
6.	$\sim \sim p \lor \sim \sim q$	DeM. 5
7.	$p \lor q$	D.N. 6
8.	$r \supset (p \lor q)$	C.P. 2–7
9.	$(\sim p \supset (\sim q \supset \sim r)) \supset (r \supset (p \lor q))$	C.P. 1–8

d.

1.	$(p \equiv \sim q) \cdot \sim (p \lor q)$	Assp. (I.P.)
2.	$p \equiv \sim q$	Simp. 1
3.	$\sim (p \lor q)$	Simp. 1
4.	$\sim p \cdot \sim q$	DeM. 3
5.	$\sim p$	Simp. 4
6.	$\sim q$	Simp. 4
7.	$(p \supset \sim q) \cdot (\sim q \supset p)$	B.E. 2
8.	$\sim q \supset p$	Simp. 7
9.	p	M.P. 6,8
10.	$p \cdot \sim p$	Conj. 5,9
11.	$\sim ((p \equiv \sim q) \cdot \sim (p \lor q))$	I.P. 1–10

7. The following assignments yield counterexamples:
   a. A = True, B = T or F, C = False, D = False, E = T or F, F = False
   (The values of B and E may be anything, but the values for A, C, D, and F *must* be those above for a counterexample.)
   c. A = True, B = False, C = True, D = False, F = True
   (These are the only values which yield a counterexample.)
   e. A = T or F, B = False, C = True, D = False, E = False, F = True, H = True

8. Symbolizations and results only.

a. $(N \lor C) \supset (A \lor P), \sim (N \lor C) \cdot S \mathbin{/\therefore} \sim P$	Invalid
b. $B \supset ((G \cdot D) \cdot S), S \supset (W \cdot \sim K), G \supset (N \cdot K) \mathbin{/\therefore} \sim B$	Valid
c. $B \supset (G \cdot D), G \supset W, D \supset E, W \cdot E \mathbin{/\therefore} B$	Invalid
d. $N \equiv (W \cdot U), W \equiv (P \cdot \sim S), \sim E \supset U, P \cdot \sim E \mathbin{/\therefore} \sim N \supset S$	Valid
e. $M \supset (I \equiv S), K \supset (I \equiv D), L \supset (I \supset F), F \supset T,$	
$(\sim D \cdot \sim T) \cdot I \mathbin{/\therefore} \sim (M \lor K \lor L)$	Invalid

Unit 10

1. a. (1) x is a master safecracker (2) Mx (3) Mr
   c. (1) x's apartment contained huge diamond rings (2) Dx (3) Dr
   e. (1) x solved the mystery (2) Sx (3) Sf

g. x gets enough sleep; x misses a lot; x is alert; x is very sharp (2) Sx; Mx; Ax; Vx (3) $(\sim Sa \supset Ma) \cdot (Aa \supset \sim Va)$

i. (1) x loves police work; x is happy with Andrew's job; x will stay in police work; x's supervisor gives x a raise; x's supervisor gives x a commendation (2) Px; Hx; Sx; Rx; Cx (3) $Pa \cdot \sim (Hj \lor Hb) \cdot (Sa \equiv (Ra \cdot Ca))$

2. a. $Pa \cdot Pb \cdot Pk$
    c. $Sp \cdot \sim Sk \cdot ((Fk \cdot \sim Rk) \supset Dk)$
    e. $Ca \cdot Ck \cdot \sim (Cb \lor Cp)$
    g. $(Rk \cdot \sim Dk) \supset Ha$
    i. $Hb \supset (Ta \cdot La \cdot \sim Fb)$

3. a. Singular   k. Categorical
    c. Categorical   m. Singular
    e. Categorical   o. Singular
    g. Identity   q. Identity
    i. Singular   s. Singular

Unit 11

1. a. Singular
    c. Singular
    e. Existential
    g. Propositional function
    i. Singular

2. a. Existential. $Cx \equiv x$ is clear. $(\exists x)Cx$
    c. Existential. $Fx \equiv x$ is a flying saucer. $(\exists x)Fx$
    e. Universal. $Px \equiv x$ has a price. $(x)Px$
    g. Existential. $Wx \equiv x$ is wrong. $(\exists x)Wx$
    i. Existential. $Ex \equiv x$ is evil. $(\exists x)Ex$

3. a. $\sim (\exists x)Lx$ or $(x) \sim Lx$
    c. $\sim (x)Ex$ or $(\exists x) \sim Ex$
    e. $\sim (\exists x)Ux$ or $(x) \sim Ux$
    g. $\sim (\exists x) \sim Gx$ or $(x)Gx$
    i. $\sim (x) \sim Bx$ or $(\exists x)Bx$
    k. $(x) \sim Cx$ or $\sim (\exists x)Cx$
    m. $(\exists x)Cx$ (or $\sim (x) \sim Cx$)
    o. $\sim (\exists x) \sim \sim Ux$ or $\sim (\exists x)Ux$ or $(x) \sim Ux$

4. a. There aren't any angels.
    c. Not everything is beautiful.
    e. Nothing is beautiful.
    g. There are devils. (Devils exist.)
    i. Not everything is not an angel.

Unit 12

1. a. E S = men, P = islands
    c. A S = dogs, P = beings that have their days
    e. E S = people, P = beings that understand modern art

    g. O S = cats, P = things that have tails
    i. E S = people, P = beings that are calling the police
    k. O S = people, P = beings that care
    m. E S = police cars, P = things that are arriving
    o. O S = jails, P = pleasant things

2. a. S = corporation executives, P = wealthy beings
      Put an × in region 1 (an O sentence)
    c. S = pine trees, P = coniferous things
      Shade region 1 (A.)
    e. S = hickory nuts, P = things that are high in fat
      Shade region 1 (A.)
    g. S = John's children, P = things that know how to ski
      Put an × in region 2 (I.)
    i. S = fruit bearers, P = coniferous things
      Shade region 2 (E.)

3. a. 1. $\sim (\exists x)(Fx \cdot \sim Gx)$
      2. $(x) \sim (Fx \cdot \sim Gx)$ Q.N. 1   (*Note*: To get the other half of the
      3. $(x)(\sim Fx \lor \sim \sim Gx)$ DeM. 2   equivalence, just do the steps
      4. $(x)(\sim Fx \lor Gx)$ D.N. 3   in reverse order.)
      5. $(x)(Fx \supset Gx)$ C.E. 4
    c. 1. $\sim (\exists x)(Fx \cdot Gx)$
      2. $(x) \sim (Fx \cdot Gx)$ Q.N. 1
      3. $(x)(\sim Fx \lor \sim Gx)$ DeM. 2
      4. $(x)(Fx \supset \sim Gx)$ C.E. 3

4. a. $Dx \equiv x$ is a dog, $Fx \equiv x$ has feelings. $(x)(Dx \supset Fx)$
    c. $Rx \equiv x$ is in the room, $Bx \equiv x$ is beautiful. $(x)(Rx \supset \sim Bx)$ or
      $\sim (\exists x)(Rx \cdot Bx)$
    e. $Dx \equiv x$ is a diamond, $Vx \equiv x$ is valuable. $\sim (x)(Dx \supset Vx)$ or
      $(\exists x)(Dx \cdot \sim Vx)$
    g. $Vx \equiv x$ is valuable, $Hx \equiv x$ is in the house. $\sim (\exists x)(Vx \cdot Hx)$ or
      $\sim (\exists x)(Hx \cdot Vx)$ or $(x)(Vx \supset \sim Hx)$ or $(x)(Hx \supset \sim Vx)$
    i. $Rx \equiv x$ is a resemblance to persons living or dead, $Cx \equiv x$ is purely coin-
      cidental. $(x)(Rx \supset Cx)$
    k. $Cx = x$ is a candidate; $Dx = x$ is candid. $(x)(Cx \supset \sim Dx)$
    m. $Kx = x$ kisses babies; $Cx = x$ is a candidate. $(x)(Kx \supset Cx)$
    o. $Px \equiv x$ is a person, $Sx \equiv x$ enjoyed the show. $\sim (\exists x)(Px \cdot \sim Sx)$ or
      $(x)(Px \supset Sx)$
    q. $Hx \equiv x$ is one of the hardiest, $Nx \equiv x$ will survive a nuclear war.
      $(x)(Nx \supset Hx)$ or $(x)(\sim Hx \supset \sim Nx)$ or $\sim (\exists x)(Nx \cdot \sim Hx)$
    s. $Px \equiv x$ is a person, $Hx \equiv x$ was happy about the decision.
      $\sim (\exists x)(Px \cdot \sim \sim Hx)$ or $\sim (\exists x)(Px \cdot Hx)$ or $(x)(Px \supset \sim Hx)$

5. a. $(x)(Bx \supset Vx)$
    c. $(\exists x)(Vx \cdot Wx)$
    e. $(x)(Wx \supset Dx)$ or $(x)(\sim Dx \supset \sim Wx)$ or $\sim (\exists x)(Wx \cdot \sim Dx)$
    g. $(x)(Ex \supset \sim Gx)$ or $\sim (\exists x)(Ex \cdot Gx)$
    i. $\sim (\exists x)(Rx \cdot Hx)$ or $(x)(Rx \supset \sim Hx)$
    k. $(x)(Mx \supset \sim Rx)$ or $\sim (\exists x)(Rx \cdot Mx)$ or $(x)(Rx \supset \sim Mx)$

m. $\sim (\exists x)(Gx \cdot \sim Ax)$ or $(x)(Gx \supset Ax)$
o. $(\exists x)(Bx \cdot \sim \sim Ax)$ or $(\exists x)(Bx \cdot Ax)$
q. $(x)(Rx \supset \sim Hx)$ or $\sim (\exists x)(Rx \cdot Hx)$
s. $\sim (\exists x)(Rx \cdot \sim Bx)$ or $(x)(Rx \supset Bx)$

6. a. $(x)(Ax \supset Bx)$
   c. $(\exists x)(Ax \cdot \sim Sx)$
   e. $\sim (x)(Px \supset Hx)$ or $(\exists x)(Px \cdot \sim Hx)$
   g. $(x)(Ux \supset Bx)$
   i. $\sim (x)(Dx \supset \sim Hx)$ or $(\exists x)(Dx \cdot Hx)$
   k. $(x)(Ex \supset Dx)$ or $(x)(\sim Dx \supset \sim Ex)$ or $\sim (\exists x)(Ex \cdot \sim Dx)$
   m. $\sim (x)(Ix \supset Ax)$ or $(\exists x)(Ix \cdot \sim Ax)$
   o. $\sim (\exists x)(Dx \cdot \sim \sim Hx)$ or $\sim (\exists x)(Dx \cdot Hx)$ or $(x)(Dx \supset \sim Hx)$
   q. $\sim (\exists x)(Bx \cdot Ex)$ or $(x)(Bx \supset \sim Ex)$
   s. $(x)(\sim Bx \supset Dx)$ or $\sim (\exists x)(\sim Bx \cdot \sim Dx)$

7. a. Not all honest people are good.
   c. No saints are politicians.
   e. No devils are saints.
   g. There are no beautiful people who are not lucky.
   i. There is nothing innocent which is not beautiful.

Unit 13

1. a. $(\exists x)(Fx \cdot \sim (Gx \lor Hx))$ C.Q.N., $(\exists x)(Fx \cdot Gx \cdot \sim Hx)$ DeM.
   c. $(\exists x)((Fx \cdot \sim Gx) \cdot \sim (Hx \cdot \sim Ix))$ C.Q.N.
       $(\exists x)((Fx \cdot \sim Gx) \cdot (\sim Hx \lor \sim \sim Ix))$ DeM.
       $(\exists x)((Fx \cdot \sim Gx) \cdot (Hx \supset Ix))$ C.E., D.N.
       $(\exists x)((Fx \cdot \sim Gx) \cdot (\sim Ix \supset \sim Hx))$ Contrap.
   e. $(\exists x)((Fx \cdot (Gx \lor Hx)) \cdot \sim ((Px \cdot Qx) \supset \sim Rx))$ C.Q.N.
       $(\exists x)((Fx \cdot (Gx \lor Hx)) \cdot \sim (\sim (Px \cdot Qx) \lor \sim Rx))$ C.E.
       $(\exists x)((Fx \cdot (Gx \lor Hx)) \cdot ((Px \cdot Qx) \cdot Rx))$ DeM., D.N.
       $(\exists x)(((Fx \cdot Gx) \lor (Fx \cdot Hx)) \cdot ((Px \cdot Qx) \cdot Rx))$ Dist.

2. a. $(\exists x)(Px \cdot Cx \cdot Sx)$
   c. $(x)((Px \cdot Sx) \supset (Cx \lor Hx))$
   e. $\sim (x)((Px \cdot Cx \cdot Hx) \supset (Ox \cdot Ex))$ or $(\exists x)((Px \cdot Cx \cdot Hx) \cdot \sim (Ox \cdot Ex))$ or $(\exists x)(Px \cdot Cx \cdot Hx \cdot (\sim Ox \lor \sim Ex))$
   g. $\sim (\exists x)(Bx \cdot Px \cdot \sim Gx \cdot Wx)$ or $(x)((Bx \cdot Px) \supset \sim (\sim Gx \cdot Wx))$ or $(x)((Bx \cdot Px) \supset (\sim Gx \supset \sim Wx))$ or $(x)((Bx \cdot Px) \supset (Wx \supset Gx))$ or $(x)((Bx \cdot Px \cdot \sim Gx) \supset \sim Wx)$ or $(x)((Bx \cdot Px \cdot Wx) \supset Gx)$
   i. $(\exists x)((Gx \cdot Px) \cdot (Ax \cdot \sim (Rx \lor Fx)))$ or $(\exists x)(Gx \cdot Px \cdot Ax \cdot \sim Rx \cdot \sim Fx)$
   k. $(x)((Px \cdot \sim Ax) \supset (\sim Rx \supset \sim Yx))$ or $(x)((Px \cdot \sim Ax) \supset (Yx \supset Rx))$ or $(x)((Px \cdot \sim Ax \cdot Rx) \supset \sim Yx)$ or $(x)((Px \cdot \sim (Ax \lor Rx)) \supset \sim Yx)$ or $\sim (\exists x)(Px \cdot \sim Ax \cdot \sim Rx \cdot Yx)$ (There are many other possibilities)
   m. $\sim (x)(Wx \supset (Rx \lor Fx))$ or $(\exists x)(Wx \cdot \sim (Rx \lor Fx))$ or $(\exists x)(Wx \cdot \sim Rx \cdot \sim Fx)$
   o. $(x)((Px \cdot \sim (Bx \lor Rx)) \supset (\sim (Yx \lor Ax) \cdot Wx))$ or $(x)((Px \cdot \sim Bx \cdot \sim Rx) \supset (\sim Yx \cdot \sim Ax \cdot Wx))$

3. a. No one who ate ham salad got sick.
 c. Not everyone who is not appreciated and not rich is unhappy.
 e. No one who is rich or famous, and beautiful but not good, will get to heaven.

4. a. $(\exists x)(Px \cdot Jx \cdot Bx \cdot \sim Vx)$
 c. $(x)[(Px \cdot Jx) \supset ((Gx \cdot Dx) \supset Vx)]$
 e. $(x)[(Px \cdot (Jx \vee Hx)) \supset (\sim Ux \supset \sim (Bx \vee Vx))]$ or
 $\sim (\exists x)(Px \cdot (Jx \vee Hx) \cdot (Bx \vee Vx) \cdot \sim Ux)$
 g. $(\exists x)(Px \cdot Jx \cdot Ux \cdot \sim Cx \cdot (Ax \equiv \sim Lx))$
 i. $(x)[(Px \cdot (Hx \vee Jx)) \supset (\sim ((Bx \cdot Ux) \vee (Vx \cdot Gx)) \supset \sim Cx)]$ or
 $\sim (\exists x)[Px \cdot (Hx \vee Jx) \cdot Cx \cdot \sim ((Bx \cdot Ux) \vee (Vx \cdot Gx))]$

5. a. $\sim (x)((Bx \cdot Cx) \supset Lx)$ or $(\exists x)(Bx \cdot Cx \cdot \sim Lx)$
 c. $(x)((Px \cdot \sim Tx) \supset \sim Ex)$ or $(x)(Px \supset (\sim Tx \supset \sim Ex))$ or
 $\sim (\exists x)(Px \cdot \sim Tx \cdot Ex)$
 e. $(x)((Px \cdot Lx \cdot Ax) \supset \sim (Rx \vee Ex))$ or $\sim (\exists x)(Px \cdot Lx \cdot Ax \cdot (Rx \vee Ex))$ or
 $(x)((Px \cdot Lx \cdot Ax) \supset (\sim Rx \cdot \sim Ex))$
 g. $(x)((Px \cdot Tx \cdot \sim Ax) \supset \sim Ix)$ or $(x)((Px \cdot Tx) \supset (\sim Ax \supset \sim Ix))$ or
 $\sim (\exists x)(Px \cdot Tx \cdot \sim Ax \cdot Ix)$ or $(x)((Px \cdot Ix) \supset (\sim Tx \vee Ax))$
 i. $(x)((Mx \cdot Px) \supset ((Tx \cdot \sim Lx \cdot \sim Ux) \supset Ex))$ or
 $(x)((Mx \cdot Px \cdot Tx \cdot \sim Lx \cdot \sim Ux) \supset Ex)$

6. a. $(x)((Px \cdot \sim (Rx \cdot Vx)) \supset \sim Ex)$ or $(x)((Px \cdot (\sim Rx \vee \sim Vx)) \supset \sim Ex)$ or
 $(x)((Px \cdot Ex) \supset (Rx \cdot Vx))$ or $\sim (\exists x)(Px \cdot Ex \cdot \sim (Rx \cdot Vx))$
 c. $(x)(Px \supset ((Tx \cdot \sim Lx) \supset \sim (Ix \vee Ux)))$ or
 $(x)((Px \cdot Tx \cdot \sim Lx) \supset (\sim Ix \cdot \sim Ux))$ or
 $\sim (\exists x)(Px \cdot Tx \cdot \sim Lx \cdot (Ix \vee Ux))$
 e. $(x)((Px \vee Bx) \supset (Cx \supset (Lx \vee Mx \vee (Ax \cdot Ux))))$ or
 $(x)((Px \vee Bx) \supset (\sim (Lx \vee Mx \vee (Ax \cdot Ux)) \supset \sim Cx))$ or
 $(x)((Px \vee Bx) \supset ((\sim Lx \cdot \sim Mx \cdot \sim (Ax \cdot Ux)) \supset \sim Cx))$ or
 $\sim (\exists x)((Px \vee Bx) \cdot \sim (Lx \vee Mx \vee (Ax \cdot Ux)) \cdot Cx)$

Unit 14

2. a. Negation
 c. Universal
 e. Conditional
 g. Universal
 i. Existential
 k. Disjunction
 m. Negation
 o. Existential
 q. Conditional
 s. Disjunction

3. a. $(\exists x)(Px \cdot \sim Ax) \cdot (\exists x)(Px \cdot Ax)$
 c. $(x)(Px \supset Vx) \supset (x)(Px \supset \sim Ix)$
 e. $\sim (\exists x)(Px \cdot Bx) \vee \sim (\exists x)(Px \cdot Rx)$
 g. $\sim Hj \supset \sim (\exists x)(Hx \cdot Px)$
 i. $\sim (Br \vee Ar) \supset (\sim Ir \cdot Rr)$
 k. $Bj \supset (x)(Px \supset Bx)$
 m. $(\exists x)(Px \cdot (Bx \vee Lx) \cdot Ix)$

4. a. If there are honest politicians, then John does not take bribes.
 c. If no politicians are respected, then some politicians take bribes.
 e. If John is not respected and not reelected, then either he abuses his powers or is not honest or is not a millionaire.
 g. Any dishonest politician who takes a bribe is a millionaire who is not respected and does not vote his conscience.

i. No lying politician is respected, but not all lying politicians who take bribes do not get reelected.

Unit 15

1. The following are correct applications of the rule cited: b, c, e, f, h, j, l, n, p, s.

2. b.

1. $(x)(Fx \supset \sim Gx)$	Pr.
2. $(\exists x)(Hx \cdot Gx)$	Pr. /∴ $\sim (x)(Hx \supset Fx)$
3. $Ha \cdot Ga$	E.I. 2, a/x (flag a)
4. $Fa \supset \sim Ga$	U.I. 1, a/x
5. Ha	Simp. 3
6. Ga	Simp. 3
7. $\sim \sim Ga$	D.N. 6
8. $\sim Fa$	M.T. 4,7
9. $Ha \cdot \sim Fa$	Conj. 5,8
10. $(\exists x)(Hx \cdot \sim Fx)$	E.G. 9, a/x
11. $\sim (x)(Hx \supset Fx)$	C.Q.N. 10

e.

1. $(x)(Tx \supset (Fx \cdot Dx))$	Pr.
2. $\sim (x)(Tx \supset \sim Bx)$	Pr. /∴ $\sim (x)(Dx \supset \sim Bx)$
3. $(\exists x)(Tx \cdot Bx)$	C.Q.N. 2
4. $Ta \cdot Ba$	E.I. 3, a/x (flag a)
5. $Ta \supset (Fa \cdot Da)$	U.I. 1, a/x
6. Ta	Simp. 4
7. Ba	Simp. 4
8. $Fa \cdot Da$	M.P. 5,6
9. Da	Simp. 8
10. $Da \cdot Ba$	Conj. 9,7
11. $(\exists x)(Dx \cdot Bx)$	E.G. 10, a/x
12. $\sim (x)(Dx \supset \sim Bx)$	C.Q.N. 11

h.

1. $(x((Bx \lor Wx) \supset ((Ax \lor Fx) \supset Sx))$	Pr. /∴ $(x)((Bx \cdot Fx) \supset Sx)$
2. ┌► flag a	F.S. (U.G.)
3. │ $(Ba \lor Wa) \supset ((Aa \lor Fa) \supset Sa)$	U.I. 1, a/x
4. │ ┌►$Ba \cdot Fa$	Assp. (C.P.)
5. │ │ Ba	Simp. 4
6. │ │ Fa	Simp. 4
7. │ │ $Ba \lor Wa$	Add. 5
8. │ │ $(Aa \lor Fa) \supset Sa$	M.P. 3,7
9. │ │ $Aa \lor Fa$	Add. 6
10. │ └ Sa	M.P. 8,9
11. └ $(Ba \cdot Fa) \supset Sa$	C.P. 4–10
12. $(x)((Bx \cdot Fx) \supset Sx)$	U.G. 11, a/x

j.

1. $(\exists x)Cx \cdot (\sim Sx \supset (Vx \lor Wx)))$	Pr.
2. $(x)(Vx \supset \sim Cx)$	Pr.
3. $\sim (\exists x)(Wx \cdot Cx)$	Pr. /∴ $(\exists x)(Sx \cdot \sim Wx)$
4. $(x)(Wx \supset \sim Cx)$	C.Q.N. 3
5. $Ca \cdot (\sim Sa \supset (Va \lor Wa))$	E.I. 1, a/x (flag a)

6. Ca	Simp. 5
7. $Wa \supset \sim Ca$	U.I. 4, a/x
8. $Va \supset \sim Ca$	U.I. 2, a/x
9. $\sim \sim Ca$	D.N. 6
10. $\sim Wa$	M.T. 7,9
11. $\sim Va$	M.T. 8,9
12. $\sim Va \cdot \sim Wa$	Conj. 11,10
13. $\sim (Va \lor Wa)$	DeM. 12
14. $\sim Sa \supset (Va \lor Wa)$	Simp. 5
15. $\sim \sim Sa$	M.T. 13,14
16. Sa	D.N. 15
17. $Sa \cdot \sim Wa$	Conj. 16,10
18. $(\exists x)(Sx \cdot \sim Wx)$	E.G. 17, a/x

k.
1. $\sim (\exists x)(Fx \cdot \sim (Gx \cdot Hx))$	Pr.
2. $(x)((Gx \lor Sx) \supset Zx)$	Pr.
3. $\sim (\exists x)(Zx \cdot Ax)$	Pr. /∴ $(x)(Fx \supset \sim Ax)$
4. $(x)(Zx \supset \sim Ax)$	C.Q.N. 3
5. $(x)(Fx \supset (Gx \cdot Hx))$	C.Q.N. 1
6. flag a	F.S. (U.G.)
7. $(Ga \lor Sa) \supset Za$	U.I. 2, a/x
8. $Za \supset \sim Aa$	U.I. 4, a/x
9. $Fa \supset (Ga \cdot Ha)$	U.I. 5, a/x
10. Fa	Assp. (C.P.)
11. $Ga \cdot Ha$	M.P. 10,9
12. Ga	Simp. 11
13. $Ga \lor Sa$	Add. 12
14. Za	M.P. 13,7
15. $\sim Aa$	M.P. 14,8
16. $Fa \supset \sim Aa$	C.P. 10–15
17. $(x)(Fx \supset \sim Ax)$	U.G. 16, a/x

o.
1. $(x)((Ax \lor Bx) \supset (Gx \cdot \sim Hx))$	Pr.
2. $(x)(Gx \supset Hx)$	Pr.
3. $(x)((Dx \cdot Ex) \supset Bx)$	Pr.
4. $\sim (\exists x)(Px \cdot \sim Ex)$	Pr. /∴ $\sim (\exists x)(Px \cdot (Ax \lor Dx))$
5. $(x)(Px \supset Ex)$	C.Q.N. 4
6. flag a	F.S. (U.G.)
7. $(Aa \lor Ba) \supset (Ga \cdot \sim Ha)$	U.I. 1, a/x
8. $Ga \supset Ha$	U.I. 2, a/x
9. $(Da \cdot Ea) \supset Ba$	U.I. 3, a/x
10. $Pa \supset Ea$	U.I. 5, a/x
11. Pa	Assp. (C.P.)
12. Ea	M.P. 10,11
13. $\sim Ga \lor Ha$	C.E. 8
14. $\sim Ga \lor \sim \sim Ha$	D.N. 13
15. $\sim (Ga \cdot \sim Ha)$	DeM. 14
16. $\sim (Aa \lor Ba)$	M.T. 7,15
17. $\sim Aa \cdot \sim Ba$	DeM. 16

18.	$\sim Ba$	Simp. 17
19.	$\sim (Da \cdot Ea)$	M.T. 9,18
20.	$\sim Da \lor \sim Ea$	DeM. 19
21.	$\sim \sim Ea$	D.N. 12
22.	$\sim Da$	D.S. 20,21
23.	$\sim Aa$	Simp. 17
24.	$\sim Aa \cdot \sim Da$	Conj. 23,22
25.	$\sim (Aa \lor Da)$	DeM. 24
26.	$Pa \supset \sim (Aa \lor Da)$	C.P. 11–25
27.	$(x)(Px \supset \sim (Ax \lor Dx))$	U.G. 26, a/x
28.	$\sim (\exists x)(Px \cdot (Ax \lor Dx))$	C.Q.N. 27

3. Symbolizations only:

a. $(x)((Dx \lor Lx) \supset (Wx \cdot Px))$, $\sim (\exists x)((Wx \cdot Dx) \cdot Mx)$, $\sim (\exists x)(Px \cdot Sx)$
   $/\therefore \sim (\exists x)(Dx \cdot (Mx \lor Sx))$

b. $(x)((Px \cdot Rx) \supset (Hx \cdot Sx))$, $(\exists x)(Px \cdot Rx \cdot Mx)$ $/\therefore (\exists x)(Px \cdot Mx \cdot Hx)$

c. $(\exists x)(Px \cdot Jx)$, $(x)(Jx \supset \sim Ax)$, $(x)((Px \cdot \sim Ax) \supset Ex)$ $/\therefore (\exists x)(Px \cdot Ex)$

d. $(\exists x)(Jx \cdot Cx \cdot Px) \cdot (x)(Jx \cdot Px) \supset Ex)$, $(x)((Jx \cdot Ex) \supset (Bx \cdot Tx))$,
   $(x)(Tx \supset (Bx \supset Mx))$ $/\therefore (\exists x)(Jx \cdot Cx \cdot Mx)$

4. e.

1.	$(\exists x)Fx \supset (x)(Gx \lor Hx)$	Pr.
2.	$(x)(Fx \supset \sim Gx)$	Pr. $/\therefore \sim (\exists x)Hx \supset (x) \sim Fx$
3.	$\sim (x) \sim Fx$	Assp. (C.P.)
4.	$(\exists x)Fx$	Q.N. 3
5.	$(x)(Gx \lor Hx)$	M.P. 1,4
6.	Fa	E.I. 4, a/x (flag a)
7.	$Fa \supset \sim Ga$	U.I. 2, a/x
8.	$\sim Ga$	M.P. 6,7
9.	$Ga \lor Ha$	U.I. 5, a/x
10.	Ha	D.S. 8,9
11.	$(\exists x)Hx$	E.G., 10, a/x
12.	$\sim (x) \sim Fx \supset (\exists x)Hx$	C.P. 3–11
13.	$\sim (x) \sim Fx \supset \sim \sim (\exists x)Hx$	D.N. 12
14.	$\sim (\exists x)Hx \supset (x) \sim Fx$	Contrap. 13

f.

1.	$(\exists x)Fx \supset (x)(Hx \supset \sim Jx)$	Pr.
2.	$\sim (\exists x)(Sx \cdot \sim Jx)$	Pr. $/\therefore (x)Fx \supset \sim (\exists x)(Hx \cdot Sx)$
3.	$(x)Fx$	Assp. (C.P.)
4.	Fa	U.I. 3, a/x
5.	$(\exists x)Fx$	E.G. 4, a/x
6.	$(x)(Hx \supset \sim Jx)$	M.P. 1,5
7.	$(x)(Sx \supset Jx)$	C.Q.N. 2
8.	flag b	F.S. (U.G.)
9.	$Sb \supset Jb$	U.I. 7, b/x
10.	$Hb \supset \sim Jb$	U.I. 6, b/x
11.	$\sim Jb \supset \sim Sb$	Contrap. 9
12.	$Hb \supset \sim Sb$	H.S. 10,11
13.	$(x)(Hx \supset \sim Sx)$	U.G. 12, b/x
14.	$\sim (\exists x)(Hx \cdot Sx)$	C.Q.N. 13
15.	$(x)Fx \supset \sim (\exists x)(Hx \cdot Sx)$	C.P. 3–14

h.
1.	$\sim ((x)Fx \lor (\exists x)Gx)$	Pr.
2.	$(x)(Hx \supset Fx)$	Pr.
3.	$((\exists x) \sim Hx \cdot (x)Sx) \supset (\exists x)Gx$	Pr. $/\therefore (\exists x) \sim Sx$
4.	$\sim (x)Fx \cdot \sim (\exists x)Gx$	DeM. 1
5.	$\sim (\exists x)Gx$	Simp. 4
6.	$\sim ((\exists x) \sim Hx \cdot (x)Sx)$	M.T. 3,5
7.	$\sim (\exists x) \sim Hx \lor \sim (x)Sx$	DeM. 6
8.	$\sim (x)Fx$	Simp. 4
9.	$(\exists x) \sim Fx$	Q.N. 8
10.	$\sim Fa$	E.I. 9, a/x (flag a)
11.	$Ha \supset Fa$	U.I. 2, a/x
12.	$\sim Ha$	M.T. 10,11
13.	$(\exists x) \sim Hx$	E.G. 12, a/x
14.	$\sim \sim (\exists x) \sim Hx$	D.N. 13
15.	$\sim (x)Sx$	D.S. 7,14
16.	$(\exists x) \sim Sx$	Q.N. 15

j.
1.	$\sim ((\exists x) \sim Fx \cdot (\exists x)(Gx \cdot \sim Hx))$	Pr.
2.	$(x)Hx \supset \sim (\exists x)(Zx \cdot Wx)$	Pr.
3.	$\sim (x)(Wx \supset Fx)$	Pr. $/\therefore \sim (\exists x) \sim Zx \supset \sim (x)Gx$
4.	$\sim (\exists x) \sim Zx$	Assp. (C.P.)
5.	$(x)Zx$	Q.N. 4
6.	$(\exists x)(Wx \cdot \sim Fx)$	C.Q.N. 3
7.	$Wa \cdot \sim Fa$	E.I. 6, a/x (flag a)
8.	Wa	Simp. 7
9.	$\sim Fa$	Simp. 7
10.	Za	U.I. 5, a/x
11.	$Za \cdot Wa$	Conj. 10,8
12.	$(\exists x)(Zx \cdot Wx)$	E.G. 11, a/x
13.	$\sim \sim (\exists x)(Zx \cdot Wx)$	D.N. 12
14.	$\sim (x)Hx$	M.T. 2,13
15.	$\sim (\exists x) \sim Fx \lor \sim (\exists x)(Gx \cdot \sim Hx)$	DeM. 1
16.	$(\exists x) \sim Fx$	E.G. 9, a/x
17.	$\sim \sim (\exists x) \sim Fx$	D.N. 16
18.	$\sim (\exists x)(Gx \cdot \sim Hx)$	D.S. 15,17
19.	$(x)(Gx \supset Hx)$	C.Q.N. 18
20.	$(\exists x) \sim Hx$	Q.N. 14
21.	$\sim Hb$	E.I. 20, b/x (flag b)
22.	$Gb \supset Hb$	U.I. 19, b/x
23.	$\sim Gb$	M.T. 21,22
24.	$(\exists x) \sim Gx$	E.G. 23, b/x
25.	$\sim (x)Gx$	Q.N. 24
26.	$\sim (\exists x) \sim Zx \supset \sim (x)Gx$	C.P. 4–25

5. a.
1.	$(x)(Fx \supset Gx)$	Assp. (C.P.)
2.	$\sim (x)Gx$	Assp. (C.P.)
3.	$(\exists x) \sim Gx$	Q.N. 2
4.	$\sim Ga$	E.I. 3, a/x (flag a)
5.	$Fa \supset Ga$	U.I. 1, a/x

6.	$\sim Fa$	M.T. 4,5
7.	$(\exists x) \sim Fx$	E.G. 6, a/x
8.	$\sim (x)Fx$	Q.N. 7
9.	$\sim (x)Gx \supset \sim (x)Fx$	C.P. 2–8
10.	$(x)(Fx \supset Gx) \supset (\sim (x)Gx \supset \sim (x)Fx)$	C.P. 1–9

c.
1.	$(x)(Fx \cdot Gx)$	Assp. (C.P.)
2.	flag a	F.S. (U.G.)
3.	$Fa \cdot Ga$	U.I. 1, a/x
4.	Fa	Simp. 3
5.	$(x)Fx$	U.G. 4, a/x
6.	flag b	F.S. (U.G.)
7.	$Fb \cdot Gb$	U.I. 1, b/x
8.	Gb	Simp. 7
9.	$(x)Gx$	U.G. 8, b/x
10.	$(x)Fx \cdot (x)Gx$	Conj. 5,9
11.	$(x)(Fx \cdot Gx) \supset ((x)Fx \cdot (x)Gx)$	C.P. 1–10
12.	$(x)Fx \cdot (x)Gx$	Assp. (C.P.)
13.	$(x)Fx$	Simp. 12
14.	$(x)Gx$	Simp. 12
15.	flag c	F.S. (U.G.)
16.	Fc	U.I. 13, c/x
17.	Gc	U.I. 14, c/x
18.	$Fc \cdot Gc$	Conj. 16,17
19.	$(x)(Fx \cdot Gx)$	U.G. 18, c/x
20.	$((x)Fx \cdot (x)Gx) \supset (x)(Fx \cdot Gx)$	C.P. 12–19
21.	$((x)(Fx \cdot Gx) \supset ((x)Fx \cdot (x)Gx))$	Conj. 11,20
	$\cdot (((x)Fx \cdot (x)Gx) \supset (x)(Fx \cdot Gx))$	B.E. 21
22.	$(x)(Fx \cdot Gx) \equiv ((x)Fx \cdot (x)Gx)$	

d.
1.	$\sim ((\exists x)Fx \lor (\exists x) \sim Gx)$	Assp. (C.P.)
2.	$\sim (\exists x)Fx \cdot \sim (\exists x) \sim Gx$	DeM. 1
3.	$\sim (\exists x)Fx$	Simp. 2
4.	$(x) \sim Fx$	Q.N. 3
5.	flag a	F.S. (U.G.)
6.	$\sim Fa$	U.I. 4, a/x
7.	$\sim Fa \lor Ha$	Add. 6
8.	$Fa \supset Ha$	C.E. 7
9.	$(x)(Fx \supset Hx)$	U.G. 8, a/x
10.	$\sim ((\exists x)Fx \lor (\exists x) \sim Gx) \supset (x)(Fx \supset Hx)$	C.P. 1–9

Unit 16

1. a. $(Fa \lor Ga) \cdot (Fa \lor Gb) \cdot (Fc \lor Gc)$
    c. $(Fa \lor (Ga \cdot Ha)) \cdot (Fb \lor (Gb \cdot Hb)) \cdot (Fc \lor (Gc \cdot Hc))$
    e. $((Fa \lor Ga) \cdot (Fb \lor Gb) \cdot (Fc \lor Gc)) \supset ((Ha \cdot Ia) \lor (Hb \cdot Ib) \lor (Hc \cdot Ic))$

2. No answers are possible here; check with your instructor or teaching assistant.

3. The domain will be given first, then the interpretation, and then the inval-

idating assignment of truth values. Answers have been abbreviated here to save space. You should, however, follow the format indicated in Unit 16.

a. $\{a,b\}$ Int. $(Aa \supset Ba) \cdot (Ab \supset Bb)$, $(Aa \vee Ab) /\therefore (Ba \cdot Bb)$ Let $Aa = F$, $Ba = F$, $Ab = T$, and $Bb = T$.

c. $\{a,b\}$ Int: $((Fa \cdot Ga) \vee (Fb \cdot Gb))$, $((Ga \supset Ha) \cdot (Gb \supset Hb))$ $/\therefore (Fa \supset Ha) \cdot (Fb \supset Hb)$. Let $Fa = T$, $Ga = F$, $Ha = F$, $Fb = T$, $Gb = T$, and $Hb = T$.

e. $\{a,b\}$ Int: $(Fa \supset \sim Ga) \cdot (Fb \supset \sim Gb)$, $((Ha \cdot Ga) \vee (Hb \cdot Gb))$ $/\therefore (Fa \supset \sim Ha) \cdot (Fb \supset \sim Hb)$. Let $Fa = T$, $Ga = F$, $Ha = T$, $Fb = F$, $Hb = T$, and $Gb = T$.

g. $\{a\}$ Int: $((Fa \cdot Ga) \supset Ha)$, $(Ha \supset Ja) /\therefore (Fa \supset Ja)$. Let $Fa = T$, $Ga = F$, $Ha = F$, and $Ja = F$.

i. $\{a,b\}$ Int: $(Fa \vee Fb) \cdot (Ga \vee Gb) \cdot (Ha \vee Hb)$ $/\therefore (Fa \cdot Ga \cdot Ha) \vee (Fb \cdot Gb \cdot Hb) \cdot$ Let $Fa = F$, $Fb = T$, $Ga = T$, $Gb = F$, $Ha = T$, $Hb = T$ or F (not the only possibilities).

k. $\{a,b\}$ Int: $(Fa \cdot Fb) \supset ((Ga \supset Ha) \cdot (Gb \supset Hb))$ $/\therefore ((Fa \cdot Ga) \supset Ha) \cdot ((Fb \cdot Gb) \supset Hb)$. Let $Fa = T$, $Ga = T$, $Ha = F$, $Fb = F$. (Other values may be anything.)

4. 2a. $(\exists x)(Rx \cdot Wx)$, $(\exists x)(Wx \cdot \sim Dx) /\therefore (\exists x)(Rx \cdot \sim Dx)$
   $\{a,b\}$ Int: $(Ra \cdot Wa) \vee (Rb \cdot Wb)$, $(Wa \cdot \sim Da) \vee (Wa \cdot \sim Db)$ $/\therefore (Ra \cdot \sim Da) \vee (Rb \cdot \sim Db)$. Let $Ra = T$, $Wa = T$, $Da = T$, $Rb = F$, $Wb = T$, $Db = F$.

   2c. $(x)(Px \supset Mx)$, $(x)(Px \supset \sim Ex) /\therefore (x)(Mx \supset \sim Ex)$
   $\{a\}$ Int: $Pa \supset Ma$, $Pa \supset \sim Ea /\therefore Ma \supset \sim Ea$. Let $Pa = F$, $Ma = T$, $Ea = T$.

   2d. $(\exists x)(Gx \cdot \sim Ax) /\therefore (\exists x)(Gx \cdot Ax)$
   $\{a\}$ Int: $Ga \cdot \sim Aa /\therefore Ga \cdot Aa$. Let $Ga = T$, $Aa = F$

   2f. $(\exists x)(Cx \cdot \sim Mx)$, $(x)(Cx \supset Wx) /\therefore (\exists x)(Wx \cdot \sim Cx)$
   $\{a\}$ Int: $Ca \cdot \sim Ma$, $Ca \supset Wa /\therefore Wa \cdot \sim Ca$. Let $Wa = T$, $Ca = T$, $Ma = F$

5. a. Symbolization: $(x)((Sx \cdot Bx) \supset Ix)$, $(x)((Sx \cdot Tx) \supset Ix) /\therefore (x)((Sx \cdot Tx) \supset Bx)$
   Domain $= \{a\}$; Int: $((Sa \cdot Ba) \supset Ia)$, $((Sa \cdot Ta) \supset Ia) /\therefore (Sa \cdot Ta) \supset Ba)$
   Invalid: let $Sa = T$, $Ba = F$, $Ta = T$, and $Ia = T$.

   c. Symbolization: $(x)(Rx \supset Sx) \cdot (\exists x)(Sx \cdot Px)$, $(x)(Tx \supset \sim Px)$ $/\therefore (x)(Tx \supset \sim Rx)$; Invalid: Let the domain $= \{a,b\}$.
   Int: $(Ra \supset Sa) \cdot (Rb \supset Sb)$, $(Sa \cdot Pa) \vee (Sb \cdot Pb)$, $(Ta \supset \sim Pa) \cdot (Tb \supset \sim Pb) /\therefore (Ta \supset \sim Ra) \cdot (Tb \supset \sim Rb)$.
   Let $Ta = T$, $Ra = T$, $Sa = T$, $Pa = F$, $Sb = T$, $Pb = T$; $Tb = F$; Rb may be either T or F.

Unit 17

1. a. $Txy \equiv x$ is taller than y; Tja
   c. $Wxy \equiv x$ wrote y; Wns
   e. $Wxy \equiv x$ works for y; Wrc
   g. $Mxy \equiv x$ spends more money than y; Mce
   i. $Axy \equiv x$ approves of y; Are

    k. *Txyz* ≡ *x* told *y* about *z*; *Trme*

    m. *Cxyz* ≡ *x* is closer to *y* than *z* is; *Cfwm*

    o. *Bxyz* ≡ *x* is between *y* and *z*; *Bfmw*

2. a. *Bcns*

    c. *Ocm*

    e. *Mjs*

    g. *Tja · Maj*

    i. *Laj*

    k. *Laj · ~ Lap*

    m. *~ Tjp · ~ Tjs*

    o. *~ (Maj · Mas)*

3. a. There is someone who is a better bridge player than Anne.

    c. Everyone respects Charles.

    e. Some people are better bridge players than others.

    g. Everyone is a better bridge player than somebody.

    i. There is someone whom everyone respects.

    k. Bob likes someone better than Charles.

    m. Some people like Dora better than Charles.

    o. There is someone who told stories about John to everyone.

    q. There is someone who told stories about John to everyone.

    s. There is someone about whom everyone told stories to everyone.

4. a. $(\exists x)Rxj$

    c. $(\exists x)(\exists y)Rxy$

    e. $(\exists x)Rxc$

    g. $(\exists x)Ixaj$

    i. $(x)Ijxa$

    k. $(x)(\exists y)Iyxa$

    m. $(x)(\exists y)(\exists z)Ixyz$

    o. $(\exists x)(\exists y)(z)Ixyz$

5. a. $(\exists x) \sim (\exists y)(\exists z)Fxyz$, $(\exists x)(y) \sim (\exists z)Fxyz$, $(\exists x)(y)(z) \sim Fxyz$

    c. $(\exists x) \sim (y)(z)(w) \sim Fxyzw$, $(\exists x)(\exists y) \sim (z)(w) \sim Fxyzw$,

      $(\exists x)(\exists y)(\exists z) \sim (w) \sim Fxyzw$, $(\exists x)(\exists y)(\exists z)(\exists w)Fxyzw$

    e. $(\exists x)(y) \sim (z) \sim (w)Fxyzw$, $(\exists x)(y)(\exists z)(w)Fxyzw$,

      or $\sim (x)(\exists y)(z) \sim (w)Fxyzw$, $\sim (x)(\exists y)(z)(\exists w) \sim Fxyzw$

6. a. $\sim (\exists x)Lax$ or $(x) \sim Lax$

    c. $\sim (x)Ljx$ or $(\exists x) \sim Ljx$

    e. $\sim (\exists x)(y)Lxy$ or $(x) \sim (y)Lxy$ or $(x)(\exists y) \sim Lxy$

    g. $\sim (\exists x) \sim (\exists y)Lxy$ or $(x)(\exists y)Lxy$

    i. $\sim (\exists x)Sxa$ or $(x) \sim Sxa$

    k. $\sim (x) \sim Bxj$ or $(\exists x)Bxj$

    m. $\sim (\exists x)Iaxc$ or $(x) \sim Iaxc$

    o. $\sim (x) \sim Ixra$ or $(\exists x)Ixra$

    q. $\sim (x)(\exists y)Ixyr$ or $(\exists x) \sim (\exists y)Ixyr$ or $(\exists x)(y) \sim Ixyr$

    s. $\sim (x)(\exists y)(\exists z)Ixyz$ or $(\exists x) \sim (\exists y)(\exists z)Ixyz$ or $(\exists x)(y) \sim (\exists z)Ixyz$ or

      $(\exists x)(y)(z) \sim Ixyz$

7. a. $(x)(Sx \supset (\exists y)(By \cdot Rxy))$
    c. $\sim (\exists x)(Sx \cdot (y)(By \supset Rxy))$ or $(x)(Sx \supset (\exists y)(By \cdot \sim Rxy))$
    e. $(\exists x)((Sx \cdot \sim (\exists y)(By \cdot Rxy)) \cdot Gx)$
    g. $\sim (x)((Sx \cdot \sim (\exists y)(By \cdot Rxy)) \supset Gx)$
    i. $\sim (x)(Sx \supset (y)((Fy \cdot Hxy) \supset Lxy))$ or $(\exists x)(Sx \cdot (\exists y)(Fy \cdot Hxy \cdot \sim Lxy))$
    k. $(x)((Sx \cdot (\exists y)(By \cdot Rxy) \cdot \sim (\exists z)(Cz \cdot Rxz)) \supset (Gx \cdot \sim Wx))$
    m. $(x)((Sx \cdot (y)(z)((Fy \cdot Hxy \cdot Bz \cdot Ayzx) \supset Rxz)) \supset (Wx \cdot Gx))$
    o. $(x)(Sx \supset (\exists y)(Py \cdot Rxy))$
    q. $\sim (\exists x)((Sx \cdot (Gx \vee (\exists y)(Py \cdot Wxy))) \cdot (z)(Cz \supset Rxz))$

8. a. Not all students read comics.
    c. Not all professors listen to all of their students.
    e. Any student who listens to all of his or her professors will read some books and will get good grades.
    g. No book is read by all students.
    i. Any student or professor who writes poetry is well rounded and reads no comics.
    k. A student who reads comics but does not read either books or poetry will not get good grades and will not be well rounded.

9. a. $(\exists x)(Dx \cdot \sim (y)(Pyx \supset Lxy))$
    c. $(x)[(Dx \cdot (y)(Pyx \supset Rxy)) \supset (\exists y)(Pyx \cdot Lxy)]$
    e. $(\exists x)[Dx \cdot (\exists y)(Pyx \cdot \sim (\exists z)(Mz \cdot Hyz) \cdot Txy)]$
    g. $(\exists x)[Px \cdot \sim (\exists y)(Ay \cdot Hxy) \cdot (\exists z)(Dz \cdot Tzx)]$
    i. $(\exists x)[Dx \cdot (y)(Pyx \supset (\exists z)(Nz \cdot Pxzy))]$
    k. $(x)[(Dx \cdot (\exists y)(Pyx \cdot \sim (\exists z)(Az \cdot Hyz) \cdot (\exists z)(Nz \cdot Pxzy))) \supset \sim (w)(Ewx \supset Rwx)]$
    m. $\sim (\exists x)[Nx \cdot (\exists y)(Sy \cdot Hxy) \cdot (\exists z)(Dz \cdot (\exists w)(Pwz \cdot \sim (\exists v)(Av \cdot Hwv)) \cdot Pzxw)]$
    o. $(\exists x)\{Nx \cdot (\exists y)(Sy \cdot Hxy) \cdot (\exists z)(Dz \cdot (w)[[Pwz \cdot \sim (\exists v)(Av \cdot Hwv) \cdot (\exists v)(Lv \cdot Hwv \cdot (\exists t)(Bt \cdot Hvt) \cdot (\exists s)(Ds \cdot Svs))] \supset Pzxw])\}$

Unit 18

1. a.
1.	$(\exists x)(Tx \cdot Sx)$	Pr.
2.	$(x)(Sx \supset \sim (\exists y)((Fy \vee By) \cdot Txy))$	Pr. /
		$\therefore (\exists x)(Tx \cdot \sim (\exists y)(By \cdot Txy))$
3.	$Ta \cdot Sa$	E.I. 1, a/x (flag a)
4.	$Sa \supset \sim (\exists y)((Fy \vee By) \cdot Tay)$	U.I. 2, a/x
5.	Sa	Simp. 3
6.	$\sim (\exists y)((Fy \vee By) \cdot Tay)$	M.P. 4,5
7.	$(y)((Fy \vee By) \supset \sim Tay)$	C.Q.N. 6
8.	flag b	F.S. (U.G.)
9.	Bb	Assp. (C.P.)
10.	$Fb \vee Bb$	Add. 9
11.	$(Fb \vee Bb) \supset \sim Tab$	U.I. 7, b/y
12.	$\sim Tab$	M.P. 10,11
13.	$Bb \supset \sim Tab$	C.P. 9–12

14. $(y)(By \supset \sim Tay)$　　　　　　　　U.G. 13, b/y
15. $\sim (\exists y)(By \cdot Tay)$　　　　　　　　C.Q.N. 14
16. Ta　　　　　　　　　　　　　　Simp. 3
17. $Ta \cdot \sim (\exists y)(By \cdot Tay)$　　　　Conj. 15,16
18. $(\exists x)(Tx \cdot \sim (\exists y)(By \cdot Txy))$　　E.G. 17, a/x

b. 1. $(x)(Rx \supset \sim (\exists y)(Uy \cdot Ty \cdot Bxy))$　　Pr.
　 2. $(\exists x)(Ux \cdot Tx \cdot Sx \cdot Bjx)$　　　Pr, $/\therefore \sim Rj$
　 3. $Ua \cdot Ta \cdot Sa \cdot Bja$　　　　　E.I. 2, a/x (flag a)
　 4. $Rj \supset \sim (\exists y)(Uy \cdot Ty \cdot Bjy)$　　U.I. 1, j/x
　 5. $Ua \cdot Ta$　　　　　　　　Simp. 3
　 6. Bja　　　　　　　　　　Simp. 3
　 7. $Ua \cdot Ta \cdot Bja$　　　　　Conj. 5,6
　 8. $(\exists y)(Uy \cdot Ty \cdot Bjy)$　　　E.G. 7, a/y
　 9. $\sim \sim (\exists y)(Uy \cdot Ty \cdot Bjy)$　　D.N. 8
　10. $\sim Rj$　　　　　　　　　　M.T. 4,9

c. 1. $(\exists x)(Bx \cdot (y)(Gy \supset Lxy))$　　Pr.
　 2. $\sim (\exists x)(Bx \cdot (\exists y)(Wy \cdot Lxy))$　　Pr. $/\therefore (x)(Gx \supset \sim Wx)$
　 3. $(x)(Bx \supset \sim (\exists y)(Wy \cdot Lxy))$　　C.Q.N. 2
　 4. $Ba \cdot (y)(Gy \supset Lay)$　　　E.I. 1, a/x (flag a)
　 5. Ba　　　　　　　　　　Simp. 4
　 6. $(y)(Gy \supset Lay)$　　　　　Simp. 4
　 7. ┌─► flag b　　　　　　　F.S. (U.G.)
　 8. │ ┌─► Gb　　　　　　　Assp. (C.P.)
　 9. │ │ $Gb \supset Lab$　　　　U.I. 6, b/y
　10. │ │ Lab　　　　　　　M.P. 8,9
　11. │ │ $Ba \supset \sim (\exists y)(Wy \cdot Lay)$　U.I. 3, a/x
　12. │ │ $\sim (\exists y)(Wy \cdot Lay)$　　M.P. 5,11
　13. │ │ $(y)(Wy \supset \sim Lay)$　　C.Q.N. 12
　14. │ │ $Wb \supset \sim Lab$　　　U.I. 13, b/y
　15. │ │ $\sim \sim Lab$　　　　　D.N. 10
　16. │ └── $\sim Wb$　　　　　　M.T. 14,15
　17. │ $Gb \supset \sim Wb$　　　　C.P. 8–16
　18. $(x)(Gx \supset \sim Wx)$　　　　U.G. 17, b/x

d. 1. $(x)((Px \cdot Lxj) \supset Gx)$　　　Pr.
　 2. $(x)(Gx \supset Sx)$　　　　　Pr.
　 3. $\sim (\exists x)((Px \cdot Hx) \cdot Sx)$　　Pr. $/\therefore (x)((Px \cdot Hx) \supset \sim Lxj)$
　 4. $(x)((Px \cdot Hx) \supset \sim Sx)$　　C.Q.N. 3
　 5. ┌─► flag a　　　　　　　F.S. (U.G.)
　 6. │ ┌─► $Pa \cdot Ha$　　　　　Assp. (C.P.)
　 7. │ │ $(Pa \cdot Ha) \supset \sim Sa$　　U.I. 4, a/x
　 8. │ │ $\sim Sa$　　　　　　　M.P. 6,7
　 9. │ │ $Ga \supset Sa$　　　　　U.I. 2, a/x
　10. │ │ $\sim Ga$　　　　　　　M.T. 8,9
　11. │ │ $(Pa \cdot Laj) \supset Ga$　　U.I. 1, a/x
　12. │ │ $\sim (Pa \cdot Laj)$　　　　M.T. 10,11
　13. │ │ $\sim Pa \lor \sim Laj$　　　DeM. 12
　14. │ │ Pa　　　　　　　　Simp. 6

	15.	$\sim \sim Pa$	D.N. 14
	16.	$\sim Laj$	D.S. 13,15
	17.	$(Pa \cdot Ha) \supset \sim Laj$	C.P. 6–16
	18.	$(x)((Px \cdot Hx) \supset \sim Lxj)$	U.G. 17, a/x

j.
1. $(x)((\exists y)(Dy \cdot Kyx) \supset (z)(Az \supset Hxz))$ — Pr.
2. $(x)(y)(Fxy \supset \sim Hxy)$ — Pr.
3. $\sim (x)(Ax \supset \sim Dx)$ — Pr.
4. $(x)(Cx \supset (y)Kyx)$ — Pr. $/\therefore (x)(Cx \supset (\exists y)(Dy \cdot \sim Fxy))$
5. $(\exists x)(Ax \cdot Dx)$ — C.Q.N. 3
6. $Ab \cdot Db$ — E.I. 5, b/x (flag b)
7. Ab — Simp. 6
8. Db — Simp. 6
9. flag a — F.S. (U.G.)
10. Ca — Assp. (C.P.)
11. $Ca \supset (y)Kya$ — U.I. 4, a/x
12. $(y)Kya$ — M.P. 10,11
13. Kba — U.I. 12, b/y
14. $Db \cdot Kba$ — Conj. 8,13
15. $(\exists y)(Dy \cdot Kya)$ — E.G. 14, b/y
16. $(\exists y)(Dy \cdot Kya) \supset (z)(Az \supset Haz)$ — U.I. 1, a/x
17. $(z)(Az \supset Haz)$ — M.P. 15,16
18. $Ab \supset Hab$ — U.I. 17, b/z
19. Hab — M.P. 7,18
20. $(y)(Fay \supset \sim Hay)$ — U.I. 2, a/x
21. $Fab \supset \sim Hab$ — U.I. 20, b/y
22. $\sim \sim Hab$ — D.N. 19
23. $\sim Fab$ — M.T. 21,22
24. $Db \cdot \sim Fab$ — Conj. 8,23
25. $(\exists y)(Dy \cdot \sim Fay)$ — E.G. 24, b/y
26. $Ca \supset (\exists y)(Dy \cdot \sim Fay)$ — C.P. 10–25
27. $(x)(Cx \supset (\exists y)(Dy \cdot \sim Fxy))$ — U.G. 26, a/x

2. c.
1. $(x)((\exists y)Fxy \supset Gx)$ — Assp. (C.P.)
2. flag a — F.S. (U.G.)
3. flag b — F.S. (U.G.)
4. Fab — Assp. (C.P.)
5. $(\exists y)Fay \supset Ga$ — U.I. 1, a/x
6. $(\exists y)Fay$ — E.G. 4, b/y
7. Ga — M.P. 5,6
8. $Fab \supset Ga$ — C.P. 4–7
9. $(y)(Fay \supset Ga)$ — U.G. 8, b/y
10. $(x)(y)(Fxy \supset Gx)$ — U.G. 9, a/x
11. $(x)((\exists y)Fxy \supset Gx) \supset (x)(y)(Fxy \supset Gx)$ — C.P. 1–10

d.
1. $(x)(\exists y)(Fxy \supset Gx)$ — Assp. (C.P.)
2. flag a — F.S. (U.G.)
3. $(y)Fay$ — Assp. (C.P.)
4. $(\exists y)(Fay \supset Ga)$ — U.I. 1, a/x
5. $Fab \supset Ga$ — E.I. 4, b/y (flag b)

6. | | | *Fab*   U.I. 3, *b/y*
7. | | *Ga*   M.P. 5,6
8. | | *(y)Fay ⊃ Ga*   C.P. 3–7
9. | *(x)((y)Fxy ⊃ Gx)*   U.G. 8, *a/x*
10. *(x)(∃ y)(Fxy ⊃ Gx) ⊃ (x)((y)Fxy ⊃ Gx)* C.P. 1–9

3. a. {*a,b*} Int: *(y)Fay ∨ (y)Fby* /∴ *(y)Fay · (y)Fby*. Let *(y)Fay* = T and
    (y)Fby = F

    c. {*a,b*} Int: *(Fa ⊃ (∃ y)Gay) · (Fb ⊃ (∃ y)Gby)*
    /∴ *(y)(Fa ⊃ Gay) ∨ (y)(Fb ⊃ Gby)*
    So, *(Fa ⊃ (Gaa ∨ Gab)) · (Fb ⊃ (Gba ∨ Gbb))*
    /∴ *((Fa ⊃ Gaa) · (Fa ⊃ Gab)) ∨ ((Fb ⊃ Gba) · (Fb ⊃ Gbb))*
    Let *Fa* = T, *Gaa* = F, *Gab* = T, *Fb* = T, *Gba* = F, *Gbb* = T

    e. {*a,b*} Int: *(y)(∃ z)Fayz · (y)(∃ z)Fbyz* /∴ *(y)(x)Fxya ∨ (y)(x)Fxyb*
    So, *(∃ z) Faaz · (∃ z) Fabz · (∃ z) Fbaz · (∃ z) Fbbz*
    /∴ *((x)Fxaa · (x)Fxba) ∨ ((x)Fxab · (x)Fxbb)*
    So, *(Faaa ∨ Faab) · (Faba ∨ Fabb) · (Fbaa ∨ Fbab) · (Fbba ∨ Fbbb)*
    /∴ *(Faaa · Fbaa · Faba · Fbba) ∨ (Faab · Fbab · Fabb · Fbbb)*
    Let *Faaa* = F, *Faab* = T, *Fbab* = F, *Fbaa* = T, *Faba* = T, *Fbbb* = T

Unit 19

1. a. Relational singular
    c. Identity
    e. Categorical
    g. Identity
    i. Categorical
    k. Relational singular
    m. Categorical
    o. Identity
    q. Categorical
    s. Identity

2. a. Singular; *Wm*
    c. Sing; *Ema*
    e. Sing; ~ Ai
    g. Id; *i ≠ g*
    i. Sing; *Dma*
    k. Sing; *(x)Fix*
    m. Sing; *Am*
    o. Sing; *(x)Max*
    q. Sing; ~ *(∃ x)Fdx*
    s. Sing; ~ *Dia*

3. a. *(x)((Px · x ≠ a) ⊃ Zxm) · Pa · ~ Zam*
    c. *(x)((Wx · x ≠ m) ⊃ ~ Ax) · Wm · Am*
    e. *(x)((Px · x ≠ d) ⊃ ~ Mxa) · Mda · Pd*
    g. *(∃ x)(Wx · x ≠ m · Zix) · Wm · Zim*
    i. *(x)((Wx · x ≠ m) ⊃ Smx) · Wm*
    k. *(x)((Nx · ~ Wx) ⊃ ~ Sx)*

4. a. $(\exists x)(\exists y)(Exa \cdot Eya \cdot x \neq y)$
    c. $(\exists x)(\exists y)(Fxm \cdot Fym \cdot x \neq y \cdot (z)(Fzm \supset (z = x \lor z = y)))$
    e. $(\exists x)(\exists y)(\exists z)(Cx \cdot Wx \cdot Cy \cdot Wy \cdot Cz \cdot Wz \cdot x \neq y \cdot y \neq z \cdot x \neq z)$
    g. $(\exists x)(\exists y)(\exists z)(Dx \cdot Dy \cdot Dz \cdot x \neq y \cdot y \neq z \cdot x \neq z \cdot$
        $(w)(Dw \supset (w = x \lor w = y \lor w = z)))$

5. a. $(\exists x)(Bx \cdot Tx \cdot (y)((By \cdot Ty) \supset x = y) \cdot Nx)$
    c. $(\exists x)(Cx \cdot (y)(Cy \supset x = y) \cdot Ixw)$
    e. $(\exists x)(Px \cdot (y)((Py \cdot x \neq y) \supset Fxy) \cdot x = h)$
    g. $(\exists x)(Mx \cdot Sx \cdot (y)((My \cdot Sy) \supset x = y) \cdot Bx \cdot Lx)$

Unit 20

1. Symbolizations
    a. $(\exists x)(Bx \cdot Mx \cdot (y)((By \cdot My) \supset x = y) \cdot \sim Ex), Bm \cdot Mm \; /\therefore \sim Em$
    b. $(\exists x)(Px \cdot Tx \cdot Fx \cdot (y)((Py \cdot Ty \cdot Fy) \supset x = y) \cdot Wx \cdot Mx), Tj \cdot Pj \cdot \sim Mj$
        $/\therefore \sim Fj$
    c. $(\exists x)(Px \cdot Fx \cdot (y)((Py \cdot Fy) \supset x = y) \cdot Ix),$
        $(\exists x)(Px \cdot Ix \cdot (y)((Py \cdot Iy) \supset x = y) \cdot Tx) \; /\therefore (\exists x)(Px \cdot Fx \cdot Tx)$
    d. $(\exists x)(Px \cdot Fx \cdot (y)((Py \cdot Fy) \supset x = y) \cdot x = j)$
        $(\exists x)(Px \cdot Vx \cdot (y)((Py \cdot Vy) \supset x = y) \cdot x = s), (x)((Px \cdot Vx) \supset \sim Fx)$
        $/\therefore s \neq j$
    e. $(x)((Px \cdot x \neq j) \supset \sim Fx) \cdot Pj \cdot Fj, (x)((Px \cdot x \neq c) \supset \sim Mx) \cdot Pc \cdot Mc,$
        $(\exists x)(Px \cdot Fx \cdot Mx) \; /\therefore j = c$
    f. $(\exists x)(Cx \cdot Mx), (x)((Rx \cdot Cx) \supset \sim Mx) \cdot (\exists x)(Rx \cdot Cx \cdot Lx),$
        $(\exists x)(Cx \cdot \sim Mx \cdot \sim Lx)$
        $/\therefore (\exists x)(\exists y)(\exists z) (Cx \cdot Cy \cdot Cz \cdot x \neq y \cdot y \neq z \cdot x \neq z)$
    g. $(\exists x)(Px \cdot (y)(Py \supset x = y)), (x)(Cx \equiv Px) \; /\therefore (\exists x)(Cx \cdot (y)(Cy \supset x = y))$
    h. $(x)((Px \cdot Hjx) \supset Sx), (\exists x)(Px \cdot Hjx) \cdot (x)(y)((Sx \cdot Hjx \cdot Sy \cdot Hjy) \supset x = y)$
        $/\therefore (\exists x)(Px \cdot Hjx \cdot (y)((Py \cdot Hjy) \supset x = y) \cdot Sx)$
    i. $(\exists x)(Px \cdot Fx \cdot (y)((Py \cdot Fy) \supset x = y)), (\exists x)(Px \cdot Bx \cdot (y)((Py \cdot By) \supset x = y)), (x)((Px \cdot Fx) \supset \sim Bx) \; /\therefore (\exists x)(\exists y)(Px \cdot Py \cdot (Fx \lor Bx) \cdot (Fy \lor By) \cdot x \neq y \cdot (z)((Pz \cdot (Fz \lor Bz)) \supset (x = z \lor y = z)))$
    j. $(x)((Px \cdot Ex) \supset (x = a \lor x = e)) \cdot (Pa \cdot Ea \cdot Pe \cdot Ee) \cdot Ta \cdot Te,$
        $(x)((Px \cdot Ex \cdot Tx) \supset Sx), (x)((Px \cdot Sx) \supset Kx) \; /\therefore (x)((Px \cdot Ex) \supset Kx)$
    k. $(\exists x)(\exists y)(Px \cdot Rx \cdot Py \cdot Ry \cdot x \neq y), (x)((Px \cdot Rx) \supset Cx),$
        $(x)(y)(z)((Cx \cdot Cy \cdot Cz \cdot Rx \cdot Ry \cdot Rz) \supset (x = y \lor y = z \lor x = z))$
        $/\therefore (\exists x)(\exists y)(Px \cdot Rx \cdot Py \cdot Ry \cdot x \neq y \cdot (z)((Pz \cdot Rz) \supset (z = x \lor z = y)))$
    l. $(\exists x)(\exists y)(\exists z)(Cx \cdot Cy \cdot Cz \cdot Rx \cdot Ry \cdot Rz \cdot x \neq y \cdot y \neq z \cdot x \neq z \cdot$
        $(w)((Cw \cdot Rw) \supset (w = x \lor w = y \lor w = z))),$
        $(\exists x)(Cx \cdot Rx \cdot Px \cdot (y)((Cy \cdot Ry \cdot Py) \supset x = y)),$
        $(x)((Cx \cdot Rx \cdot \sim Px) \supset Ox) \; /\therefore (\exists x)(\exists y)(Ox \cdot Oy \cdot Rx \cdot Ry \cdot x \neq y)$
    m. $(\exists x)(Ax \cdot Tx \cdot (y)((Ay \cdot Ty \cdot x \neq y) \supset Fxy) \cdot Dx)$
        $/\therefore (x)((Ax \cdot Tx \cdot \sim Dx) \supset (\exists y)(Dy \cdot Fyx))$

1. Proofs for some problems.
    b. 1. $(\exists x)(Px \cdot Tx \cdot Fx \cdot$ Pr.
            $(y)((Py \cdot Ty \cdot Fy) \supset x = y) \cdot Wx \cdot Mx)$
        2. $Tj \cdot Pj \cdot \sim Mj$ Pr. $/\therefore \sim Fj$

   3. $Pa \cdot Ta \cdot Fa \cdot$

      $(y)((Py \cdot Ty \cdot Fy) \supset a = y) \cdot Wa \cdot Ma$ E.I. 1, a/x (flag a)

   4. $(y)((Py \cdot Ty \cdot Fy) \supset a = y)$ Simp. 3

   5. $(Pj \cdot Tj \cdot Fj) \supset a = j$ U.I. 4, j/y

   6. $a = j$ Assp. (I.P.)

   7. Ma Simp. 3

   8. Mj I Sub. 6,7

   9. $\sim Mj$ Simp. 2

 10. $Mj \cdot \sim Mj$ Conj. 8,9

 11. $a \neq j$ I.P. 6–10

 12. $\sim (Pj \cdot Tj \cdot Fj)$ M.T. 5,11

 13. $\sim (Tj \cdot Pj \cdot Fj)$ Com. 12

 14. $\sim (Tj \cdot Pj) \vee \sim Fj$ DeM. 13

 15. $\sim \sim (Tj \cdot Pj)$ Simp. D.N. 2

 16. $\sim Fj$ D.S. 14,15

e.   1. $(x)((Px \cdot x \neq j) \supset \sim Fx) \cdot Pj \cdot Fj$ Pr.

    2. $(x)((Px \cdot x \neq c) \supset \sim Mx) \cdot Pc \cdot Mc$ Pr.

    3. $(\exists x)(Px \cdot Fx \cdot Mx)$ Pr. $/\therefore j = c$

    4. $Pa \cdot Fa \cdot Ma$ E.I. 3, a/x (flag a)

    5. $(Pa \cdot a \neq j) \supset \sim Fa$ Simp., U.I. 1, a/x

    6. $(Pa \cdot a \neq c) \supset \sim Ma$ Simp. U.I. 2, a/x

    7. $\sim \sim Fa$ Simp., D.N. 4

    8. $\sim \sim Ma$ Simp., D.N. 4

    9. $\sim (Pa \cdot a \neq j)$ M.T. 5,7

  10. $\sim (Pa \cdot a \neq c)$ M.T. 6,8

  11. $\sim Pa \vee \sim (a \neq j)$ DeM. 9

  12. $\sim Pa \vee \sim (a \neq c)$ DeM. 10

  13. $\sim \sim Pa$ Simp., D.N. 4

  14. $\sim (a \neq j)$ D.S. 11,13

  15. $\sim (a \neq c)$ D.S. 12,13

  16. $a = j$ D.N. 14

  17. $a = c$ D.N. 15

  18. $j = a$ I Sym. 16

  19. $j = c$ I Sub. 17,18

f.   1. $(\exists x)(Cx \cdot Mx)$ Pr.

    2. $(x)((Rx \cdot Cx) \supset \sim Mx)$ Pr.

    3. $(\exists x)(Rx \cdot Cx \cdot Lx)$ Pr.

    4. $(\exists x)(Cx \cdot \sim Mx \cdot \sim Lx)$ Pr.

    5. $Ca \cdot Ma$ E.I. 1, a/x (flag a)

    6. $Rb \cdot Cb \cdot Lb$ E.I. 3, b/x (flag b)

    7. $Cc \cdot \sim Mc \cdot \sim Lc$ E.I. 4, c/x (flag c)

    8. Ca Simp. 5

    9. Ma Simp. 5

  10. $(Ra \cdot Ca) \supset \sim Ma$ U.I. 2, a/x

  11. $\sim (Ra \cdot Ca)$ D.N., M.T. 9,10

  12. $\sim Ra \vee \sim Ca$ DeM. 11

  13. $\sim Ra$ D.N., D.S. 8,12

14.	Rb	Simp. 6
15.	$\llcorner a = b$	Assp. (I.P.)
16.	Ra	I Sub. 15,14
17.	$Ra \cdot \sim Ra$	Conj. 16,13
18.	$a \neq b$	I.P. 15–17
19.	$\llcorner b = c$	Assp. (I.P.)
20.	Lb	Simp. 6
21.	Lc	I. Sub. 19,20
22.	$\sim Lc$	Simp. 7
23.	$Lc \cdot \sim Lc$	Conj. 21,22
24.	$b \neq c$	I.P. 19–23
25.	$\llcorner a = c$	Assp. (I.P.)
26.	Mc	I. Sub. 25,9
27.	$\sim Mc$	Simp. 7
28.	$Mc \cdot \sim Mc$	Conj. 26,27
29.	$a \neq c$	I.P. 25–28
30.	$Cb \cdot Cc$	Simp., Conj., 6,7
31.	$Ca \cdot Cb \cdot Cc \cdot a \neq b \cdot b \neq c \cdot a \neq c$	Conj. 8,30,18,24,29
32.	$(\exists z)(Ca \cdot Cb \cdot Cz \cdot a \neq b \cdot b \neq z \cdot a \neq z)$	E.G. 31, c/z
33.	$(\exists y)(\exists z)(Ca \cdot Cy \cdot Cz \cdot a \neq y \cdot$ $y \neq z \cdot a \neq z)$	E.G., 32, b/y
34.	$(\exists x)(\exists y)(\exists z)(Cx \cdot Cy \cdot Cz \cdot x \neq y \cdot$ $y \neq z \cdot x \neq z)$	E.G., 33, a/x

i.

1.	$(\exists x)(Px \cdot Fx \cdot (y)((Py \cdot Fy) \supset x = y))$	Pr
2.	$(\exists x)(Px \cdot Bx \cdot (y)((Py \cdot By) \supset x = y))$	Pr.
3.	$(x)((Px \cdot Fx) \supset \sim Bx)$	Pr.
4.	$Pa \cdot Fa \cdot (y)((Py \cdot Fy) \supset a = y)$	E.I. 1, a/x (flag a)
5.	$Pb \cdot Bb \cdot (y)((Py \cdot By) \supset b = y)$	E.I. 2, b/y (flag b)
6.	$(Pa \cdot Fa) \supset \sim Ba$	U.I. 3, a/x
7.	$Pa \cdot Fa$	Simp. 4
8.	$\sim Ba$	M.P. 6,7
9.	$\llcorner a = b$	Assp. (I.P.)
10.	Bb	Simp. 5
11.	Ba	I. Sub. 9,10
12.	$Ba \cdot \sim Ba$	Conj. 11,8
13.	$a \neq b$	I.P. 9–12
14.	$\llcorner$ flag c	F.S. (U.G.)
15.	$\llcorner Pc \cdot (Fc \lor Bc)$	Assp. (C.P.)
16.	$(Pc \cdot Fc) \lor (Pc \cdot Bc)$	Dist. 15
17.	$(Pc \cdot Fc) \supset a = c$	Simp., U.I. 4, c/y
18.	$(Pc \cdot Bc) \supset b = c$	Simp., U.I. 5, c/y
19.	$a = c \lor b = c$	Dil. 16,17,18
20.	$(Pc \cdot (Fc \lor Bc)) \supset (a = c \lor b = c)$	C.P. 15–19
21.	$(z)((Pz \cdot (Fz \lor Bz)) \supset (a = z \lor b = z))$	U.G. 20, c/z
22.	Pa	Simp. 4
23.	Fa	Simp. 4
24.	$Fa \lor Ba$	Add. 23
25.	Pb	Simp. 5

26. Bb Simp. 5
27. $Fb \lor Bb$ Add. 26
28. $Pa \cdot Pb \cdot (Fa \lor Ba) \cdot (Fb \lor Bb) \cdot a \neq b \cdot$
    $(z)((Pz \cdot (Fz \lor Bz)) \supset (a = z \lor b = z))$ Conj. 22,25,24,27,13,21
29. $(\exists y)(Pa \cdot Py \cdot (Fa \lor Ba) \cdot (Fy \lor By) \cdot$
    $a \neq y \cdot (z)((Pz \cdot (Fz \lor Bz)) \supset$
    $(a = z \lor y = z)))$ E.G. 28, b/y
30. $(\exists x)(\exists y)(Px \cdot Py \cdot (Fx \lor Bx) \cdot$
    $(Fy \lor By) \cdot x \neq y \cdot (z)((Pz \cdot (Fz \lor Bz)) \supset$
    $(x = z \lor y = z)))$ E.G. 29, a/x

2. b. 1. $(\exists x)(Ax \cdot (y)(Ay \supset x = y) \cdot x = a)$ Pr.
        2. $Ab \lor Ac$ Pr.
        3. $Ad \cdot (y)(Ay \supset d = y) \cdot d = a$ E.I. 1, d/x (flag d)
        4. $(y)(Ay \supset d = y)$ Simp. 3
        5. ┌─ Ab Assp. (C.P.)
        6. │ $Ab \supset d = b$ U.I. 4, b/y
        7. │ $d = b$ M.P. 5,6
        8. │ $b = d$ I. Sym. 7
        9. │ $d = a$ Simp. 3
        10.│ $b = a$ I. Sub. 8,9
        11.└─ $a = b$ I. Sym. 10
        12. $Ab \supset a = b$ C.P. 5–11
        13. ┌─ Ac Assp. (C.P.)
        14. │ $Ac \supset d = c$ U.I. 4, c/y
        15. │ $d = c$ M.P. 13,14
        16. │ $d = a$ Simp. 3
        17. │ $a = d$ I. Sym. 16
        18. └─ $a = c$ I. Sub. 17,15
        19. $Ac \supset a = c$ C.P. 13–18
        20. $a = b \lor a = c$ Dil., 2,12,19

    e. 1. $(\exists x)(\exists y)(\exists z)(Fx \cdot Fy \cdot Fz \cdot x \neq y \cdot y \neq z \cdot x \neq z))$ Pr.
        2. $(x)(Fx \supset Gx)$ Pr.
        3. $(x)(y)(z)(w)((Gx \cdot Gy \cdot Gz \cdot Gw) \supset$
        $(x = y \lor x = z \lor x = w \lor y = z \lor y = w \lor z = w))$ Pr.
        4. $(\exists y)(\exists z)(Fa \cdot Fy \cdot Fz \cdot a \neq y \cdot y \neq z \cdot a \neq z)$ E.I. 1, a/x (flag a)
        5. $(\exists z)(Fa \cdot Fb \cdot Fz \cdot a \neq b \cdot b \neq z \cdot a \neq z)$ E.I. 4, b/y (flag b)
        6. $Fa \cdot Fb \cdot Fc \cdot a \neq b \cdot b \neq c \cdot a \neq c$ E.I. 5, c/z (flag c)
        7. $Fa \supset Ga$ U.I. 2, a/x
        8. $Fb \supset Gb$ U.I. 2, b/x
        9. $Fc \supset Gc$ U.I. 2, c/x
        10. Fa Simp. 6
        11. Fb Simp. 6
        12. Fc Simp. 6
        13. Ga M.P. 7,10
        14. Gb M.P. 8,11
        15. Gc M.P. 9,12

16.	┌─ flag d	F.S. (U.G.)
17.	│ ┌─ Fd	Assp. (C.P.)
18.	│ │ $Fd \supset Gd$	U.I. 2, d/x
19.	│ │ Gd	M.P. 17,18
20.	│ │ $Ga \cdot Gb \cdot Gc \cdot Gd$	Conj. 13, 14, 15, 19
21–24.	│ │ $(Ga \cdot Gb \cdot Gc \cdot Gd) \supset$	
	│ │ $(a = b \vee a = c \vee a = d \vee b = c \vee$	
	│ │ $b = d \vee c = d)$	U.I. 3, a/x, b/y, c/z, d/w
25.	│ │ $a = b \vee a = c \vee a = d \vee b = c \vee$	
	│ │ $b = d \vee c = d$	M.P. 20,24
26.	│ │ $a \neq b$	Simp. 6
27.	│ │ $a = c \vee a = d \vee b = c \vee b = d \vee c = d$ D.S. 25,26	
28.	│ │ $b \neq c$	Simp. 6
29.	│ │ $a = c \vee a = d \vee b = d \vee c = d$	D.S. 27,28
30.	│ │ $a \neq c$	Simp. 6
31.	│ └─ $a = d \vee b = d \vee c = d$	D.S. 29,30
32.	└─ $Fd \supset (a = d \vee b = d \vee c = d)$	C.P. 17–31
33.	$(w)(Fw \supset (a = w \vee b = w \vee c = w))$	U.G. 32, d/w
34.	$Fa \cdot Fb \cdot Fc \cdot a \neq b \cdot b \neq c \cdot a \neq c \cdot$	
	$(w)(Fw \supset (a = w \vee b = w \vee c = w))$	Conj. 6,33
35–36.	$(\exists y)(\exists z)(Fa \cdot Fy \cdot Fz \cdot a \neq y \cdot y \neq z \cdot a \neq z)$	
	$\cdot (w)(Fw \supset (a = w \vee y = w \vee z = w)))$	E.G. 34, a/x, b/y
37.	$(\exists x)(\exists y)(\exists z)(Fx \cdot Fy \cdot Fz \cdot x \neq y \cdot y \neq z \cdot x \neq z$	
	$\cdot (w)(Fw \supset (x = w \vee y = w \vee z = w)))$	E.G. 36, a/x

3. a.

1.	┌─ Fa	Assp. (C.P.)
2.	│ $a = a$	I. Ref. 1
3.	│ $Fa \cdot a = a$	Conj. 1,2
4.	└─ $(\exists x)(Fx \cdot x = a)$	E.G. 3, a/x
5.	$Fa \supset (\exists x)(Fx \cdot x = a)$	C.P, 1–4
6.	┌─ $(\exists x)(Fx \cdot x = a)$	Assp. (C.P.)
7.	│ $Fb \cdot b = a$	E.I. 6, b/x (flag b)
8.	│ Fb	Simp. 7
9.	│ $b = a$	Simp. 7
10.	└─ Fa	I. Sub. 8,9
11.	$(\exists x)(Fx \cdot x = a) \supset Fa$	C.P. 6–10
12.	$(Fa \supset (\exists x)(Fx \cdot x = a)) \cdot$	
	$((\exists x)(Fx \cdot x = a) \supset Fa)$	Conj. 5,11
13.	$Fa \equiv (\exists x)(Fx \cdot x = a)$	B.E. 12

b.

1.	┌─ $(\exists x)(y)(Fy \equiv x = y)$	Assp. (C.P.)
2.	│ $(y)(Fy \equiv a = y)$	E.I. 1, a/x (flag a)
3.	│ $a = a$	I. Ref. 2
4.	│ $Fa \equiv a = a$	U.I. 2, a/y
5.	│ $(Fa \supset a = a) \cdot (a = a \supset Fa)$	B.E. 4
6.	│ $a = a \supset Fa$	Simp. 5
7.	│ Fa	M.P. 3,6
8.	│ $Fa \cdot (y)(Fy \equiv a = y)$	Conj. 7,2
9.	└─ $(\exists x)(Fx \cdot (y)(Fy \equiv x = y))$	E.G. 8, a/x

10. $(\exists x)(y)(Fy \equiv x = y) \supset$
    $(\exists x)(Fx \cdot (y)(Fy \equiv x = y))$ C.P. 1–9
11. $(\exists x)(Fx \cdot (y)(Fy \equiv x = y))$ Assp. (C.P.)
12. $Fb \cdot (y)(Fy \equiv b = y)$ E.I. 11, b/x (flag b)
13. $(y)(Fy \equiv b = y)$ Simp. 12
14. $(\exists x)(y)(Fy \equiv x = y)$ E.G. 13, b/x
15. $(\exists x)(Fx \cdot (y)(Fy \equiv x = y)) \supset$
    $(\exists x)(y)(Fy \equiv x = y)$ C.P. 11–14
16. Step 10 $\cdot$ Step 15   Conj. 10,15
17. $(\exists x)(y)(Fy \equiv x = y) \equiv$
    $(\exists x)(Fx \cdot (y)(Fy \equiv x = y))$ B.E. 16

Unit 21

The following are all Wffs: 1, 4, 5, 7, 8, 11, 12, 14, 15, 16, 18, 20. 2 and 19 lack outside parentheses; 3 has no operators; 6 has an extra pair of parentheses, as does 9; 10 lacks inner parentheses; 13 has an extra left parenthesis; 17 has misplaced parentheses.

Unit 22

1. a, c, d, e, f, i are all Wffs. b is not because it has too many C's. g and h are not because they do not have enough operators.

2. a. a. $(p \cdot (q \supset r))$
      c. $((\sim p \supset \sim q) \vee (\sim p \cdot q))$
      d. $\sim \sim \sim \sim \sim \sim \sim (p \supset q)$
      e. $((\sim p \cdot \sim q) \supset \sim (p \vee q))$
      f. $((((p \supset q) \supset r) \supset s) \supset t)$
      i. $((p \supset (q \supset r)) \supset ((p \supset q) \supset (p \supset r)))$
   b. $r \supset (s \equiv t)$
   c. $r \supset (\sim (s \cdot u) \vee p)$
   d. $\sim (p \vee s) \cdot (((s \cdot s) \supset s) \vee s)$
   e. $(r \vee t) \supset (r \equiv s)$
   f. $\sim ((p \equiv (\sim t \vee s)) \equiv s) \cdot s$
   g. $((p \equiv o) \equiv u) \cdot t$
   h. $((p \equiv t) \equiv r) \cdot (((s \cdot s) \supset s) \vee s)$

3. a. $CKpqp$ f. $CCApqrCpr$
   b. $CKCpqpq$ g. $EKpAqrAKpqKpr$
   c. $EKpKqrKKpqr$ h. $EApKqrKApqApr$
   d. $ENKpqANpNq$ i. $CNCpKqrKpANqNr$
   e. $CpNNp$ j. $EpAKpqKpNq$

Unit 23

1. Unit 5; 1f (valid)

  1. $p \equiv\, \sim q$
  2. $q \lor p$
  3. $\sim (\sim q \lor \sim p)$ (neg. concl.)
  4. $\sim\sim q$ (from 3)

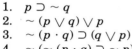

  5.   (from 1)

2. Unit 5; 1g (invalid)

  1. $p \supset\, \sim q$
  2. $\sim (p \lor q) \lor p$
  3. $\sim (p \cdot q) \supset (q \lor p)$
  4. $\sim (\sim (p \cdot q) \supset\, \sim p)$ (negated conclusion)
  5. $\sim (p \cdot q)$
         $\sim\sim p$ (from 4)

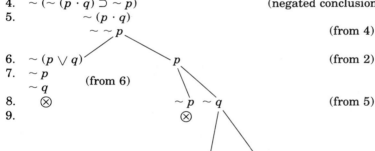

  6. $\sim (p \lor q)$ p (from 2)
  7. $\sim p$ (from 6)
     $\sim q$
  8. $\otimes$ $\sim p$ $\sim q$ (from 5)
  9. $\otimes$

  10. $\sim\sim (p \cdot q)$ $q \lor p$ (from 3)
             p
             q
  11. $\otimes$ q $ⓟ$
                 $\otimes$ (from 10)

Note: Circled formulas are those which occur last in an open branch. Thus, any tree with a circled formula is a tree for an invalid argument form.

Unit 5; 1h   (invalid)

1. $(p \supset q) \vee (q \supset r)$
2. $\sim r \supset \sim (p \cdot q)$
3. $\sim (q \supset \sim p)$ (neg. concl.)
4. q

4. $\sim \sim p$ (from 3)

5. $\sim \sim r$ $\sim (p \cdot q)$ (from 2)

6. $\sim p$ $\sim q$

6. $\otimes$ $\otimes$ (from 5)

7. $(p \supset q)$ $(q \supset r)$ (from 1)

8. $\sim q$ $\widehat{r}$ (from 7)

8. $\otimes$

9. $\sim p$ $\widehat{q}$ (from 7)

9. $\otimes$

Unit 5; 1i   (valid)

1. $p \supset (q \cdot r)$
2. $\sim p \supset \sim (q \cdot r)$
3. $q \cdot \sim p$
4. $\sim \sim r$ (negated conclusion)
5. q

5. $\sim p$ (from 3)

6. $\sim \sim p$ $\sim (q \cdot r)$ (from 2)

6. $\otimes$

7. $\sim q$ $\sim r$

7. $\otimes$ $\otimes$ (from 6)

Unit 24

1. VALID

Trespassers · People prosecuted · People shot

2. INVALID

Dogs · Things that have fleas · Things that make good pets

3. VALID

Politicians · Dishonest people · Trustworthy people

4. VALID

Students · Good teachers · People who go to heaven

5. VALID

Nuclear plants · Unsafe things · Things that should be funded by the government

6. INVALID

Lovers · People whom everyone loves · Happy people

7. INVALID

Coffee drinkers · Nervous things · Those with large savings accounts

8. VALID

Those with large savings accounts · Nervous things · Coffee drinkers

9. INVALID

Criminals · Nervous things · Coffee drinkers

10. VALID

Hijackings · Crimes · Political acts

Unit 25

1. a. $p \,/\, q$
   c. $((p \,/\, p) \,/\, (q \,/\, q)) \,/\, ((p \,/\, p) \,/\, (q \,/\, q))$
   e. $p \,/\, ((q \,/\, r) \,/\, (q \,/\, r))$
   g. $(((p \,/\, p) \,/\, (q \,/\, q)) \,/\, ((p \,/\, p) \,/\, (q \,/\, q))) \,/\, (r \,/\, r)$
   i. $((p \,/\, p) \,/\, q) \,/\, (q \,/\, q)$
   k. $p \,/\, ((q \,/\, (p \,/\, q)) \,/\, (q \,/\, (p \,/\, q)))$

2. a. $((p \downarrow p) \downarrow (q \downarrow q)) \downarrow ((p \downarrow p) \downarrow (q \downarrow q))$
   c. $p \downarrow q$
   e. $[(p \downarrow p) \downarrow (((p \downarrow p) \downarrow (q \downarrow q)) \downarrow ((p \downarrow p) \downarrow (q \downarrow q)))] \downarrow$
   $[(p \downarrow p) \downarrow (((p \downarrow p) \downarrow (q \downarrow q)) \downarrow ((p \downarrow p) \downarrow (q \downarrow q)))]$
   g. $(((p \downarrow q) \downarrow (p \downarrow q)) \downarrow r) \downarrow (((p \downarrow q) \downarrow (p \downarrow q)) \downarrow r)$
   i. $((p \downarrow (q \downarrow q)) \downarrow q) \downarrow ((p \downarrow (q \downarrow q)) \downarrow q)$
   k. $[(p \downarrow p) \downarrow (((q \downarrow q) \downarrow ((p \downarrow p) \downarrow (q \downarrow q))) \downarrow ((q \downarrow q) \downarrow ((p \downarrow p) \downarrow (q \downarrow q))))]$
   $\downarrow [(p \downarrow p) \downarrow (((q \downarrow q) \downarrow ((p \downarrow p) \downarrow (q \downarrow q))) \downarrow ((q \downarrow q) \downarrow ((p \downarrow p) \downarrow (q \downarrow q))))]$

3. a. $\sim (p \cdot q) \lor \sim (r \cdot s)$
   b. $(q \supset p) \lor \sim (q \supset p)$
   c. $((\sim p \lor \sim q) \cdot (\sim r \lor \sim s)) \lor ((p \lor q) \cdot (r \lor s))$
   d. $((p \cdot q) \lor (r \cdot s)) \lor ((p \lor q) \cdot (r \lor s))$

4. a. $\sim (p \lor q) \cdot \sim (r \lor s)$
   b. $\sim (p \supset q) \cdot (p \supset q)$
   c. $((\sim p \cdot \sim q) \lor (\sim r \cdot \sim s)) \cdot ((p \cdot q) \lor (r \cdot s))$
   d. $((p \lor q) \cdot (r \lor s)) \cdot ((p \cdot q) \lor (r \cdot s))$

5. a. $\sim (p \supset q)$ e. $(p \supset r) \cdot (\sim p \supset \sim r)$ or $p \equiv \sim r$
   c. $\sim p$ g. $(p \supset \sim q) \cdot (\sim p \supset (q \equiv r))$

Index

STATEMENT OF THE QUANTIFIER RULES, WITH ALL
NECESSARY RESTRICTIONS

A. Preliminary Definitions

1. φx is a propositional function on x, simple or complex. If complex, it is assumed that it is enclosed in parentheses, so that the scope of any prefixed quantifier extends to the end of the formula.
2. φa is a formula just like φx, except that every occurrence of x in φx has been replaced by an a.
3. An instance of a general formula is the result of deleting the initial quantifier and replacing each variable bound by that quantifier uniformly with some name.
4. An a-flagged subproof is a subproof that begins with the words "flag a," and ends with some instance containing a.

B. The Four Quantifier Rules

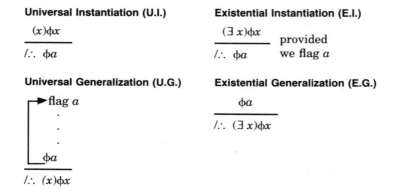

Universal Instantiation (U.I.)

$(x)\phi x$

/∴ φa

Existential Instantiation (E.I.)

$(\exists x)\phi x$

/∴ φa   provided we flag a

Universal Generalization (U.G.)

➤ flag a

.
.
.

φa

/∴ (x)φx

Existential Generalization (E.G.)

φa

/∴ $(\exists x)\phi x$

C. Flagging Restrictions

R_1 A letter being flagged must be new to the proof, that is, it may not appear, either in a formula or as a letter being flagged, previous to the step in which it gets flagged.

R_2 A flagged letter may not appear either in the premises or in the conclusion of a proof.

R_3 A flagged letter may not appear outside the subproof in which it gets flagged.